Lecture Notes in Physics

For information about Vols. 1–172, please contact your bookseller or Springer-Verlag.

Vol. 173: Stochastic Processes in Quantum Theory and Statistical Physics. Proceedings, 1981. Edited by S. Albeverio, Ph. Combe, and M. Sirugue-Collin. VIII, 337 pages. 1982.

Vol. 174: A. Kadić, D.G.B. Edelen, A Gauge Theory of Dislocations and Disclinations. VII, 290 pages. 1983.

Vol. 175: Defect Complexes in Semiconductor Structures. Proceedings, 1982. Edited by J. Giber, F. Beleznay, J. C. Szép, and J. László. VI, 308 pages. 1983.

Vol. 176: Gauge Theory and Gravitation. Proceedings, 1982. Edited by K. Kikkawa, N. Nakanishi, and H. Nariai. X, 316 pages. 1983.

Vol. 177: Application of High Magnetic Fields in Semiconductor Physics. Proceedings, 1982. Edited by G. Landwehr. XII, 552 pages. 1983.

Vol. 178: Detectors in Heavy-Ion Reactions. Proceedings, 1982. Edited by W. von Oertzen. VIII, 258 pages. 1983.

Vol. 179: Dynamical Systems and Chaos. Proceedings, 1982. Edited by L. Garrido. XIV, 298 pages. 1983.

Vol. 180: Group Theoretical Methods in Physics. Proceedings, 1982. Edited by M. Serdaroğlu and E. İnönü. XI, 569 pages. 1983.

Vol. 181: Gauge Theories of the Eighties. Proceedings, 1982. Edited by R. Raitio and J. Lindfors. V, 644 pages. 1983.

Vol. 182: Laser Physics. Proceedings, 1983. Edited by J. D. Harvey and D. F. Walls. V, 263 pages. 1983.

Vol. 183: J. D. Gunton, M. Droz, Introduction to the Theory of Metastable and Unstable States. VI, 140 pages. 1983.

Vol. 184: Stochastic Processes – Formalism and Applications. Proceedings, 1982. Edited by G. S. Agarwal and S. Dattagupta. VI, 324 pages. 1983.

Vol. 185: H. N. Shirer, R. Wells, Mathematical Structure of the Singularities at the Transitions between Steady States in Hydrodynamic Systems. XI, 276 pages. 1983.

Vol. 186: Critical Phenomena. Proceedings, 1982. Edited by F.J.W. Hahne. VII, 353 pages. 1983.

Vol. 187: Density Functional Theory. Edited by J. Keller and J.L. Gázquez. V, 301 pages. 1983.

Vol. 188: A. P. Balachandran, G. Marmo, B.-S. Skagerstam, A. Stern, Gauge Symmetries and Fibre Bundles. IV, 140 pages. 1983.

Vol. 189: Nonlinear Phenomena. Proceedings, 1982. Edited by K. B. Wolf. XII, 453 pages. 1983.

Vol. 190: K. Kraus, States, Effects, and Operations. Edited by A. Böhm, J. W. Dollard and W. H. Wootters. IX, 151 pages. 1983.

Vol. 191: Photon Photon Collisions. Proceedings, 1983. Edited by Ch. Berger. V, 417 pages. 1983.

Vol. 192: Heidelberg Colloquium on Spin Glasses. Proceedings, 1983. Edited by J. L. van Hemmen and I. Morgenstern. VII, 356 pages. 1983.

Vol. 193: Cool Stars, Stellar Systems, and the Sun. Proceedings, 1983. Edited by S. L. Balliunas and L. Hartmann. VII, 364 pages. 1984.

Vol. 194: P. Pascual, R. Tarrach, QCD: Renormalization for the Practitioner. V, 277 pages. 1984.

Vol. 195: Trends and Applications of Pure Mathematics to Mechanics. Proceedings, 1983. Edited by P. G. Ciarlet and M. Roseau. V, 422 pages. 1984.

Vol. 196: WOPPLOT 83. Parallel Processing: Logic, Organization and Technology. Proceedings, 1983. Edited by J. Becker and I. Eisele. V, 189 pages. 1984.

Vol. 197: Quarks and Nuclear Structure. Proceedings, 1983. Edited by K. Bleuler. VIII, 414 pages. 1984.

Vol. 198: Recent Progress in Many-Body Theories. Proceedings, 1983. Edited by H. Kümmel and M. L. Ristig. IX, 422 pages. 1984.

Vol. 199: Recent Developments in Nonequilibrium Thermodynamics. Proceedings, 1983. Edited by J. Casas-Vázquez, D. Jou and G. Lebon. XIII, 485 pages. 1984.

Vol. 200: H. D. Zeh, Die Physik der Zeitrichtung. V, 86 Seiten. 1984.

Vol. 201: Group Theoretical Methods in Physics. Proceedings, 1983. Edited by G. Denardo, G. Ghirardi and T. Weber. XXXVII, 518 pages. 1984.

Vol. 202: Asymptotic Behavior of Mass and Spacetime Geometry. Proceedings, 1983. Edited by F. J. Flaherty. VI, 213 pages. 1984.

Vol. 203: C. Marchioro, M. Pulvirenti, Vortex Methods in Two-Dimensional Fluid Dynamics. III, 137 pages. 1984.

Vol. 204: Y. Waseda, Novel Application of Anomalous (Resonance) X-Ray Scattering for Structural Characterization of Disordered Materials. VI, 183 pages. 1984.

Vol. 205: Solutions of Einstein's Equations: Techniques and Results. Proceedings, 1983. Edited by C. Hoenselaers and W. Dietz. VI, 439 pages. 1984.

Vol. 206: Static Critical Phenomena in Inhomogeneous Systems. Edited by A. Pękalski and J. Sznajd. Proceedings, 1984. VIII, 358 pages. 1984.

Vol. 207: S. W. Koch, Dynamics of First-Order Phase Transitions in Equilibrium and Nonequilibrium Systems. III, 148 pages. 1984.

Vol. 208: Supersymmetry and Supergravity/Nonperturbative QCD. Proceedings, 1984. Edited by P. Roy and V. Singh. V, 389 pages. 1984.

Vol. 209: Mathematical and Computational Methods in Nuclear Physics. Proceedings, 1983. Edited by J. S. Dehesa, J. M. G. Gomez and A. Polls. V, 276 pages. 1984.

Vol. 210: Cellular Structures in Instabilities. Proceedings, 1983. Edited by J. E. Wesfreid and S. Zaleski. VI, 389 pages. 1984.

Vol. 211: Resonances – Models and Phenomena. Proceedings, 1984. Edited by S. Albeverio, L. S. Ferreira and L. Streit. VI, 369 pages. 1984.

Vol. 212: Gravitation, Geometry and Relativistic Physics. Proceedings, 1984. Edited by Laboratoire "Gravitation et Cosmologie Relativistes", Université Pierre et Marie Curie et C.N.R.S., Institut Henri Poincaré, Paris. VI, 336 pages. 1984.

Vol. 213: Forward Electron Ejection in Ion Collisions. Proceedings, 1984. Edited by K. O. Groeneveld, W. Meckbach and I. A. Sellin. VII, 165 pages. 1984.

Lecture Notes in Physics

246

Field Theory, Quantum Gravity and Strings

Proceedings of a Seminar Series Held at DAPHE, Observatoire de Meudon, and LPTHE, Université Pierre et Marie Curie, Paris, Between October 1984 and October 1985

Edited by H. J. de Vega and N. Sánchez

Springer-Verlag Berlin Heidelberg GmbH

Editors

H. J. de Vega
Université Pierre et Marie Curie, L.P.T.H.E.
Tour 16, 1er étage, 4, place Jussieu, F-75230 Paris Cedex, France

N. Sánchez
Observatoire de Paris, Section d'Astrophysique de Meudon
5, place Jules Janssen, F-92195 Meudon Principal Cedex, France

ISBN 978-3-540-16452-4 ISBN 978-3-540-39789-2 (eBook)
DOI 10.1007/978-3-540-39789-2

Originally published by Springer-Verlag Berlin Heidelberg New York in 1986
Softcover reprint of the hardcover 1st edition 1986

2153/3140-543210

PREFACE

Perhaps the main challenge in theoretical physics today is the quantum unification of all interactions, including gravity. Such a unification is strongly suggested by the beautiful non-Abelian gauge theory of strong, electromagnetic and weak interactions, and, in addition, is required for a conceptual unification of general relativity and quantum theory.

The revival of interest in string theory since 1984 has arisen in this context. Superstring models appear to be candidates for the achievement of such unification. A consistent description of primordial cosmology ($t \lesssim t$ Planck) requires a quantum theory of gravity. Since a full quantum theory of gravity is not yet available, different types of approximations and models are used, in particular, the wave function of the Universe approach and semiclassical treatments of gravity. A nice possibility for a geometrical unification of gravity and gauge theories arises from higher-dimensional theories through dimensional reduction following Kaluza and Klein's proposal. Perturbative schemes are not sufficient to elucidate the physical content of different field theories of interest in different contexts. Exactly solvable theories can be helpful for understanding more realistic models; they can be important in four (or more) dimensions or else as models in the two-dimensional sheet of a string. In addition, the development of powerful methods for solving non-linear problems is of conceptual and practical importance.

A seminar series "Seminaires sur les équations non-linéaires en théorie des champs" intended to follow current developments in mathematical physics, particularly in the above-mentioned areas, was started in the Paris region in October 1983. The seminars take place alternately at DAPHE-Observatoire de Meudon and LPTHE-Université Pierre et Marie Curie (Paris VI), and they encourage regular meetings between theoretical physicists of different disciplines and a number of mathematicians. Participants come from Paris VI and VII, IHP, ENS, Collège de France, CPT-Marseille, DAPHE-Meudon, IHES and LPTHE-Orsay. The first volume "Non-Linear Equations in Classical and Quantum Field Theory", comprising the twenty-two lectures delivered in this series up to October 1984, has already been published by Springer-Verlag as **Lecture Notes in Physics**, Vol.226. The present volume "Field Theory, Quantum Gravity and Strings" accounts for the next twenty-two lectures delivered up to October 1985.

It is a pleasure to thank all the speakers for accepting our invitations and for their interesting contributions. We thank all the participants for their interest and for their stimulating discussions. We also thank M. Dubois-Violette at Orsay and J.L. Richard at Marseille, and B. Carter and B. Whiting at Meudon for their cooperation and encouragement. We acknowledge Mrs. C. Rosolen and Mrs. D. Lopes for their typing of part of these proceedings.

We particularly thank the Scientific Direction "Mathématiques-Physique de Base" of C.N.R.S. and the "Observatoire de Paris-Meudon" for the financial support which has made this series possible. We extend our appreciation to Springer-Verlag for their co-operation and efficiency in publishing these proceedings and hope that the possibility of making our seminars more widely available in this way will continue in the future.

Paris-Meudon
December 1985

H.J. de Vega
N. Sánchez

TABLE OF CONTENTS

(*) Lecture given by this author

Lectures on Quantum Cosmology

S. W. Hawking

Department of Applied Mathematics & Theoretical Physics,
Silver Street,
Cambridge CB3 9EW.

1. Introduction.

The aim of cosmology is to describe the Universe and to explain why it should be the way it is. For this purpose one constructs a mathematical model of the universe and a set of rules which relate elements of the model to observable quantities. This model normally consists of two parts:

[1] Local Laws which govern the physical fields in the model. In classical physics, these Laws are normally expressed as differential equations which can be derived from an action I. In quantum physics the Laws can be obtained from a path integral over all field configurations weighted with $\exp(iI)$.

[2] Boundary Conditions which pick out one particular state from among the set of those allowed by the Local Laws. The classical state can be specified by the boundary conditions for the differential equations at some initial time and the quantum state can be determined by the asymptotic conditions on the class C of field configurations that are summed over in the path integral.

Many people would say that the boundary conditions for the universe were not a question for science but for metaphysics or religion. However, in classical general relativity one cannot avoid the problem of boundary conditions because there are a number of theorems [1] which show that the universe must have started out with a spacetime singularity of infinite density and spacetime curvature. At this singularity all the Laws of physics would break down. Thus one could not predict how the universe would emerge from the Big Bang singularity but would have to impose it as a boundary condition. One can, however, interpret the singularity theorems in a different way: namely, that they indicate that the gravitational field was so strong in the very early universe that classical general relativity breaks down and that quantum gravitational effects have to be taken into account. There does not seem to be any necessity for singularities in quantum gravity and, as I shall show, one can avoid the problem of boundary conditions.

I shall adopt what is called the Euclidean approach to quantum gravity. In this one performs a path integral over Euclidean i.e. positive definite metrics rather than over metrics with Lorentzian signature (− + + +) and then analytically continues the result to the Lorentzian regime. The basic assumption of the Euclidean approach is that the "probability" of a positive definite 4-metric $g_{\mu\nu}$ and matter field configuration Φ is proportional to

$$\exp(-\tilde{I}[g_{\mu\nu},\Phi]) \tag{1.1}$$

where $\tilde{I} = -iI$ is the Euclidean action.

$$\tilde{I}[g_{\mu\nu},\Phi] = \frac{m_p^2}{16\pi}\left[-\int_{\partial M} 2Kh^{1/2}d^3x \right. \tag{1.2}$$

$$\left. -\int_M \left(R - 2\Lambda - \frac{16\pi}{m_p^2}L(g_{\mu\nu},\Phi)\right)g^{1/2}d^4x\right]$$

where h_{ij} is the 3-metric on the boundary ∂M and K is the trace of the second fundamental form of the boundary. The surface term in the action is necessary because the curvature scalar R contains second derivatives of the metric. The physics of the universe is governed by probabilities of the form above for all 4-metrics $g_{\mu\nu}$ and matter field configurations belonging to a certain class C. The specification of this class determines the quantum state of the universe.

There seem to be two and only two natural choices of the class C:

a) Compact Metrics

b) Non-compact metrics which are asymptotic to metrics of maximal symmetry, i.e. flat Euclidean space or Euclidean anti-de Sitter space

Boundary conditions of type b) define the usual vacuum state. In this state the expectation values of most quantities are defined to be zero so the vacuum state is not of as much interest as the quantum state of the universe. In particle scattering calculations one starts with the vacuum state and one changes the state by creating particles by the action of field operators at infinity in the infinite past. One lets the particles interact and then annihilates the resultant particles by the action of other field operators at future infinity. This gets one back to the vacuum state. If one supposed that the quantum state of the universe was some such particle scattering state, one one would loose all ability to predict the state of the universe because one would have no idea what was coming in. One would also expect that the matter in the universe would become concentrated in a certain region and that it would decrease to zero at large distances instead of the roughly homogeneous universe that we observe.

In particle scattering problems, one is interested in observables at infinity. One is therefore concerned only with metrics which are connected to infinity: any disconnected compact parts of the metric would not contribute to the scattering of particles from infinity. In cosmology, on the other hand, one is concerned with observables in a finite region in the middle of the space and it does not matter whether this region is connected to an infinite asymptotic region. Suppose that the class C which defines the quantum state of the universe consists of metrics of type b). The expectation value of an observable in a finite region will be given by a path integral which contains contributions from two kinds of metric.

i) Connected asymptotically Euclidean or anti-de Sitter metrics

ii) Disconnected metrics which consist of a compact part which contains the region of observation and an asymptotically Euclidean or anti-de Sitter part

One cannot exclude disconnected metrics from the path integral because they can be approximated by connected metrics in which the different parts were joined by thin tubes. The tubes could be chosen to have negligible action. Similarly, topologically non-trivial metrics cannot be excluded because they can be approximated by topologically trivial metrics. It turns out that the dominant contribution to the path integral comes from disconnected metrics of the second kind. Thus, as far as observations in a finite region are concerned, the result of choosing the class C that defines the quantum state to be non-compact metrics of type b) would be almost the same as choosing it to be compact metrics of type a). It would therefore seem more natural to choose C to be the class of all compact non-singular metrics. This would mean that the universe would be completely self-contained without any singularities at which the laws of physics break down and without any edges at which boundary conditions would have to be set. It should be emphasised, however, that this is only a proposal for the quantum state of the universe. One cannot derive it from some other principle but merely show that it is a natural choice. The ultimate test is not whether it is aesthetically appealing but whether it enables one to make predictions that agree with observations. I shall endeavour to do this for a simple model.

2. The Wavefunction of the Universe

In practice, one is normally interested in the probability, not of the entire 4-metric, but of a more restricted set of observables. Such a probability can be derived from the basic probability (1.1) by integrating over the unobserved quantities. A particularly important case is the probability $P[h_{ij},\Phi_0]$ of finding a closed compact 3-submanifold S which divides the 4-manifold M into two parts $M_\pm$ and on which the induced 3-metric is h_{ij} and the matter field configuration is Φ_0 is

$$P[h_{ij},\Phi_0] = \int d[g_{\mu\nu}]d[\Phi]\exp(-\tilde{I}[g_{\mu\nu},\Phi]) \tag{2.1}$$

where the integral is taken over all 4-metrics and matter field configurations belonging to the class C which contain the submanifold S on which the induced 3-metric is h_{ij} and the matter field configuration is Φ_0. This probability can be factorized into the product of two amplitudes or wave functions $\Psi_\pm[h_{ij},\Phi_0]$. $P[h_{ij},\Phi_0] = \Psi_+[h_{ij},\Phi_0]\Psi_-[h_{ij},\Phi_0]$ where

$$\Psi_\pm[h_{ij},\Phi_0] = \int_{C_\pm} d[g_{\mu\nu}]d[\Phi]\exp(-\tilde{I}[g_{\mu\nu},\Phi]) \tag{2.2}$$

The path integral is over the classes $C_\pm$ of metrics on the compact manifolds $M_\pm$ with boundary S. With the choice of compact metrics for C, $\Psi_+ = \Psi_-$ and both are real. I shall therefore drop the subscripts + and - and refer to Ψ as the "Wavefunction of the Universe".

In a neighbourhood of S in M, one can introduce a time coordinate t, which is zero on S, and three space coordinates x^i and one can write the metric in the 3 + 1 form

$$ds^2 = -(N^2 - N_iN^i)dt^2 + 2N_idx^idt + h_{ij}dx^idx^j \tag{2.3}$$

A Lorentzian metric corresponds to the lapse N being real and a Euclidean metric corresponds to N negative imaginary. The shift vector N_i is real in both cases. In the Lorentzian case the classical action is

$$I = \int (L_g + L_m) d^3x dt \tag{2.4}$$

where

$$L_g = \frac{m_p^2}{16\pi} N (G^{ijkl} K_{ij} K_{kl} + h^{1/2}\, {}^3R) \tag{2.5}$$

$$K_{ij} = \frac{1}{2N}\left[- \frac{\partial h_{ij}}{\partial t} + 2N_{(i|j)} \right] \tag{2.6}$$

is the second fundamental form of S and

$$G^{ijkl} = \tfrac{1}{2} h^{1/2} (h^{ik} h^{jl} + h^{il} h^{jk} - 2h^{ij} h^{kl}) \tag{2.7}$$

In the case of a massive scalar field Φ

$$L_m = \tfrac{1}{2} N h^{1/2} \left[N^{-2} \left(\frac{\partial \Phi}{\partial t} \right)^2 - \frac{2N^i}{N^2} \frac{\partial \Phi}{\partial t} \frac{\partial \Phi}{\partial x^i} - \left(h^{ij} - \frac{N^i N^j}{N^2} \right) \frac{\partial \Phi}{\partial x^i} \frac{\partial \Phi}{\partial x^j} - m^2 \Phi^2 \right] \tag{2.8}$$

In the Hamiltonian treatment of General Relativity one regards the components h_{ij} of the 3-metric and the field Φ as the canonical coordinates. The canonically conjugate momenta are

$$\pi^{ij} = \frac{\partial L_g}{\partial \dot{h}_{ij}} = - \frac{h^{1/2} m_p^2}{16\pi} (K^{ij} - h^{ij} K) \tag{2.9}$$

$$\pi_\Phi = \frac{\partial L_m}{\partial \dot{\Phi}} = N^{-1} h^{1/2} \left[\dot{\Phi} - N^i \frac{\partial \Phi}{\partial x^i} \right] \tag{2.10}$$

The Hamiltonian is

$$H = \int(\pi^{ij}\dot{h}_{ij} + \pi_\Phi \dot{\Phi} - L_g - L_m)d^3x \tag{2.11}$$

$$= \int(NH_0 + N_i H^i)d^3x$$

where

$$H_0 = 16\pi m_p^{-2} G_{ijkl}\pi^{ij}\pi^{kl} - \frac{m_p^2}{16\pi}h^{1/2}\,{}^3R \tag{2.12}$$

$$+ \tfrac{1}{2}h^{1/2}\left[\frac{\pi_\Phi^2}{h} + h^{ij}\frac{\partial\Phi}{\partial x^i}\frac{\partial\Phi}{\partial x^j} + m^2\Phi^2\right]$$

$$H^i = -2\pi^{ij}{}_{|j} + h^{ij}\frac{\partial\Phi}{\partial x^j}\pi_\Phi \tag{2.13}$$

and

$$G_{ijkl} = \tfrac{1}{2}h^{-1/2}(h_{ik}h_{jl} + h_{il}h_{jk} - h_{ij}h_{kl}) \tag{2.14}$$

From its path integral definition, the wavefunction Ψ is a function only of the 3-metric h_{ij} and the matter field configuration Φ_0 on S but it is not a function of t, which is merely a coordinate that can be given any value. It therefore follows that Ψ will be unchanged if the surface S is displaced a distance N along the normals and shifted an amount N^i along itself. The change in Ψ under that displacement will be the quantum Hamiltonian operator acting on Ψ. Thus Ψ will obey the zero energy Schroedinger equation.

$$H\Psi = 0 \tag{2.15}$$

where the Hamiltonian operator is obtained from the classical Hamiltonian by the replacements

$$\pi^{ij}(x) \to -i\frac{\delta}{\delta h_{ij}(x)}, \qquad \pi_{\phi}(x) \to -i\frac{\delta}{\delta\Phi(x)} \tag{2.16}$$

3 Quantization

The wavefunction Ψ can be regarded as a function on the infinite dimensional manifold W of all 3-metrics h_{ij} and matter fields Φ on S. A tangent vector to W is a pair of fields (γ_{ij},μ) on S where γ_{ij} can be regarded as a small change of the metric h_{ij} and μ can be regarded as a small change of Φ. For each choice of N on S there is a natural metric $\Gamma(N)$ on W [2].

$$ds^2 = \int N^{-1}\left[\frac{m_p^2}{32\pi}G^{ijkl}\gamma_{ij}\gamma_{kl} + \tfrac{1}{2}h^{1/2}\mu^2\right]d^3x \tag{3.1}$$

The zero energy Schrodinger equation

$$H\Psi = 0 \tag{3.2}$$

can be decomposed into the momentum constraint

$$\underline{H}\,\Psi = \int N_i H^i d^3x\,\Psi \tag{3.3}$$

$$= \int h^{1/2}N_i\left[2\left[\frac{\delta}{\delta h_{ij}(x)}\right]_{|j} - h^{ij}\frac{\partial\Phi}{\partial x^j}\frac{\delta}{\delta\Phi(x)}\right]d^3x\,\Psi = 0$$

This implies that Ψ is the same on 3-metrics and matter field configurations that are related by coordinate transformations in S. The other part of the Schroedinger equation, corresponding to

$$H_I \Psi = 0 \tag{3.4}$$

where $H_I = \int N H_0 d^3x$ is called the Wheeler-DeWitt equation. There is one Wheeler-DeWitt equation for each choice of N on S. One can regard them as a system of second order partial differential equations for Ψ on W. There is some ambiguity in the choice of operator ordering in these equations but this will not affect the results of this paper. We shall assume that H_I has the form[2]

$$(- \tfrac{1}{2}\nabla^2 + \xi \mathbb{R} + V)\Psi = 0 \tag{3.5}$$

where ∇^2 is the Laplacian in the metric $\Gamma(N)$, $\mathbb{R}$ is the curvature scalar of this metric and the potential V is

$$V = \int h^{1/2} N \left[- \frac{m_p^2}{16\pi} \, {}^3R + \epsilon + U \right] d^3x \tag{3.6}$$

where $U = T^{00} - \tfrac{1}{2}\pi_\Phi^2$. The constant ϵ can be regarded as a renormalization of the cosmological constant Λ. We shall assume that the renormalized Λ is zero. We shall also assume that the coefficient ξ of the scalar curvature $\mathbb{R}$ of W is zero.

Any wavefunction Ψ which satisfies the momentum constraint and the Wheeler-DeWitt equation for each choice of N and N_i on S describes a possible quantum state of the Universe. We shall be concerned with the particular solution which represents the quantum state defined by a path integral over compact 4-metrics without boundary. In this case

$$\Psi = \int d[g_{\mu\nu}] d[\Phi] \exp(- \tilde{I}(g_{\mu\nu}, \Phi)) \tag{3.7}$$

where $\tilde{I}$ is the Euclidean action obtained by setting N negative imaginary. One can regard (3.7) as a boundary condition on the Wheeler-DeWitt equations. It implies that Ψ tends to a constant, which can be normalized to one, as h_{ij} goes to zero.

4 Unperturbed Friedman Model

References [3,4,5] considered the Minisuperspace model which consisted of a Friedman model with metric

$$ds^2 = \sigma^2(-N^2dt^2 + a^2d\Omega_3^2) \tag{4.1}$$

where $d\Omega_3^2$ is the metric of the unit 3-sphere. The normalization factor $\sigma^2 = \frac{2}{3\pi m_p^2}$ has been included for convenience. The model contains a scalar field $(2^{1/2}\pi\sigma)^{-1}\phi$ with mass $\sigma^{-1}m$ which is constant on surfaces of constant t. One can easily generalize this to the case of a scalar field with a potential $V(\phi)$. Such generalizations include models with higher derivative quantum corrections [6]. The action is

$$I = -\tfrac{1}{2}\int dtNa^3\left[\frac{1}{N^2a^2}\left(\frac{da}{dt}\right)^2 - \frac{1}{a^2} - \frac{1}{N^2}\left(\frac{d\phi}{dt}\right)^2 + m^2\phi^2\right] \tag{4.2}$$

The classical Hamiltonian is

$$H = \tfrac{1}{2}N(-a^{-1}\pi_a^2 + a^{-3}\pi_\phi^2 - a + a^3m^2\phi^2) \tag{4.3}$$

where

$$\pi_a = -\frac{ada}{Ndt} \qquad \pi_\phi = \frac{a^3d\phi}{Ndt} \tag{4.4}$$

The classical Hamiltonian constraint is $H = 0$. The classical field equations are

$$N\frac{d}{dt}\left(\frac{1}{N}\frac{d\phi}{dt}\right) + \frac{3}{a}\frac{da}{dt}\frac{d\phi}{dt} + N^2m^2\phi = 0 \tag{4.5}$$

$$N\frac{d}{dt}\left(\frac{1}{N}\frac{da}{dt}\right) = N^2am^2\phi^2 - 2a\left(\frac{d\phi}{dt}\right)^2 \tag{4.6}$$

The Wheeler-DeWitt equation is

$$\tfrac{1}{2}Ne^{-3\alpha}\left[\frac{\partial^2}{\partial\alpha^2} - \frac{\partial^2}{\partial\phi^2} + 2V\right]\Psi(\alpha,\phi) = 0 \tag{4.7}$$

where

$$V = \tfrac{1}{2}(e^{6\alpha}m^2\phi^2 - e^{4\alpha}) \tag{4.8}$$

and $\alpha = \ln a$. One can regard equation (4.7) as a hyperbolic equation for Ψ in the flat space with coordinates (α,ϕ) with α as the time coordinate. The boundary condition that gives the quantum state defined by a path integral over compact 4-metrics is $\Psi \to 1$ as $\alpha \to -\infty$. If one integrates equation (4.7) with this boundary condition, one finds that the wavefunction starts oscillating in the region $V > 0$, $|\phi| > 1$ (this has been confirmed numerically [5]). One can interpret the oscillatory component of the wavefunction by the WKB approximation:

$$\Psi = \mathrm{Re}\,(\,C\,e^{iS}\,) \tag{4.9}$$

where C is a slowly varying amplitude and S is a rapidly varying phase. One chooses S to satisfy the classical Hamilton-Jacobi equation:

$$H(\pi_\alpha,\pi_\phi,\alpha,\phi) = 0 \tag{4.10}$$

where

$$\pi_\alpha = \frac{\partial S}{\partial\alpha}, \quad \pi_\phi = \frac{\partial S}{\partial\phi} \tag{4.11}$$

One can write (4.10) in the form

$$\tfrac{1}{2}f^{ab}\frac{\partial S}{\partial q^a}\frac{\partial S}{\partial q^b} + e^{-3\alpha}V = 0 \tag{4.12}$$

where f^{ab} is the inverse to the metric $\Gamma(1)$:

$$f^{ab} = e^{-3\alpha}\mathrm{diag}(\,-1,1) \tag{4.13}$$

The wavefunction (4.9) will then satisfy the Wheeler-DeWitt equation if

$$\nabla^2 C + 2if^{ab}\frac{\partial C}{\partial q^a}\frac{\partial S}{\partial q^b} + iC\nabla^2 S = 0 \tag{4.14}$$

where ∇^2 is the Laplacian in the metric f_{ab}. One can ignore the first term in equation (4.14) and can integrate the equation along the trajectories of the vector field $X^a = \frac{dq^a}{dt} = f^{ab}\frac{\partial S}{\partial q^b}$ and so determine the amplitude C. These trajectories correspond to classical solutions of the field equations. They are parameterized by the coordinate time t of the classical solutions.

The solutions that correspond to the oscillating part of the wavefunction of the Minisuperspace model start out at $V = 0$, $|\phi| > 1$ with $\frac{d\alpha}{dt} = \frac{d\phi}{dt} = 0$. They expand exponentially with

$$S = -\tfrac{1}{3}e^{3\alpha}m|\phi|(1 - m^{-2}e^{-2\alpha}\phi^{-2}) \sim -\tfrac{1}{3}e^{3\alpha}m|\phi| \tag{4.15}$$

$$\frac{d\alpha}{dt} = m|\phi|, \qquad \frac{d|\phi|}{dt} = -\tfrac{1}{3}m \tag{4.16}$$

After a time of order $3m^{-1}(|\phi_1| - 1)$, where ϕ_1 is the initial value of ϕ, the field ϕ starts to oscillate with frequency m. The solution then becomes matter dominated and expands with e^{α} proportional to $t^{2/3}$. If there were other fields present, the massive scalar particles would decay into light particles and then the solution would expand with e^{α} proportional to $t^{1/2}$. Eventually the solution would reach a maximum radius of order $\exp(\frac{9\phi_1^2}{2})$ or $\exp(9\phi_1^2)$ depending on whether it is radiation or matter dominated for most of the expansion. The solution would then recollapse in a similar manner.

5 The Perturbed Friedman Model

We assume that the metric is of the form (2.3) except the right hand side has been multiplied by a normalization factor σ^2. The 3-metric h_{ij} has the form

$$h_{ij} = a^2(\Omega_{ij} + \epsilon_{ij}) \tag{5.1}$$

where Ω_{ij} is the metric on the unit 3-sphere and ϵ_{ij} is a perturbation on this metric and may be expanded in harmonics:

$$\epsilon_{ij} = \sum_{n,\ell,m} \Big[6^{1/2}\, a_{n\ell m}\, \tfrac{1}{3}\Omega_{ij} Q^n_{\ell m} + 6^{1/2}\, b_{n\ell m}\, (P_{ij})^n_{\ell m} + 2^{1/2}\, c^o_{n\ell m}\, (S^o_{ij})^n_{\ell m}$$

$$+ 2^{1/2}\, c^e_{n\ell m}\, (S^e_{ij})^n_{\ell m} + 2\, d^o_{n\ell m}\, (G^o_{ij})^n_{\ell m} + 2\, d^e_{n\ell m}\, (G^e_{ij})^n_{\ell m}\Big] \tag{5.2}$$

The coefficients $a_{n\ell m}, b_{n\ell m}, c^o_{n\ell m}, c^e_{n\ell m}, d^o_{n\ell m}, d^e_{n\ell m}$ are functions of the time coordinate t but not the three spatial coordinates x^i.

The $Q(x^i)$ are the standard scalar harmonics on the 3-sphere. The $P_{ij}(x^i)$ are given by (suppressing all but the i,j indices)

$$P_{ij} = \frac{1}{(n^2 - 1)}\, Q_{|ij} + \tfrac{1}{3}\Omega_{ij} Q \tag{5.3}$$

They are traceless, $P_i{}^i = 0$. The S_{ij} are defined by

$$S_{ij} = S_{i|j} + S_{j|i} \tag{5.4}$$

where S_i are the transverse vector harmonics, $S_i{}^{|i} = 0$. The G_{ij} are the transverse traceless tensor harmonics, $G_i{}^i = G_{ij}{}^{|j} = 0$. Further details about the harmonics and their normalization can be found in appendix A.

The lapse, shift and the scalar field $\Phi(x^i,t)$ can be expanded in terms of harmonics:

$$N = N_0\left[1 + 6^{-1/2} \sum_{n,\ell,m} g_{n\ell m} Q^n_{\ell m}\right] \qquad (5.5)$$

$$N_i = e^\alpha \sum_{n,\ell,m}\left[6^{-1/2} k_{n\ell m} (P_i)^n_{\ell m} + 2^{1/2} j_{n\ell m} (S_i)^n_{\ell m}\right] \qquad (5.6)$$

$$\Phi = \sigma^{-1}\left[\frac{1}{2^{1/2}\pi}\phi(t) + \sum_{n,\ell,m} f_{n\ell m} Q^n_{\ell m}\right] \qquad (5.7)$$

where $P_i = \frac{1}{(n^2-1)} Q_{|i}$. Hereafter, the labels n,ℓ,m,o and e will be denoted simply by n. One can then expand the action to all orders in terms of the "background" quantities a,ϕ,N_0 but only to second order in the "perturbations" $a_n,b_n,c_n,d_n,f_n,g_n,k_n,j_n$:

$$I = I_0(a,\phi,N_0) + \sum_n I_n \qquad (5.8)$$

where I_0 is the action of the unperturbed model (4.2) and I_n is quadratic in the perturbations and is given in appendix B.

One can define conjugate momenta in the usual manner. They are:

$$\pi_\alpha = -N_0^{-1}e^{3\alpha}\dot{\alpha} + \text{quadratic terms} \qquad (5.9)$$

$$\pi_\phi = N_0^{-1}e^{3\alpha}\dot{\phi} + \text{quadratic terms} \qquad (5.10)$$

$$\pi_{a_n} = -N_0^{-1}e^{3\alpha}\left[\dot{a}_n + \dot{\alpha}(a_n - g_n) + \tfrac{1}{3}e^{-\alpha}k_n\right] \qquad (5.11)$$

$$\pi_{b_n} = N_0^{-1}e^{3\alpha}\frac{(n^2-4)}{(n^2-1)}\left[\dot{b}_n + 4\dot{\alpha}b_n - \tfrac{1}{3}e^{-\alpha}k_n\right] \qquad (5.12)$$

$$\pi_{c_n} = N_0^{-1} e^{3\alpha} (n^2 - 4) \left[\dot{c}_n + 4\dot{\alpha} c_n - e^{-\alpha} j_n\right] \tag{5.13}$$

$$\pi_{d_n} = N_0^{-1} e^{3\alpha} \left[\dot{d}_n + 4\dot{\alpha} d_n\right] \tag{5.14}$$

$$\pi_{f_n} = N_0^{-1} e^{3\alpha} \left[\dot{f}_n + \dot{\phi}(3a_n - g_n)\right] \tag{5.15}$$

The quadratic terms in equations (5.9) and (5.10) are given in appendix B. The Hamiltonian can then be expressed in terms of these momenta and the other quantities:

$$H = N_0\left[H_{|0} + \sum_n H^n_{|2} + \sum_n g_n H^n_{|1}\right] + \sum_n \left[k_n\, {}^S H^n_{\_1} + j_n\, {}^V H^n_{\_1}\right] \tag{5.16}$$

The subscripts 0,1,2 on the $H_|$ and $H_\_$ denote the orders of the quantities in the perturbations and S and V denote the scalar and vector parts of the shift part of the Hamiltonian. $H_{|0}$ is the Hamiltonian of the unperturbed model with $N = 1$:

$$H_{|0} = \tfrac{1}{2} e^{-3\alpha}\left[-\pi_\alpha^2 + \pi_\phi^2 + e^{6\alpha} m^2 \phi^2 - e^{4\alpha}\right] \tag{5.17}$$

The second order Hamiltonian is given by $H_{|2} = \sum_n H^n_{|2} = \sum_n ({}^S H^n_{|2} + {}^V H^n_{|2} + {}^T H^n_{|2})$ where

$$\begin{aligned}
{}^S H^n_{|2} = \tfrac{1}{2} e^{-3\alpha}\Bigg[&\left[\tfrac{1}{2} a_n^2 + \frac{10(n^2-4)}{(n^2-1)} b_n^2\right]\pi_\alpha^2 + \left[\frac{15}{2} a_n^2 + \frac{6(n^2-4)}{(n^2-1)} b_n^2\right]\pi_\phi^2 \\
&- \pi_{a_n}^2 + \frac{(n^2-1)}{(n^2-4)} \pi_{b_n}^2 + \pi_{f_n}^2 + 2a_n \pi_{a_n} \pi_\alpha + 8b_n \pi_{b_n} \pi_\alpha - 6a_n \pi_{f_n} \pi_\phi \\
&- e^{4\alpha}\left[\tfrac{1}{3}(n^2 - \tfrac{5}{2}) a_n^2 + \frac{(n^2-7)}{3} \frac{(n^2-4)}{(n^2-1)} b_n^2 + \tfrac{2}{3}(n^2-4)\, a_n b_n - (n^2-1) f_n^2\right]
\end{aligned}$$

$$+ e^{6\alpha}m^2\left[f_n^2 + 6a_n f_n\phi\right] + e^{6\alpha}m^2\phi^2\left[\frac{3}{2}a_n^2 - \frac{6(n^2-4)}{(n^2-1)}b_n^2\right]\Bigg] \quad (5.18)$$

$${}^{V}H^n_{|2} = \tfrac{1}{2}e^{-3\alpha}\Bigg[(n^2-4)c_n^2\left[10\pi_\alpha^2 + 6\pi_\phi^2\right] + \frac{1}{(n^2-4)}\pi_{c_n}^2 + 8c_n\pi_{c_n}\pi_\alpha +$$

$$(n^2-4)c_n^2\left[2e^{4\alpha} - 6e^{6\alpha}m^2\phi^2\right]\Bigg] \quad (5.19)$$

$${}^{T}H^n_{|2} = \tfrac{1}{2}e^{-3\alpha}\Bigg[d_n^2\left[10\pi_\alpha^2 + 6\pi_\phi^2\right] + \pi_{d_n}^2 + 8d_n\pi_{d_n}\pi_\alpha +$$

$$d_n^2\left[(n^2+1)e^{4\alpha} - 6e^{6\alpha}m^2\phi^2\right]\Bigg] \quad (5.20)$$

The first order Hamiltonians are

$$H^n_{|1} = \tfrac{1}{2}e^{-3\alpha}\Bigg[- a_n\left[\pi_\alpha^2 + 3\pi_\phi^2\right] + 2\left[\pi_\phi\pi_{f_n} - \pi_\alpha\pi_{a_n}\right]$$

$$+ m^2e^{6\alpha}\left[2f_n\phi + 3a_n\phi^2\right] - \tfrac{2}{3}e^{4\alpha}\left[(n^2-4)b_n + (n^2+\tfrac{1}{2})a_n\right]\Bigg] \quad (5.21)$$

The shift parts of the Hamiltonian are

$${}^{S}H^n_{\_1} = \tfrac{1}{3}e^{-3\alpha}\left[- \pi_{a_n} + \pi_{b_n} + \left[a_n + \frac{4(n^2-4)}{(n^2-1)}b_n\right]\pi_\alpha + 3f_n\pi_\phi\right] \quad (5.22)$$

$${}^{V}H^n_{\_1} = e^{-\alpha}\left[\pi_{c_n} + 4(n^2-4)\, c_n\pi_\alpha\right] \quad (5.23)$$

The classical field equations are given in appendix B.

Because the Lagrange multipliers N_0, g_n, k_n, j_n are independent, the zero energy Schroedinger equation

$$H\Psi = 0 \quad (5.24)$$

can be decomposed as before into momentum constraints and Wheeler-DeWitt equations. As the momentum constraints are linear in the momenta, there is no ambiguity in the operator ordering. One therefore has

$$ {}^{S}H^{n}_{\_1}\Psi = -\frac{1}{3}e^{-3\alpha}\left[\frac{\partial}{\partial a_n} - \left(a_n + \frac{4(n^2-4)}{(n^2-1)}b_n\right)\frac{\partial}{\partial\alpha}\right. $$

$$ \left. - \frac{\partial}{\partial b_n} - 3f_n\frac{\partial}{\partial\phi}\right]\Psi = 0 \qquad (5.25) $$

$$ {}^{V}H^{n}_{\_1}\Psi = e^{-\alpha}\left[\frac{\partial}{\partial c_n} + 4(n^2-4)\,c_n\frac{\partial}{\partial\alpha}\right]\Psi = 0 \qquad (5.26) $$

The first order Hamiltonians $H^{n}_{|1}$ give a series of finite dimensional second order differential equations, one for each n. In the order of approximation that we are using, the ambiguity in the operator ordering will consist of the possible addition of terms linear in $\frac{\partial}{\partial\alpha}$. The effect of such terms can be compensated for by multiplying the wavefunction by powers of e^{α}. This will not affect the relative probabilities of different observations at a given value of α. We shall therefore ignore such ambiguities and terms.

$$ \tfrac{1}{2}e^{-3\alpha}\left[a_n\left(\frac{\partial^2}{\partial\alpha^2} + 3\frac{\partial^2}{\partial\phi^2}\right) - 2\left(\frac{\partial}{\partial f_n}\frac{\partial}{\partial\phi} - \frac{\partial}{\partial a_n}\frac{\partial}{\partial\alpha}\right)\right. $$

$$ \left. + m^2e^{6\alpha}\left[2\phi f_n + 3a_n\phi^2\right] - \frac{2}{3}e^{4\alpha}\left[(n^2-4)b_n + (n^2+\tfrac{1}{2})a_n\right]\right]\Psi = 0 \qquad (5.27) $$

Finally, one has an infinite dimensional second order differential equation

$$ \left[H_{|0} + \sum_n\left({}^{S}H^{n}_{|2} + {}^{V}H^{n}_{|2} + {}^{T}H^{n}_{|2}\right)\right]\Psi = 0 \qquad (5.28) $$

where $H_{|0}$ is the operator in the Wheeler-DeWitt equation of the unperturbed

Friedman Minisuperspace model:

$$H_{|0} = \tfrac{1}{2} e^{-3\alpha}\left[\frac{\partial^2}{\partial\alpha^2} - \frac{\partial^2}{\partial\phi^2} + e^{6\alpha} m^2\phi^2 - e^{4\alpha}\right] \tag{5.29}$$

and

$$\begin{aligned}{}^{S}H^{n}_{|2} = \tfrac{1}{2} e^{-3\alpha}\Bigg[&- \left[\tfrac{1}{2}a_n^2 + \frac{10(n^2-4)}{(n^2-1)} b_n^2\right]\frac{\partial^2}{\partial\alpha^2} - \left[\frac{15}{2}a_n^2 + \frac{6(n^2-4)}{(n^2-1)} b_n^2\right]\frac{\partial^2}{\partial\phi^2} \\ &+ \frac{\partial^2}{\partial a_n^2} - \frac{(n^2-1)}{(n^2-4)}\frac{\partial^2}{\partial b_n^2} - \frac{\partial^2}{\partial f_n^2} - 2a_n\frac{\partial}{\partial a_n}\frac{\partial}{\partial\alpha} - 8b_n\frac{\partial}{\partial b_n}\frac{\partial}{\partial\alpha} + 6a_n\frac{\partial}{\partial f_n}\frac{\partial}{\partial\phi} \\ &- e^{4\alpha}\left[\frac{1}{3}\left(n^2-\frac{5}{2}\right)a_n^2 + \frac{(n^2-7)}{3}\frac{(n^2-4)}{(n^2-1)} b_n^2 + \frac{2}{3}(n^2-4)\, a_n b_n - (n^2-1)f_n^2\right] \\ &+ e^{6\alpha}m^2\left[f_n^2 + 6a_n f_n\phi\right] + e^{6\alpha}m^2\phi^2\left[\frac{3}{2}a_n^2 - \frac{6(n^2-4)}{(n^2-1)}b_n^2\right]\Bigg]\end{aligned} \tag{5.30}$$

$$\begin{aligned}{}^{V}H^{n}_{|2} = \tfrac{1}{2} e^{-3\alpha}\Bigg[&- (n^2-4)c_n^2\left[10\frac{\partial^2}{\partial\alpha^2} + 6\frac{\partial^2}{\partial\phi^2}\right] - \frac{1}{(n^2-4)}\frac{\partial^2}{\partial c_n^2} - 8c_n\frac{\partial}{\partial c_n}\frac{\partial}{\partial\alpha} + \\ &(n^2-4)c_n^2\left[2e^{4\alpha} - 6e^{6\alpha}m^2\phi^2\right]\Bigg]\end{aligned} \tag{5.31}$$

$$\begin{aligned}{}^{T}H^{n}_{|2} = \tfrac{1}{2} e^{-3\alpha}\Bigg[&- d_n^2\left[10\frac{\partial^2}{\partial\alpha^2} + 6\frac{\partial^2}{\partial\phi^2}\right] - \frac{\partial^2}{\partial d_n^2} - 8d_n\frac{\partial}{\partial d_n}\frac{\partial}{\partial\alpha} + \\ &d_n^2\left[(n^2+1)e^{4\alpha} - 6e^{6\alpha}m^2\phi^2\right]\Bigg]\end{aligned} \tag{5.32}$$

We shall call equation (5.28) the master equation. It is not hyperbolic because, as well as the positive second derivatives $\frac{\partial^2}{\partial\alpha^2}$ in $H_{|0}$, there are the positive second derivatives $\frac{\partial^2}{\partial a_n^2}$ in each ${}^S H^n_{|2}$. However, one can use the momentum constraint (5.25) to substitute for the partial derivatives with respect to a_n and then solve the resultant differential equation on $a_n = 0$. Similarly, one can use the momentum constraint (5.26) to substitute for the partial derivatives with respect to c_n and then solve on $c_n = 0$. One thus obtains a modified equation which is hyperbolic for small f_n. If one knows the wavefunction on $a_n = 0 = c_n$, one can use the momentum constraints to calculate the wavefunction at other values of a_n and c_n.

6 The Wavefunction

Because the perturbation modes are not coupled to each other, the wavefunction can be expressed as a sum of terms of the form

$$\Psi = \mathrm{Re}\,(\ \Psi_0(\alpha,\phi)\prod_n \Psi^{(n)}(\alpha,\phi,a_n,b_n,c_n,d_n,f_n)) \tag{6.1}$$

$$= \mathrm{Re}\,(\ Ce^{iS})$$

where S is a rapidly varying function of α and ϕ and C is a slowly varying function of all the variables. If one substitutes (6.1) into the master equation and divides by Ψ, one obtains

$$-\frac{\nabla_2^2\Psi_0}{2\Psi_0} - \sum_n \frac{\nabla_2^2\Psi^{(n)}}{2\Psi^{(n)}} - \sum_{n\leq m}\frac{(\nabla_2\Psi^{(n)}).(\nabla_2\Psi^{(m)})}{2\Psi^{(n)}\Psi^{(m)}} - \frac{(\nabla_2\Psi_0)}{\Psi_0}.\left[\sum_n \frac{\nabla_2\Psi^{(n)}}{\Psi^{(n)}}\right] \tag{6.2}$$

$$+\sum_n \frac{H^n_{|2}\Psi}{\Psi} + e^{-3\alpha}\,V(\alpha,\phi) = 0$$

where ∇_2^2 is the Laplacian in the Minisuperspace metric $f_{ab} = e^{3\alpha}\text{diag}(-1,1)$ and the dot product is with respect to this metric.

An individual perturbation mode does not contribute a significant fraction of the sums in the third and fourth terms in equation (6.2). Thus these terms can be replaced by

$$-\frac{(\nabla_2\Psi)}{\Psi}.\sum_n\frac{(\nabla_2\Psi^{(n)})}{\Psi^{(n)}} + \tfrac{1}{2}\left[\sum_n\frac{\nabla_2\Psi^{(n)}}{\Psi^{(n)}}\right]^2 \tag{6.3}$$

$$\approx \; -\,i(\nabla_2 S).\sum_n\frac{(\nabla_2\Psi^{(n)})}{\Psi^{(n)}} + \tfrac{1}{2}\left[\sum_n\frac{\nabla_2\Psi^{(n)}}{\Psi^{(n)}}\right]^2$$

In order that the ansatz (6.1) be valid, the terms in (6.2) that depend on a_n, b_n, c_n, d_n, f_n have to cancel out. This implies

$$\frac{(\nabla_2\Psi)}{\Psi}.(\nabla_2\Psi^{(n)}) + \tfrac{1}{2}\nabla_2^2\Psi^{(n)} = \frac{H^n_{|2}\Psi}{\Psi}\Psi^{(n)} \tag{6.4}$$

$$(-\tfrac{1}{2}\nabla_2^2 + e^{-3\alpha}\,V + \tfrac{1}{2}J.J)\Psi_0 = 0 \tag{6.5}$$

where $J = \sum_n\frac{\nabla_2\Psi^{(n)}}{\Psi^{(n)}}$

In regions in which the phase S is a rapidly varying function of α and ϕ, one can neglect the second term in (6.4) in comparison with the first term. One can also replace the π_α and π_ϕ which appear in $H^n_{|2}$ by $\frac{\partial S}{\partial\alpha}$ and $\frac{\partial S}{\partial\phi}$ respectively. The vector $X^a = f^{ab}\frac{\partial S}{\partial q^b}$ obtained by raising the covector $\nabla_2 S$ by the inverse minisuperspace metric f^{ab} can be regarded as $\frac{\partial}{\partial t}$ where t is the time parameter of the classical Friedman metric that corresponds to Ψ by the WKB approximation. One then obtains a time dependent Schroedinger equation for each mode along a trajectory of the vector field X^a:

$$i\frac{\partial\Psi^{(n)}}{\partial t} = H^n_{|2}\Psi^{(n)} \tag{6.6}$$

Equation (6.5) can be interpreted as the Wheeler–DeWitt equation for a two dimensional minisuperspace model with an extra term ½J.J arising from the perturbations. In order to make J finite, one will have to make subtractions. Subtracting out the ground state energies of the $H^n_{|2}$ corresponds to a renormalization of the cosmological constant Λ. There is a second subtraction which corresponds to a renormalization of the Planck mass m_p and a third one which corresponds to a curvature squared counterterm. The effect of such higher derivative terms in the action has been considered elsewhere [6].

One can write Ψ^n as

$$\Psi_n = {}^S\Psi^{(n)}(\alpha,\phi,a_n,b_n,f_n)\ {}^V\Psi^{(n)}(\alpha,\phi,c_n)\ {}^T\Psi^{(n)}(\alpha,\phi,d_n) \tag{6.7}$$

where ${}^S\Psi^{(n)}$, ${}^V\Psi^{(n)}$ and ${}^T\Psi^{(n)}$ obey independent Schroedinger equations with ${}^SH^n_{|2}$, ${}^VH^n_{|2}$ and ${}^TH^n_{|2}$ respectively.

7 The Boundary Conditions

We want to find the solution of the master equation that corresponds to

$$\Psi[h_{ij},\Phi] = \int d[g_{\mu\nu}]d[\Phi] \exp(-\tilde{I}) \qquad (7.1)$$

where the integral is taken over all compact 4-metrics and matter fields which are bounded by the 3-surface S. If one takes the scale parameter α to be very negative but keeps the other parameters fixed, the Euclidean action $\tilde{I}$ tends to zero like $e^{2\alpha}$. Thus one would expect Ψ to tend to one as α tends to minus infinity.

One can estimate the form of the scalar, vector and tensor parts $S_\Psi(n), V_\Psi(n), T_\Psi(n)$ of the perturbation $\Psi^{(n)}$ from the path integral (7.1). One takes the 4-metric $g_{\mu\nu}$ and the scalar field Φ to be of the background form

$$ds^2 = \sigma^2(- N^2dt^2 + e^{2\alpha(t)}d\Omega_3^2) \qquad (7.2)$$

and $\phi(t)$ respectively plus a small perturbation described by the variables (a_n, b_n, f_n), c_n and d_n as functions of t. In order for the background 4-metric to be compact, it has to be Euclidean when $\alpha = -\infty$ ie N has to be purely negative imaginary at $\alpha = -\infty$, which we shall take to be $t = 0$. In regions in which the metric is Lorentzian, N will be real and positive. In order to allow a smooth transition from Euclidean to Lorentzian, we shall take N to be of the form $-ie^{i\mu}$ where $\mu = 0$ at $t = 0$. In order that the 4-metric and the scalar field be regular at $t = 0$, a_n, b_n, c_n, d_n, f_n have to vanish there.

The tensor perturbations d_n have the Euclidean action

$${}^T\tilde{I}_n = \tfrac{1}{2}\int dt\ d_n\ {}^TD\ d_n + \text{boundary term} \qquad (7.3)$$

where

$$ {}^{T}D = \left[-\frac{d}{dt}\left[\frac{e^{3\alpha}}{iN_0}\frac{d}{dt}\right] + iN_0 e^{\alpha}(n^2-1)\right] \tag{7.4}$$

$$ + 4iN_0 e^{3\alpha}\left[+ \tfrac{1}{2}e^{-2\alpha} - \frac{3}{2}m^2\phi^2 - \frac{3\dot{\phi}^2}{2(iN_0)^2} - \frac{3\dot{\alpha}^2}{2(iN_0)^2} - \frac{1}{iN_0}\frac{d}{dt}\left[\frac{\dot{\alpha}}{iN_0}\right]\right] $$

The last term in (7.4) vanishes if the background metric satisfies the background field equations. The action is extremized when d_n satisfies the equation

$$ {}^{T}D\, d_n = 0 \tag{7.5}$$

For a d_n that satisfies (7.5), the action is just the boundary term

$$ {}^{T}\tilde{I}_n^{cl} = \frac{1}{2iN_0} e^{3\alpha}\left[d_n\dot{d}_n + 4\dot{\alpha}d_n^2\right] \tag{7.6}$$

The path integral over d_n will be

$$ \int d[d_n]\, \exp(-\,{}^{T}\tilde{I}_n) = (\det {}^{T}D)^{-1/2} \exp(-\,{}^{T}\tilde{I}_n^{cl}) \tag{7.7}$$

One now has to integrate (7.7) over different background metrics to obtain the wavefunction ${}^{T}\Psi(n)$. One expects the dominant contribution to come from background metrics that are near a solution of the classical background field equations. For such metrics one can employ the adiabatic approximation in which one regards α to be a slowly varying function of t. Then the solution of (7.5) which obeys the boundary condition $d_n = 0$ at $t = 0$ is

$$ d_n = A(e^{\nu\tau} - e^{-\nu\tau}) \tag{7.8}$$

where $\nu = e^{-\alpha}(n^2-1)^{1/2}$ and $\tau = \int iN_0\, dt$. This approximation will be valid for background fields which are near a solution of the background field equations and for which

$$|\frac{\dot{\alpha}}{N_0}| \ll n e^{-\alpha} \tag{7.9}$$

For a regular Euclidean metric, $|\frac{\dot{\alpha}}{N_0}| = e^{-\alpha}$ near $t = 0$. If the metric is a Euclidean solution of the background field equations, then $|\frac{\dot{\alpha}}{N_0}| < e^{-\alpha}$. Thus the adiabatic approximation should hold for large values of n into the region in which the solution of the background field equations becomes Lorentzian and the WKB approximation can be used. The wavefunction $T_\Psi(n)$ will then be

$$T_\Psi(n) = B \exp\left[- \left[\tfrac{1}{2} n e^{2\alpha} \coth(\nu\tau) + \frac{2}{iN_0} \dot{\alpha} e^{3\alpha} \right] d_n^2 \right] \tag{7.10}$$

In the Euclidean region, τ will be real and positive. For large values of n, $\coth(\nu\tau) \approx 1$. In the Lorentzian region where the WKB approximation applies, τ will be complex but it will still have a positive real part and $\coth(\nu\tau)$ will still be approximately 1 for large n. Thus

$$T_\Psi(n) = B \exp\left[- 2i \frac{\partial S}{\partial \alpha} d_n^2 - \tfrac{1}{2} n e^{2\alpha} d_n^2 \right] \tag{7.11}$$

The normalization constant B can be chosen to be 1. Thus, apart from a phase factor, the gravitational wave modes enter the WKB region in their ground state.

We now consider the vector part $V_\Psi(n)$ of the wavefunction. This is pure gauge as the quantities c_n can be given any values by gauge transformations parameterized by the j_n. The freedom to make gauge transformations is reflected quantum mechanically in the constraint

$$e^{-\alpha}\left[\frac{\partial}{\partial c_n} + 4(n^2-4) c_n \frac{\partial}{\partial \alpha} \right] \Psi = 0 \tag{7.12}$$

One can integrate (7.12) to give

$$\Psi(\alpha,\{c_n\}) = \Psi(\alpha - 2\sum_n (n^2-4)c_n^2, 0) \tag{7.13}$$

where the dependence on the other variables has been suppressed. One can also replace $\frac{\partial\Psi}{\partial\alpha}$ by $i\frac{\partial S}{\partial\alpha}\Psi$. One can then solve for $V_\Psi(n)$:

$$V_\Psi(n) = \exp\left[2i\,(n^2-4)c_n^2\,\frac{\partial S}{\partial\alpha}\right] \tag{7.14}$$

The scalar perturbation modes a_n, b_n and f_n involve a combination of the behaviour of the tensor and vector perturbations. The scalar part of the action is given in appendix B. The action is extremized by solutions of the classical equations

$$N_0\frac{d}{dt}\left(e^{3\alpha}\frac{\dot a_n}{N_0}\right) + \frac{1}{3}(n^2 - 4)N_0^2e^{\alpha}(a_n + b_n) + 3e^{3\alpha}(\dot\phi\dot f_n - N_0^2m^2\phi f_n) =$$

$$N_0^2\left[3e^{3\alpha}m^2\phi^2 - \frac{1}{3}(n^2+2)e^{\alpha}\right]g_n + e^{3\alpha}\dot\alpha\dot g_n - \frac{1}{3}N_0\frac{d}{dt}\left[e^{2\alpha}\frac{k_n}{N_0}\right] \tag{7.15}$$

$$N_0\frac{d}{dt}\left(e^{3\alpha}\frac{\dot b_n}{N_0}\right) - \frac{1}{3}(n^2 - 1)N_0^2e^{\alpha}(a_n + b_n) = \frac{1}{3}(n^2 - 1)N_0^2e^{\alpha}g_n$$

$$+ \frac{1}{3}N_0\frac{d}{dt}\left[e^{2\alpha}\frac{k_n}{N_0}\right] \tag{7.16}$$

$$N_0\frac{d}{dt}\left(e^{3\alpha}\frac{\dot f_n}{N_0}\right) + 3e^{3\alpha}\dot\phi\dot a_n + N_0^2\left[m^2e^{3\alpha} + (n^2-1)e^{\alpha}\right]f_n =$$

$$e^{3\alpha}\left[-2N_0^2m^2\phi g_n + \dot\phi\dot g_n - e^{-\alpha}\dot\phi k_n\right] \tag{7.17}$$

There is a three parameter family of solutions to (7.15) to (7.17) which obey the boundary condition $a_n = b_n = f_n = 0$ at $t = 0$. There are however, two constraint equations:

$$\dot{a}_n + \frac{(n^2-4)}{(n^2-1)}\dot{b}_n + 3f_n\dot{\phi} = \dot{\alpha}g_n - \frac{e^{-\alpha}}{(n^2-1)}k_n \tag{7.18}$$

$$\begin{aligned} &3a_n\left(-\dot{\alpha}^2 + \dot{\phi}^2\right) + 2\left(\dot{\phi}\dot{f}_n - \dot{\alpha}\dot{a}_n\right) \\ &+ N_0^2\, m^2\left(2f_n\phi + 3a_n\phi^2\right) - \tfrac{2}{3}N_0^2 e^{-2\alpha}\left[(n^2-4)b_n + (n^2 + \tfrac{1}{2})a_n\right] \\ &= \tfrac{2}{3}\dot{\alpha}e^{-\alpha}k_n + 2g_n\left(-\dot{\alpha}^2 + \dot{\phi}^2\right) \end{aligned} \tag{7.19}$$

These correspond to the two gauge degrees of freedom parameterized by k_n and g_n respectively. The Euclidean action for a solution to equations (7.15) to (7.19) is

$$\begin{aligned} S_{I_n}^{cl} = \frac{1}{2iN_0}e^{3\alpha}\Bigg[&-a_n\dot{a}_n + \frac{(n^2-4)}{(n^2-1)}b_n\dot{b}_n + f_n\dot{f}_n \\ &+ \dot{\alpha}\left[-a_n^2 + \frac{4(n^2-4)}{(n^2-1)}b_n^2\right] + 3\dot{\phi}a_n f_n \\ &+ g_n\left[\dot{\alpha}a_n - \dot{\phi}f_n\right] - \frac{1}{3}e^{-\alpha}k_n\left[a_n + \frac{(n^2-4)}{(n^2-1)}b_n\right]\Bigg] \end{aligned} \tag{7.20}$$

where the background field equations have been used.

In many ways the simplest gauge to work in is that with $g_n = k_n = 0$. However, this gauge does not allow one to find a compact 4-metric which is bounded by a 3-surface with arbitrary values of a_n, b_n and f_n and which is a solution of the equations (7.15) to (7.17) and the constraint equations. Instead, we shall use the gauge $a_n = b_n = 0$ and shall solve the constraint equations (7.18) and (7.19) to find g_n and k_n:

$$g_n = 3\frac{\left[(n^2-1)\dot{\alpha}\dot{\phi}f_n + \dot{\phi}\dot{f}_n + N_0^2m^2\phi f_n\right]}{\left[(n^2-4)\dot{\alpha}^2 + 3\dot{\phi}^2\right]} \tag{7.21}$$

$$k_n = 3(n^2-1)e^{\alpha}\frac{\left[\dot{\alpha}\dot{\phi}\dot{f}_n + N_0^2m^2\phi f_n\dot{\alpha} - 3f_n\dot{\phi}(- \dot{\alpha}^2 + \dot{\phi}^2)\right]}{\left[(n^2-4)\dot{\alpha}^2 + 3\dot{\phi}^2\right]} \tag{7.22}$$

With these substituted, (7.17) becomes a second order equation for f_n

$$N_0\frac{d}{dt}\left[e^{3\alpha}\frac{\dot{f}_n}{N_0}\right] + N_0^2\left[m^2e^{3\alpha} + (n^2-1)e^{\alpha}\right]f_n =$$

$$e^{3\alpha}\left[- 2N_0^2m^2\phi g_n + \dot{\phi}\dot{g}_n - e^{-\alpha}\dot{\phi}k_n\right] \tag{7.23}$$

For large n we can again use the adiabatic approximation to estimate the solution of (7.23) when $|\phi| > 1$:

$$f_n = A\sinh(\nu\tau) \tag{7.24}$$

where $\nu^2 = e^{-2\alpha}(n^2-1)$. Thus for these modes

$$S_\Psi(n)(\alpha,\phi,\ 0\ ,\ 0\ ,f_n) \approx \exp\left[- ½ne^{2\alpha}f_n^2 - ½i\frac{\partial S}{\partial\phi}g_nf_n\right] \tag{7.25}$$

This is of the ground state form apart from a small phase factor. The value of $S_\Psi(n)$ at non-zero values of a_n and b_n can be found by integrating the constraint equations (5.25) and (5.27).

The tensor and scalar modes start off in their ground states, apart possibly from the modes at low n. The vector modes are pure gauge and can be neglected. Thus the total energy $E = \sum_n \frac{H_{|2}^{(n)}\Psi^{(n)}}{\Psi^{(n)}}$ of the perturbations will be small when the ground state energies are subtracted. But $E = i(\nabla_2 S).J$ where $J = \sum_n \frac{\nabla_2 \Psi^{(n)}}{\Psi^{(n)}}$. Thus J is small. This means that the wavefunction Ψ_0 will obey the Wheeler-DeWitt equation of the unperturbed minisuperspace model and the phase factor S will be approximately $-i\ln\Psi_0$. However the homogeneous scalar field mode ϕ will not start out in its ground state. There are two reasons for this: first, regularity at $t = 0$ requires $a_n = b_n = c_n = d_n = f_n = 0$, but does not require $\phi = 0$. Second, the classical field equation for ϕ is of the form for a damped harmonic oscillator with a constant frequency m rather than a decreasing frequency $e^{-\alpha}n$. This means that the adiabatic approximation is not valid at small t and that the solution of the classical field equation is ϕ approximately constant. The action of such solutions is small, so large values of $|\phi|$ are not damped as they are for the other variables. Thus the WKB trajectories which start out from large values of $|\phi|$ have high probability. They will correspond to classical solutions which have a long inflationary period and then go over to a matter dominated expansion. In a realistic model which included other fields of low rest mass, the matter energy in the oscillations of the massive scalar field would decay into light particles with a thermal spectrum. The model would then expand as a radiation dominated universe.

8 Growth of Perturbations

The tensor modes will obey the Schroedinger equation

$$i\frac{\partial\, {}^T\Psi^{(n)}}{\partial t} = {}^T H^n_{|2}\, {}^T\Psi^{(n)} \tag{8.1}$$

$$= \tfrac{1}{2}e^{-3\alpha}\left(+ d_n^2\left[10\left(\frac{\partial S}{\partial\alpha}\right)^2 + 6\left(\frac{\partial S}{\partial\phi}\right)^2\right] - \frac{\partial^2}{\partial d_n{}^2} - 8d_n i\frac{\partial S}{\partial\alpha}\frac{\partial}{\partial d_n} + \right.$$

$$d_n^2\left[(n^2+1)e^{4\alpha} - 6e^{6\alpha}m^2\phi^2\right]\Big) \qquad (8.2)$$

One can write

$$T_\Psi(n) = \exp(-2\alpha)\,\exp\left[-2i\,\frac{\partial S}{\partial \alpha}\,d_n^2\right]\,T_{\Psi_0}(n) \qquad (8.3)$$

then

$$i\frac{\partial T_{\Psi_0}(n)}{\partial t} = \tfrac{1}{2}e^{-3\alpha}\left[-\frac{\partial^2}{\partial d_n^2} + d_n^2(n^2-1)e^{4\alpha}\right]T_{\Psi_0}(n) \qquad (8.4)$$

The WKB approximation to the background Wheeler-DeWitt equation has been used in deriving (8.4). Then (8.4) has the form of the Schroedinger equation for an oscillator with a time dependent frequency $\nu = (n^2-1)^{1/2}e^{-\alpha}$. Initially the wavefunction $T_{\Psi_0}(n)$ will be in the ground state (apart from a normalization factor) and the frequency ν will be large compared to $\dot{\alpha}$. In this case one can use the adiabatic approximation to show that $T_{\Psi_0}(n)$ remains in the ground state

$$T_{\Psi_0}(n) \sim \exp\left[-\tfrac{1}{2}ne^{2\alpha}d_n^2\right] \qquad (8.5)$$

The adiabatic approximation will break down when $\nu \sim \dot{\alpha}$ ie the wavelength of the gravitational mode becomes equal to the horizon scale in the inflationary period. The wavefunction $T_{\Psi_0}(n)$ will then " freeze " :

$$T_{\Psi_0}(n) \sim \exp\left[-\tfrac{1}{2}ne^{2\alpha_*}d_n^2\right] \qquad (8.6)$$

where α_* is the value of α at which the mode goes outside the horizon. The wavefunction $T_{\Psi_0}(n)$ will remain of the form (8.6) until the mode re-enters the horizon in the matter or radiation dominated era at the much greater value α_e of α.

One can then apply the adiabatic approximation again to (8.4) but ${}^{T}\Psi_0^{(n)}$ will no longer be in the ground state; it will be a superposition of a number of highly excited states. This is the phenomenon of the amplification of the ground state fluctuations in the gravitational wave modes that was discussed in references [7,8,9].

The behaviour of the scalar modes is rather similar but their description is more complicated because of the gauge degrees of freedom. In the previous section we evaluated the wavefunction ${}^{S}\Psi^{(n)}$ on $a_n = b_n = 0$ by the path integral prescription. The ground state form (in f_n) that we found will be valid until the adiabatic approximation breaks down ie until the wavelength of the mode excedes the horizon distance during the inflationary period. In order to discuss the subsequent behaviour of the wavefunction, it is convenient to use the first order Hamiltonian constraint (5.27) to evaluate ${}^{S}\Psi^{(n)}$ on $a_n \neq 0, b_n = f_n = 0$. One finds that

$$ {}^{S}\Psi^{(n)}(\alpha,\phi,a_n, 0, 0) = B \exp\left[iCa_n^2\right] {}^{S}\Psi_0^{(n)}(\alpha,\phi,a_n) \tag{8.7} $$

The normalization and phase factors B and C depend on α and ϕ but not a_n.

$$ C = \frac{1}{2}\left(\frac{\partial S}{\partial \alpha}\right)^{-1}\left[\left(\frac{\partial S}{\partial \alpha}\right)^2 - \frac{1}{3}(n^2-4)e^{4\alpha}\right] \tag{8.8} $$

At the time the wavelength of the mode equals the horizon distance during the inflationary period, the wavefunction ${}^{S}\Psi_0^{(n)}$ has the form

$$ {}^{S}\Psi_0^{(n)} = \exp\left[- ½\, n\, y_*^{-2}\, e^{2\alpha_*}\, a_n^2\right] \tag{8.9} $$

where y_* is the value of $y = \frac{\partial S}{\partial \alpha}\left(\frac{\partial S}{\partial \phi}\right)^{-1}$ when the mode leaves the horizon. $y_* = 3\phi_*$. More generally, in the case of a scalar field with a potential $V(\phi)$, $y = 6V\left(\frac{\partial V}{\partial \phi}\right)^{-1}$.

One can obtain a Schroedinger equation for ${}^{S}\Psi_0^{(n)}$ by putting $b_n = f_n = 0$ in the scalar Hamiltonian ${}^{S}H^{n}_{|2}$ and substituting for $\frac{\partial}{\partial b_n}$ and $\frac{\partial}{\partial f_n}$ from the momentum constraint (5.25) and the first order Hamiltonian constraint (5.27) respectively. This gives

$$i\frac{\partial\, {}^{S}\Psi_0^{(n)}}{\partial t} = \tfrac{1}{2}e^{-3\alpha}\left[-y^2 \frac{\partial^2}{\partial a_n^2} + e^{4\alpha}(n^2-4)\left[\frac{1}{y^2} - \frac{1}{3}e^{4\alpha}\left[\frac{\partial S}{\partial \alpha}\right]^{-2}\right] a_n^2 \right] {}^{S}\Psi_0^{(n)} \tag{8.10}$$

where terms of order $\frac{1}{n^2}$ have been neglected. The term $e^{4\alpha}\left[\frac{\partial S}{\partial \alpha}\right]^{-2}$ will be small compared to $\frac{1}{y^2}$ except near the time of maximum radius of the background solution. The Schroedinger equation for ${}^{S}\Psi_0^{(n)}(a_n)$ is very similar to the equation for ${}^{T}\Psi_0^{(n)}(d_n)$, (8.4), except that the kinetic term is multiplied by a factor y^2 and the potential term is divided by a factor y^2. One would therefore expect that for wavelengths within the horizon, ${}^{S}\Psi_0^{(n)}$ would have the ground state form $\exp(-\tfrac{1}{2}ny^{-2}e^{2\alpha}a_n^2)$ and this is bourne out by (8.9). On the other hand, when the wavelength becomes larger than the horizon, the Schroedinger equation (8.10) indicates that ${}^{T}\Psi_0^{(n)}$ will freeze in the form (8.9) until the mode re-enters the horizon in the matter dominated era. Even if the equation of state of the Universe changes to radiation dominated during the period that the wavelength of the mode is greater than the horizon size, it will still be true that ${}^{S}\Psi_0^{(n)}$ is frozen in the form (8.9). The ground state fluctuations in the scalar modes will therefore be amplified in a similar manner to the tensor modes. At the time of re-entry of the horizon the rms fluctuation in the scalar modes, in the gauge in which $b_n = f_n = 0$, will be greater by the factor y_* than the rms fluctuation in the tensor modes of the same wavelength.

9 Comparison with Observation

From a knowledge of ${}^{T}\Psi_0^{(n)}$ and ${}^{S}\Psi_0^{(n)}$ one can calculate the relative probabilities of observing different values of d_n and a_n at a given point on a trajectory of the vector field X^i ie at a given value of α and ϕ in a background metric which is a solution of the classical field equations. In fact, the dependence on ϕ will be unimportant and we shall neglect it. One can then calculate the probabilities of observing different amounts of anisotropy in the microwave background and can compare these predictions with the upper limits set by observation.

The tensor and scalar perturbation modes will be in highly excited states at large values of α. This means that we can treat their development as an ensemble evolving according to the classical equations of motion with initial distributions in d_n and a_n proportional to $|{}^{T}\Psi_0^{(n)}|^2$ and $|{}^{S}\Psi_0^{(n)}|^2$ respectively. The initial distributions in $\dot{d}_n$ and $\dot{a}_n$ will be proportional to $|{}^{T}\Psi_0^{(n)}\pi_{d_n}$ TPn0$|$ and $|{}^{S}\Psi_0^{(n)}\pi_{a_n}{}^{S}\Psi_0^{(n)}|$ respectively. In fact, at the time that the modes re-enter the horizon, the distributions will be concentrated at $\dot{d}_n = \dot{a}_n = 0$.

The surfaces with $b_n = f_n = 0$ will be surfaces of constant energy density in the classical solution during the inflationary period. By local conservation of energy, they will remain surfaces of constant energy density in the era after the inflationary period when the energy is dominated by the coherent oscillations of the homogeneous background scalar field ϕ. If the scalar particles decay into light particles and heat up the universe, the surfaces with $b_n = f_n = 0$ will be surfaces of constant temperature. The surface of last scattering of the microwave background will be such a surface with temperature T_s. The microwave radiation can be considered to have propagated freely to us from this surface. Thus the observed temperature will be

$$T_o = \frac{T_s}{1 + z} \tag{9.1}$$

where z is the redshift of the surface of last scattering. Variations in the observed temperature will arise from variations in z in different directions of observation. These are given by

$$1 + z = \ell^{\mu} n_{\mu} \tag{9.2}$$

evaluated at the surface of last scattering where n_{μ} is the unit normal to the surfaces of constant t in the gauge $g_n = k_n = j_n = 0$ and $b_n = f_n = 0$ on the surface of last scattering and ℓ^{μ} is the parallelly propagated tangent vector to the null geodesic from the observer normalized by $\ell^{\mu} n_{\mu} = 1$ at the present time. One can calculate the evolution of $\ell^{\mu} n_{\mu}$ down the past light cone of the observer:

$$\frac{d}{d\lambda}\left[\ell^{\mu} n_{\mu}\right] = n_{\mu;\nu}\ell^{\mu}\ell^{\nu} \tag{9.3}$$

where λ is the affine parameter on the null geodesic. The only non-zero components of $n_{\mu;\nu}$ are

$$n_{i;j} = e^{2\alpha}\Big[\dot{\alpha}\Omega_{ij} + \sum_n(\dot{a}_n + \dot{\alpha}a_n)\tfrac{1}{3}\Omega_{ij}Q + \sum_n(\dot{b}_n + \dot{\alpha}b_n)P_{ij}$$

$$+ \sum_n(\dot{d}_n + \dot{\alpha}d_n)G_{ij}\Big] \tag{9.4}$$

In the gauge that we are using, the dominant anisotropic terms in (9.4) on the scale of the horizon, will be those involving $\dot{\alpha}a_n$ and $\dot{\alpha}d_n$. These will give temperature anisotropies of the form

$$\langle(\Delta T/T)^2\rangle \sim \langle a_n^2\rangle \quad \text{or} \quad \sim \langle d_n^2\rangle \tag{9.5}$$

The number of modes that contribute to anisotropies on the scale of the horizon is of the order of n^3. From the results of the last section

$$\langle a_n^2 \rangle = y_*^2 n^{-1} e^{-2\alpha_*} \quad (9.6)$$

$$\langle d_n^2 \rangle = n^{-1} e^{-2\alpha_*} \quad (9.7)$$

The dominant contribution comes from the scalar modes which give

$$\langle (\Delta T/T)^2 \rangle \sim y_*^2 n^2 e^{-2\alpha_*} \quad (9.8)$$

But $ne^{-\alpha_*} \sim \dot{\alpha}_*$, the value of the Hubble constant at the time that the present horizon size left the horizon during the inflationary period. The observational upper limit of about 10^{-8} on $\langle (\Delta T/T)^2 \rangle$ restricts this Hubble constant to be less than about $5.10^{-5} m_p$ (Ref. 10) which in turn restricts the mass of the scalar field to be less than 10^{14} GeV .

10 Conclusion and Summary

We started from the proposal that the quantum state of the Universe is defined by a path integral over compact 4-metrics. This can be regarded as a boundary condition for the Wheeler-DeWitt equation for the wavefunction of the Universe on the infinite dimensional manifold, superspace, the space of all 3-metrics and matter field configurations on a 3-surface S. Previous papers had considered finite dimensional approximations to superspace and had shown that the boundary condition led to a wavefunction which could be interpreted as corresponding to a family of classical solutions which were homogeneous and isotropic and which had a period of exponential or inflationary expansion. In the present paper we extended this work to the full superspace without restrictions. We treated the two basic homogeneous and isotropic degrees of freedom exactly and the other degrees of freedom to second order. We justified this approximation by showing that the inhomogeneous or anisotropic modes started out in their ground states.

We derived time dependent Schroedinger equations for each mode. We showed that they remained in the ground state until their wavelength exceded the horizon size during the inflationary period. In the subsequent expansion the ground state fluctuations got frozen until the wavelength re-entered the horizon during the radiation or matter dominated era. This part of the calculation is similar to earlier work on the development of gravitational waves [7] and density perturbations [11,12] in the inflationary universe but it has the advantage that the assumptions of a period of exponential expansion and of an initial ground state for the perturbations are justified. The perturbations would be compatible with the upper limits set by observations of the microwave background if the scalar field that drives the inflation has a mass of 10^{14} GeV or less.

In section 8 we calculated the scalar perturbations in a gauge in which the surfaces of constant time are surfaces of constant density. There are thus no density fluctuations in this gauge. However, one can make a transformation to a gauge in which $a_n = b_n = 0$. In this gauge the density fluctuation at the time that the wavelength comes within the horizon is

$$\langle(\Delta\rho/\rho)^2\rangle \approx y^2 \frac{\dot{\rho}_e^2}{\dot{\alpha}_e^2 \rho_e^2} \dot{\alpha}_*^2 \qquad (10.1)$$

Because y and $\dot{\alpha}_*$ depend only logarithmically on the wavelength of the perturbations, this gives an almost scale free spectrum of density fluctuations. These fluctuations can evolve according to the classical field equations to give rise to the formation of galaxies and all the other structure that we observe in the Universe. Thus all the complexities of the present state of the Universe have their origin in the ground state fluctuations in the inhomogeneous modes and so arise from the Heisenberg Uncertainty Principle.

References

1 S. W. Hawking & G. F. R. Ellis, "The Large-Scale Structure of Space-Time", (Cambridge University Press, 1973).

2 S. W. Hawking and D. N. Page, "Operator Ordering and the Flatness of the Universe", DAMTP preprint (1985)

3 S. W. Hawking in: "Relativity, Groups and Topology II", Les Houches 1983, Session XL, edited by B. S. DeWitt & R. Stora (North Holland Amsterdam, 1984)

4 S. W. Hawking, Nucl. Phys. B239 257 (1984).

5 S. W. Hawking and Z. C. Wu, Phys. Lett 151B 15 (1985)

6 S. W. Hawking & J. C. Luttrell, Nucl. Phys. B247 250 (1984)

7 V. A. Rubakov, M. V. Sazhin & A. V. Veryaskin, Phys. Lett. 115B 189 (1982)

8 L. P. Grischuk, Zh. Eksp. Teor. Fiz 67 825 (1974) [Sov. Phys. JETP 40 409 (1975)]; Ann. N. Y. Sci 302 439 (1977)

9 A. A. Starobinsky, Pis'ma Zh. Eksp. Teor. Fiz 30 719 (1979)

10 S. W. Hawking, Phys. Lett. 150B 339 (1985)

11 S. W. Hawking, Phys. Lett. 115B 295 (1982)

12 A. H. Guth & S. Y. Pi, Phys. Rev. Lett. 49 1110 (1982) [JETP Lett. 30 682 (1979)]

13 E. M. Lifshitz and I. M. Khalatnikov, Adv. Phys. 12 185 (1963)

14 U. H. Gerlach and U. K. Sengupta, Phys. Rev D18 1773 (1978)

Appendix A: Harmonics on the 3-sphere

In this appendix we describe the properties of the scalar, vector and tensor harmonics on the 3-sphere S^3. The metric on S^3 is Ω_{ij} and so the line element is

$$d\ell^2 = \Omega_{ij}dx^i dx^j = d\chi^2 + \sin^2\chi(d\theta^2 + \sin^2\theta d\phi^2) \qquad \text{(A1)}$$

A vertical stroke will denote covariant differentiation with respect to the metric Ω_{ij}. Indices i,j,k are raised and lowered using Ω_{ij}.

(1) **Scalar Harmonics**

The scalar spherical harmonics $Q^n_{\ell m}(\chi,\theta,\phi)$ are scalar eigenfunctions of the Laplacian operator on S^3. Thus, they satisfy the eigenvalue equation

$$Q^{(n)}{}_{|k}{}^{|k} = -(n^2-1)\,Q^{(n)} \qquad n = 1,2,3... \qquad \text{(A2)}$$

The most general solution to (A2), for given n, is a sum of solutions

$$Q^{(n)}(\chi,\theta,\phi) = \sum_{\ell=0}^{n-1}\sum_{m=-\ell}^{\ell} A^n_{\ell m}\,Q^n_{\ell m}(\chi,\theta,\phi) \qquad \text{(A3)}$$

where $A^n_{\ell m}$ are a set of arbitrary constants. The $Q^n_{\ell m}$ are given explicitly by

$$Q^n_{\ell m}(\chi,\theta,\phi) = \Pi^n_\ell(\chi)Y_{\ell m}(\theta,\phi) \qquad \text{(A4)}$$

where $Y_{\ell m}(\theta,\phi)$ are the usual harmonics on the 2-sphere, S^2, and $\Pi^n_\ell(\chi)$ are the Fock harmonics[13,14] The spherical harmonics $Q^n_{\ell m}$ constitute a complete orthogonal set for the expansion of any scalar field on S^3.

(2) Vector Harmonics

The transverse vector harmonics $(S_i)^n_{\ell m}(\chi,\theta,\phi)$ are vector eigenfunctions of the Laplacian operator on S^3 which are transverse. That is, they satisfy the eigenvalue equation

$$S_i^{(n)}{}_{|k}{}^{|k} = -(n^2-2)\,S_i^{(n)} \qquad n = 2,3,4\ldots \tag{A5}$$

and the transverse condition

$$S_i^{(n)|i} = 0 \tag{A6}$$

The most general solution to (A5) and (A6) is a sum of solutions

$$S_i^{(n)}(\chi,\theta,\phi) = \sum_{\ell=1}^{n-1}\sum_{m=-\ell}^{\ell} B^n_{\ell m}\,(S_i)^n_{\ell m}(\chi,\theta,\phi) \tag{A7}$$

where $B^n_{\ell m}$ are a set of arbitrary constants. Explicit expressions for the $(S_i)^n_{\ell m}$ are given in reference 14 where it is also explained how they are classified as odd (o) or even (e) using a parity transformation. We thus have two linearly independent transverse vector harmonics S_i^o and S_i^e (n, ℓ, m suppressed).

Using the scalar harmonics $Q^n_{\ell m}$ we may construct a third vector harmonic $(P_i)^n_{\ell m}$, defined by (n, ℓ, m suppressed)

$$P_i = \frac{1}{(n^2-1)}\,Q_{|i} \qquad n = 2,3,4\ldots \tag{A8}$$

It may be shown to satisfy

$$P_{i|k}{}^{|k} = -(n^2-3)\,P_i \qquad \text{and} \qquad P_i{}^{|i} = -Q \tag{A9}$$

The three vector harmonics S_i^o, S_i^e and P_i constitute a complete orthogonal set for the expansion of any vector field on S^3.

(3) Tensor Harmonics

The transverse traceless tensor harmonics $(G_{ij})^n_{\ell m}(\chi,\theta,\phi)$ are tensor eigenfunctions of the Laplacian operator on S^3 which are transverse and traceless. That is, they satisfy the eigenvalue equation

$$G^{(n)}_{ij\ |k}{}^{|k} = -(n^2-3)\, G^{(n)}_{ij} \qquad n = 3,4,5... \tag{A10}$$

and the transverse and traceless conditions

$$G^{(n)|i}_{ij} = o \quad , \quad G^{(n)i}_{i} = o \tag{A11}$$

The most general solution to (A11) and (A12) is a sum of solutions

$$G^{(n)}_{ij}(\chi,\theta,\phi) = \sum_{\ell=2}^{n-1} \sum_{m=-\ell}^{\ell} C^n_{\ell m}\, (G_{ij})^n_{\ell m}(\chi,\theta,\phi) \tag{A12}$$

where $C^n_{\ell m}$ are a set of arbitrary constants. As in the vector case they may be classified as odd or even. Explicit expressions for $(G^o_{ij})^n_{\ell m}$ and $(G^e_{ij})^n_{\ell m}$ are given in reference 14

Using the transverse vector harmonics $(S^o_i)^n_{\ell m}$ and $(S^e_i)^n_{\ell m}$, we may construct traceless tensor harmonics $(S^o_{ij})^n_{\ell m}$ and $(S^e_{ij})^n_{\ell m}$ defined, both for odd and even, by (n, ℓ, m suppressed)

$$S_{ij} = S_{i|j} + S_{j|i} \tag{A13}$$

and thus $S_i{}^i = o$ since S_i is transverse. In addition, the S_{ij} may be shown to satisfy

$$S_{ij}{}^{|j} = -(n^2 - 4)\, S_i \tag{A14}$$

$$S_{ij}{}^{|ij} = 0 \tag{A15}$$

$$S_{ij|k}{}^{|k} = -(n^2 - 6)\, S_{ij} \tag{A16}$$

Using the scalar harmonics $Q^n_{\ell m}$, we may construct two tensors $(Q_{ij})^n_{\ell m}$ and $(P_{ij})^n_{\ell m}$ defined by $(n, \ell, m$ suppressed)

$$Q_{ij} = \tfrac{1}{3}\Omega_{ij} Q \qquad n = 1,2,3 \tag{A17}$$

$$\text{and } P_{ij} = \frac{1}{(n^2 - 1)}\, Q_{|ij} + \tfrac{1}{3}\Omega_{ij} Q \qquad n = 2,3,4 \tag{A18}$$

The P_{ij} are traceless, $P_i{}^i = 0$, and in addition, may be shown to satisfy

$$P_{ij}{}^{|j} = -\tfrac{2}{3}(n^2 - 4)\, P_i \tag{A19}$$

$$P_{ij|k}{}^{|k} = -(n^2 - 7)\, P_{ij} \tag{A20}$$

$$P_{ij}{}^{|ij} = \tfrac{2}{3}(n^2 - 4)\, Q \tag{A21}$$

The six tensor harmonics $Q_{ij}, P_{ij}, S^o_{ij}, S^e_{ij}, G^o_{ij}$ and G^e_{ij} constitute a complete orthogonal set for the expansion of any symmetric second rank tensor field on S^3.

(4) Orthogonality and Normalization

The normalization of the scalar, vector and tensor harmonics is fixed by the orthogonality relations. We denote the integration measure on S^3 by $d\mu$. Thus

$$d\mu = d^3x\,(\det\Omega_{ij})^{1/2} = \sin^2\chi\,\sin\theta\,d\chi d\theta d\phi \tag{A22}$$

The $Q^n_{\ell m}$ are normalized so that

$$\int d\mu\, Q^n_{\ell m}\, Q^{n'}_{\ell' m'} = \delta^{nn'}\,\delta_{\ell\ell'}\,\delta_{mm'} \tag{A23}$$

This implies

$$\int d\mu\,(P_i)^n_{\ell m}\,(P^i)^{n'}_{\ell' m'} = \frac{1}{(n^2-1)}\,\delta^{nn'}\,\delta_{\ell\ell'}\,\delta_{mm'} \tag{A24}$$

and

$$\int d\mu\,(P_{ij})^n_{\ell m}\,(P^{ij})^{n'}_{\ell' m'} = \frac{2(n^2-4)}{3(n^2-1)}\,\delta^{nn'}\,\delta_{\ell\ell'}\,\delta_{mm'} \tag{A25}$$

The $(S_i)^n_{\ell m}$, both odd and even, are normalized so that

$$\int d\mu\,(S_i)^n_{\ell m}\,(S^i)^{n'}_{\ell' m'} = \delta^{nn'}\,\delta_{\ell\ell'}\,\delta_{mm'} \tag{A26}$$

This implies

$$\int d\mu\,(S_{ij})^n_{\ell m}\,(S^{ij})^{n'}_{\ell' m'} = 2(n^2-4)\,\delta^{nn'}\,\delta_{\ell\ell'}\,\delta_{mm'} \tag{A27}$$

Finally, the $(G_{ij})^n_{\ell m}$, both odd and even, are normalized so that

$$\int d\mu\,(G_{ij})^n_{\ell m}\,(G^{ij})^{n'}_{\ell' m'} = \delta^{nn'}\,\delta_{\ell\ell'}\,\delta_{mm'} \tag{A28}$$

The information given in this appendix about the spherical harmonics is all that is needed to perform the derivations presented in the main text. Further details may be found in references 13 & 14

Appendix B: Action and Field Equations of the Infinite Dimensional Model

The action

$$I = I_0(\alpha,\phi,N_0) + \sum_n I_n \tag{B1}$$

where I_0 is the action of the unperturbed model:

$$I_0 = -\tfrac{1}{2}\int dt\, N_0 e^{3\alpha}\left[\frac{\dot{\alpha}^2}{N_0^2} - e^{-2\alpha} - \frac{\dot{\phi}^2}{N_0^2} + m^2\phi^2\right] \tag{B2}$$

I_n is quadratic in the perturbations and may be written

$$I_n = \int dt(L_g^n + L_m^n) \tag{B3}$$

where

$$L_g^n = \tfrac{1}{2}e^{\alpha}N_0\left[\tfrac{1}{3}(n^2-\tfrac{5}{2})a_n^2 + \frac{(n^2-7)}{3}\frac{(n^2-4)}{(n^2-1)}b_n^2 - 2(n^2-4)c_n^2 - (n^2+1)d_n^2 + \tfrac{2}{3}(n^2-4)a_nb_n\right.$$

$$\left. g_n\left[\tfrac{2}{3}(n^2-4)b_n + \tfrac{2}{3}(n^2+\tfrac{1}{2})a_n\right] + \frac{1}{N_0^2}\left[-\frac{1}{3(n^2-1)}k_n^2 + (n^2-4)j_n^2\right]\right]$$

$$+ \tfrac{1}{2}\frac{e^{3\alpha}}{N_0}\left[-\dot{a}_n^2 + \frac{(n^2-4)}{(n^2-1)}\dot{b}_n^2 + (n^2-4)\dot{c}_n^2 + \dot{d}_n^2\right.$$

$$+ \dot{\alpha}\left[-2a_n\dot{a}_n + 8\frac{(n^2-4)}{(n^2-1)}b_n\dot{b}_n + 8(n^2-4)c_n\dot{c}_n + 8d_n\dot{d}_n\right]$$

$$+ \dot{\alpha}^2\left[-\tfrac{3}{2}a_n^2 + 6\frac{(n^2-4)}{(n^2-1)}b_n^2 + 6(n^2-4)c_n^2 + 6d_n^2\right]$$

$$\left. + g_n\left[2\dot{\alpha}\dot{a}_n + \dot{\alpha}^2(3a_n - g_n)\right]\right.$$

$$+ e^{-\alpha}\left[k_n\left[-\frac{2}{3}\dot{a}_n - \frac{2(n^2-4)}{3(n^2-1)}\dot{b}_n + \frac{2}{3}\dot{\alpha}g_n\right] - 2(n^2-4)\dot{c}_n j_n\right]\Bigg] \quad (B4)$$

and

$$L_m^n = \tfrac{1}{2}N_0 e^{3\alpha}\left[\frac{1}{N_0^2}\left[\dot{f}_n^2 + 6a_n\dot{f}_n\dot{\phi}\right] - m^2\left[f_n^2 + 6a_n f_n\phi\right] - e^{-2\alpha}(n^2-1)f_n^2\right.$$

$$+ \frac{3}{2}\left[\frac{\dot{\phi}^2}{N_0^2} - m^2\phi^2\right]\left[a_n^2 - \frac{4(n^2-4)}{(n^2-1)}b_n^2 - 4(n^2-4)c_n^2 - 4d_n^2\right] + \frac{\dot{\phi}^2}{N_0^2}g_n^2$$

$$\left. - g_n\left[2m^2 f_n\phi + 3m^2 a_n\phi^2 + 2\frac{\dot{f}_n\dot{\phi}}{N_0^2} + 3\frac{\dot{a}_n\dot{\phi}^2}{N_0^2}\right] - 2\frac{e^{-\alpha}}{N_0^2}k_n f_n\dot{\phi}\right] \quad (B5)$$

The full expressions for π_α and π_ϕ are

$$\pi_\alpha = \frac{e^{3\alpha}}{N_0}\left[-\dot{\alpha} + \sum_n \left[-a_n\dot{a}_n + \frac{4(n^2-4)}{(n^2-1)}b_n\dot{b}_n + 4(n^2-4)c_n\dot{c}_n + 4d_n\dot{d}_n\right]\right.$$

$$\dot{\alpha}\sum_n \left[-\frac{3}{2}a_n^2 + \frac{6(n^2-4)}{(n^2-1)}b_n^2 + 6(n^2-4)c_n^2 + 6d_n^2\right]$$

$$\left.\sum_n g_n\left[\dot{a}_n + \dot{\alpha}(3a_n - g_n) + \frac{1}{3}e^{-\alpha}k_n\right]\right] \quad (B7)$$

$$\pi_\phi = \frac{e^{3\alpha}}{N_0}\left[\dot{\phi} + \sum_n \left[3a_n\dot{f}_n + \frac{3}{2}\dot{\phi}\left[a_n^2 - \frac{4(n^2-4)}{(n^2-1)}b_n^2 - 4(n^2-4)c_n^2 - 4d_n^2\right]\right] + \right.$$

$$\left.\sum_n \left[\dot{\phi}g_n^2 - g_n(\dot{f}_n + 3a_n\dot{\phi}) - e^{-\alpha}k_n f_n\right]\right] \quad (B7)$$

The classical field equations may be obtained from the action (B1) by varying with respect to each of the fields in turn. Variation with respect to α and ϕ gives two field equations, similar to those obtained in section II.4, but modified by terms quadratic in the perturbations:

$$N_0 \frac{d}{dt}\left[\frac{1}{N_0}\frac{d\phi}{dt}\right] + 3\frac{d\alpha}{dt}\frac{d\phi}{dt} + N_0^2 m^2 \phi = \text{quadratic terms} \qquad (B8)$$

$$N_0 \frac{d}{dt}\left[\frac{\dot\alpha}{N_0}\right] + 3\,\dot\phi^2 - N_0^2 e^{-2\alpha}$$

$$- \frac{3}{2}\left[- \dot\alpha^2 + \dot\phi^2 - N_0^2 e^{-2\alpha} + N_0^2 m^2 \phi^2\right] = \text{quadratic terms} \qquad (B9)$$

Variation with respect to the perturbations a_n, b_n, c_n, d_n and f_n leads to five field equations:

$$N_0 \frac{d}{dt}\left(e^{3\alpha}\frac{\dot a_n}{N_0}\right) + \frac{1}{3}(n^2 - 4)N_0^2 e^{\alpha}(a_n + b_n) + 3e^{3\alpha}(\dot\phi \dot f_n - N_0^2 m^2 \phi f_n) =$$

$$N_0^2\left[3e^{3\alpha}m^2\phi^2 - \frac{1}{3}(n^2+2)e^{\alpha}\right]g_n + e^{3\alpha}\dot\alpha \dot g_n - \frac{1}{3}N_0 \frac{d}{dt}\left[e^{2\alpha}\frac{k_n}{N_0}\right] \qquad (B10)$$

$$N_0 \frac{d}{dt}\left(e^{3\alpha}\frac{\dot b_n}{N_0}\right) - \frac{1}{3}(n^2 - 1)N_0^2 e^{\alpha}(a_n + b_n) = \frac{1}{3}(n^2 - 1)N_0^2 e^{\alpha} g_n$$

$$+ \frac{1}{3}N_0 \frac{d}{dt}\left[e^{2\alpha}\frac{k_n}{N_0}\right] \qquad (B11)$$

$$\frac{d}{dt}\left(e^{3\alpha}\frac{\dot c_n}{N_0}\right) = \frac{d}{dt}\left[e^{2\alpha}\frac{j_n}{N_0}\right] \qquad (B12)$$

$$N_0 \frac{d}{dt}\left(e^{3\alpha}\frac{\dot d_n}{N_0}\right) + (n^2 - 1)N_0^2 e^{\alpha} d_n = 0 \qquad (B13)$$

$$N_0 \frac{d}{dt}\left(e^{3\alpha}\frac{\dot f_n}{N_0}\right) + 3e^{3\alpha}\dot\phi \dot a_n + N_0^2\left[m^2 e^{3\alpha} + (n^2-1)e^{\alpha}\right]f_n =$$

$$e^{3\alpha}\left[-2N_0^2 m^2 \phi g_n + \dot{\phi}\dot{g}_n - e^{-\alpha}\dot{\phi}k_n \right] \qquad (B14)$$

In obtaining (B10) – (B14), the field equations (B8) and (B9)have been used and terms cubic in the perturbations have been droppped.

Variation with respect to the Lagrange multipliers k_n, j_n, g_n and N_0 leads to a set of constraints. Variation with respect to k_n and j_n leads to the momentum constraints:

$$\dot{a}_n + \frac{(n^2-4)}{(n^2-1)}\dot{b}_n + 3f_n\dot{\phi} = \dot{\alpha}g_n - \frac{e^{-\alpha}}{(n^2-1)}k_n \qquad (B15)$$

$$\dot{c}_n = e^{-\alpha}j_n \qquad (B16)$$

Variation with respect to g_n gives the linear Hamiltonian constraint:

$$3a_n(-\dot{\alpha}^2 + \dot{\phi}^2) + 2(\dot{\phi}\dot{f}_n - \dot{\alpha}\dot{a}_n)$$

$$+ N_0^2 m^2(2f_n\phi + 3a_n\phi^2) - \tfrac{2}{3}N_0^2 e^{-2\alpha}\left[(n^2-4)b_n + (n^2 + \tfrac{1}{2})a_n\right]$$

$$= \tfrac{2}{3}\dot{\alpha}e^{-\alpha}k_n + 2g_n(-\dot{\alpha}^2 + \dot{\phi}^2) \qquad (B17)$$

Finally, variation with respect to N_0 yields the Hamiltonian constraint, which we write as

$$\tfrac{1}{2}e^{3\alpha}\left[-\frac{\dot{\alpha}^2}{N_0^2} + \frac{\dot{\phi}^2}{N_0^2} - e^{-2\alpha} + m^2\phi^2 \right] = \text{quadratic terms} \qquad (B18)$$

SOLITONS AND BLACK HOLES IN 4,5 DIMENSIONS

G.W. Gibbons
Department of Applied Mathematics and Theoretical Physics,
University of Cambridge, Silver Street, Cambridge CB3 9EW
U.K.

Contents

1. Introduction

This is the written version of two lectures given in Paris in February 1985. Since the material as given has now appeared elsewhere [1,2] I have decided not to repeat the lectures verbatim but rather to comment on the general problem of solitons in gravity, in particular on the importance or otherwise of spatial and spacetime topology contrasting the situation in 4 and in 5 spacetime dimensions. My main point will be that while there are many similarities with the situation in Yang-Mills-Higgs theory there are significant differences. In particular the apparently inevitable occurrence of spacetime singularities and their conjectured shielding by event horizons (Cosmic Censorship) means that one cannot assume that the time evolution of initial data is continuous. This substantially alters ones views of the importance of topology in the classical theory. It is highly likely that the quantum theory - should it make mathematical sense - will be similarly affected. The plan of the article is as follows: in section 2 I will discuss some topological aspects of the initial value problem. In section 3 I will describe why I don't feel one can regard black holes as solitons except in the extreme Reissner-Nordstrom case, and the relation of this to supergravity. In section 4 I will contrast the situation with that in 5-dimensions and I will argue that the true analogue of magnetic monopoles in Yang-Mills theory are the multi-Taub NUT solutions whose importance for Kaluza-Klein theory was first stressed by Gross, Perry and Sorkin. Their relation to black holes will also be described. In section 5 I will describe a duality conjecture analogous to that of Olive & Montonen in the Yang-Mills case.

2. Topology and the Initial Data

It is an attractive idea that the way to study solitons and other topological features in General Relativity is to start with an initial data set $\{\Sigma, g_{ij}, K_{ij}\}$ where Σ is a 3-dimensional manifold, g_{ij} a Riemannian metric and K_{ij} the second fundamental form. The metric and second fundamental form just satisfy certain constraints and be asymptotically flat. Indeed one could imagine more than one asymptotic region, just as there is in the Schwarzschild vacuum solution. The k asymptotic regions may be imagined to be compactified to give a compact manifold $\tilde{\Sigma}$, Σ being diffeomorphic to $\tilde{\Sigma}$ with k points removed. There is no complete topological classification of 3 manifolds but it is known [3] that for orientable manifolds and factors $\tilde{\Sigma}$ may be expressed uniquely as the connected sum of a number of "prime manifolds" Σ_i

$$\tilde{\Sigma} \cong \Sigma_1 \# \Sigma_2 \ldots \# \Sigma_n \qquad (1)$$

A complete list of prime manifolds is not known but it is known that for instance $S^2 \times S^1$ and elliptic spaces S^3/Γ where Γ is a suitable discrete subgroup of SO(4) with free action on S^3 are prime. Initial data satisfying the constraints which are orientable are, according to Schoen and Yau [4] probably limited to a sum of $S^2 \times S^1$'s and elliptic spaces.

The existence of a unique factorization has led Witten [5] to argue that there are no solitons in 4-dimensional gravity because if there were one would expect an antisoliton 3-metric such that one could write:

$$S^3 = \Sigma_s \# \bar{\Sigma}_s \qquad (2)$$

where Σ_s is the soliton 3-space topology and $\bar{\Sigma}_s$ that of the anti-soliton. If Σ_s is prime this is ruled out by the uniqueness. One now seems to have a problem with CPT since (2) implies that the soliton antisoliton pair cannot have the quantum numbers of the vacuum. The way out of this particular difficulty would seem to be that topology is not a "good quantum number". This seems reasonable because it appears that any topologically non-trivial initial data set must evolve to give spacetime singularities in its future [6].

According to the widely believed but still as yet unproved Cosmic Censorship Hypothesis [7] these singularities will be shielded inside event horizons. Furthermore it is also widely believed that the final state (in the classical theory) will consist of one or more time independent black holes. These black holes will have the metric of the Kerr solution.

The consequences of this are rather disappointing as far as spatial topology is concerned. Suppose one started with for instance one of Sorkin's non-orientable wormholes [8]. That is $\tilde{\Sigma} = P$ the non-orientable S^2 bundle over S^1. It is not difficult to construct initial data with this topology [9]. This has a number of fascinating topological properties [3]. For instance, topologically:

$$P \# (S^2 \times S^1) \cong P \# P \tag{3}$$

which one might interpret as saying that two non-orientable wormholes could turn into a non-orientable wormhole and a conventional orientable wormhole. All of this however will be invisible from infinity since presumably each or maybe both will be surrounded by event horizons and the fact that they are topologically non-trivial will play no role in the exterior dynamics. The final black hole solution will be a Schwarzschild or Kerr metric and no hint of the interior topology will show up in that.

Very much the same applies to the significance of the θ-vacuum structure of the initial data. One might view the configuration space Q for gravity as the space of Riemannian metrics on $\tilde{\Sigma}$ factored by the set of diffeomorphisms $\mathrm{Diff}_*(\Sigma)$ having a point on $\tilde{\Sigma}$ (the point at infinity) and its tangent space invariant. If $\mathrm{Diff}_*(\tilde{\Sigma})$ is not connected the configuration space Q will not be simply connected and θ-vacuum analogous to those in Yang-Mills theory are possible [10]. A particular instance of this is the beautiful work of Sorkin and Friedman [11] on spin ½ from gravity. Because Q is not simply connected a rotation of the spacetime relative to infinity may result in one moving around a closed loop in Q which is not homotopic to the constant path. The quantum wave function could in principle change sign under such a rotation. As an example consider as they do $\tilde{\Sigma}$ to be S^3/Γ where Γ is the 8 element group consisting of the quaternions and their negatives together with ± 1. It is quite easy to construct time symmetric initial data corresponding to this space. The resulting space Σ can be thought of as containing 7 black holes suitably identified [9]. Despite the exotic topology it seems rather likely that the end result will be just one large black hole. Again there will be no sign in the external metric of the initial exotic topology.

Finally as a final argument against the significance of 3-space topology let me remind the reader of the well known theorem of Serini, Einstein, Pauli and Lichnerowicz which I like to paraphrase as "No solitons without horizonts". The theorem states that there are no regular globally static solutions of the vacuum Einstein equations other than the flat one. The argument depends on the fact that if $g_{oo} = V^2$, with $V \neq 0$ on Σ and $V \to 1$ at infinity the field equations imply that

$$\nabla_i \nabla^i V = 0 \tag{4}$$

where ∇_i is covariant differentiation with respect to the spatial metric g_{ij}. The maximum principle immediately shows that $V = 1$. The remaining field equation now reads

$$R_{ij} = 0 \tag{5}$$

where R_{ij} is the Ricci tensor of g_{ij} which in 3-dimensions shows that g_{ij} and hence the 4-dimensional metric must be flat.

3. The Black Hole as Soliton

The remarks in section 2 have been intended to convince the reader of the importance of the 4-dimensional dynamics of the theory as opposed to that of 3-dimensional initial data. This does not mean that one can necessarily regard black holes as solitons. Far from it. They have no fixed mass or angular momentum even in the classical theory. Indeed the non-decreasing property of the event horizon area is anything but solitonic. The situation is even worse in the quantum theory since we know from the work of Hawking [12] that black holes are unstable against thermal evaporation. We are still ignorant of the final outcome of this process which may not be calculable in Einstein theory but may require a consistent quantum theory of gravity. A plausible guess is that the hole simply disappears in a puff of radiation. If this is true the black hole should be regarded in the quantum theory as an unstable "intermediate state", rather than a stable particle-like state.

The exception to this would be if the hole carried a "central" charge. By central I mean completely conserved and not carried by any of the fundamental fields of the theory. For example in N=2 ungauged extended supergravity [13] there is a Maxwell field. The fields of the N=2 supergravity multiplets are the graviton, the photon and the gravitino. These are all electrically neutral with respect to the Maxwell field - that is why the theory is "ungauged". It is quite possible for black holes to carry this charge - essentially because the lines of flux are "trapped in the topology" as people used to say in the days of "Geometrodynamics". The metric of such holes (if non-rotating) is that of Reissner and Nordstrom. It is parameterized by the mass M and charge Q. Because of the duality, invariance of the theory of any magnetic charge may be rotated to zero by a suitable duality rotation. The singularity is clothed by an event horizon if

$$M \geq |Q|/\kappa \tag{6}$$

where $\kappa^2 = 4\pi G$ and G is Newton's constant. I have described in more detail elsewhere [1,2] how one may view (6) as a Bogomolny type inequality [see also 14,15,16,17]. The electric charge Q is truly central in the sense of the supersymmetry algebra and the inequality in (6) is saturated by extreme black holes which are "supersymmetric" in that they possess "Killing spinors". There exist a whole family of multi-black hole metrics [17] satisfying (6). These are the Papapetrou-Majumdar metrics [18] which are included in the general class of Israel-Wilson metrics [19]. Tod [20] has shown that the Israel-Wilson metrics exhaust all the metrics with Killing spinors in N=2 supergravity. It has been known for some time that the throat of the extreme Reissner-Nordstrom metric has the geometry of the Robinson-Bertotti solution, i.e. the product metric on $S^2 \times (AdS)_2$ when $(AdS)_2$ is 2-dimensional anti-de Sitter space. The Robinson-Bertotti metric shares with flat space the property of being maximally supersymmetric - i.e. of having the largest possible number of Killing spinors. Thus the extreme Reissner-Nordstrom metrics spatially interpolate between the 2 possible "vacua" of N=2 ungauged supergravity. The possible relevance of this remark for spontaneous compactification is intriguing. For the present let me remark that this is typically soliton-like behaviour.

Since the charge is central it cannot be lost during Hawking evaporation and so a hole with an initial charge must settle down to the lowest mass state with that charge. This is the extreme (zero temperature) state. This extreme Reissner-Nordstrom holes seem to behave just like solitons. The hole with the opposite charge is clearly the antisoliton and it seems extremely plausible that a soliton-anti-soliton might completely annihilate one another. They cannot do this classically if Cosmic Censorship holds since by Hawking's area theorem the final event horizon must have non-vanishing area but the resultant Schwarzschild black hole can then evaporate thermally.

The main way in which the extreme holes differ from solitons is that there seems to be no way of fixing their mass or charge - i.e. no quantization rule.

Since the extreme holes (which need not all have the same mass), can remain in equilibrium it is reasonable to consider departures from equilibrium perturbatively. To lowest order they should move on geodesics on a suitable "moduli space", that is to lowest order the parameters specifying the solution should change slowly. This is the same approximation as has been used successfully in Yang-Mills theory [21,22]. In the present case the Papapetrou-Majumdar solution (representing N black holes) is specified by giving the positions of N points in $\mathbb{R}^3$. In principal the points could coincide though I will argue in a short while that this doesn't happen. If the holes, having equal masses, were identical one would factor by the action of the permutation group S_N on the N positions. Thus we know the moduli space. The metric is not known. However if one makes the approximation that one hole is very much smaller than all the others one can anticipate that the motion of the small hole in the field of the others should be given by the standard equation for a charged geodesic (with charge = mass $\times\,\kappa$). In the slow motion limit this does indeed give non-relativistic geodesic motion in the metric

$$ds^2 = U^3 d\underline{x}^2 \tag{7}$$

where

$$U = 1 + \sum_{i=1}^{i=N-1} \frac{GM_i}{|\underline{x} - \underline{x}_i|} \tag{8}$$

This metric is complete on $\mathbb{R}^3 - \{\underline{x}_i\}$. In this approximation the holes would take an infinite time to merge or coalesce.

The quantum scattering of extreme holes could be studied in the non-relativistic limit by looking at the Schrödinger equation on the moduli space. This would presumably correspond to the scalar Laplacian with respect to the metric on the moduli space, though it is also possible that potential terms might appear due to one loop effects. In the case that the holes all had equal mass one should divide out by the permutation group. The moduli space would have fundamental group S_N. The wave function could in principle then be even or odd under permutation. Thus one could imagine "fermionic" black holes! This is the analogue of the effect of Sorkin and Friedman I described above.

It is possible to find extreme black holes in the N=4 ungauged extended supergravity theory as well [23]. They should also probably be thought of as solitons. Like the extreme holes in N=2 they also have no natural mass quantization - at least as far as classical or semi-classical considerations are concerned. To get a satisfactory quantization rule one seems forced to turn to Kaluza-Klein theory.

4. Solitons in 5-dimensions

Much of the discussion about the relevance of topology in section 2 could be repeated here with 4 replacing 3. The details of the topological discussion would differ and we certainly don't have detailed singularity theorems and black hole uniqueness theorems in higher dimensions - indeed we know very little about black holes in higher dimensions. However in higher dimensions gravity is even more attractive (having a force inversely as distance to the power of the dimension of spacetime minus 2) than in 4-dimensions. In 5-dimensions it depends on distance in the same way as the repulsive centrifugal force (which is inversely as distance cubed in all dimensions). In higher dimensions it rises even more rapidly than the centrifugal repulsion. This would seem to make gravitational collapse and spacetime singularities even more likely in higher dimensions.

However there is an important difference. We are no longer obliged nor would we wish to confine ourselves to initial data which are asymptotically Euclidean. If we do so the argument that the vanishing of the Ricci tensor implies that the 4-space is flat still goes through according to Schoen and Yau's Positive Action Theorem [25]. If we don't require that the 4-metric be asymptotically Euclidean there are many complete Ricci flat 4-metrics, including one - that on the K3 surface - which is compact. Any gravitational instanton will give a static 5-metric with no horizons. Note that if we have no horizon we are still forced to have $V = 1$, that is the metric must be a product on $\mathbb{R} \times M$, where M is the 4-manifold. In the older language the spacetime would be said to be "ultrastatic".

Not all of these objects will be classically stable. The stability will be governed by spectrum of the Lichnerowicz Laplacian acting on symmetric tensors on M. If M has a self-dual metric this is known to be positive and hence the corresponding static lump will be classically stable. If M has a metric which is not self-dual the spectrum is not likely to have a positive spectrum and the corresponding lump will be unstable. Examples of this are the Euclidean Schwarzschild solution [26] and the "Taub-Bolt solution" [27].

The evolution of these objects in the full non-linear theory is unclear. The Euclidean Schwarzschild solution has the same asymptotics as the flat metric on $\mathbb{R}^3 \times S^1$ so presumably it loses energy to gravitational radiation and attempts to settle down to the flat metric but it can't do this without forming some sort of singularity since this would involve a spatial topology change. It seems likely that a black hole will be formed but this is not known. The same remarks apply to the Taub-Bolt metric which presumably tries to settle down to the Taub-NUT metric. Again black hole formation seems likely. It is possible that these black holes appear regular when viewed from a 4-dimensional stand-point in which case they should be included with those described in [24] and [28]. The Hawking effect may then cause these black holes to evolve to the flat or the Taub-NUT solution.

The boundary conditions of interest for Kaluza-Klein theory is that the metric be what has been called in this context asymptotically flat - i.e. that it approach the flat product metric on $R^3 \times S^1$ at infinity or that it be asymptotically locally flat. The typical example of the latter is the self-dual Taub-NUT metric, or multi-Taub-NUT with N centres. The topology at infinity in this case is $\mathbb{R} \times B_N$ when B_N is the S^1 bundle over S^2 with Hopf invariant N - i.e. the lens space $L(N,1)$.

Gross, Perry and Sorkin [29] have pointed out that the Taub-NUT solution plays the role of a magnetic monopole in Kaluza-Klein theory. Perry and myself [30] have shown that the monopole moment P of any asymptotically locally flat solution should satisfy the Bogomolny type inequality

$$\frac{|P|}{2\kappa} \leq M \tag{9}$$

with equality in the supersymmetric self-dual case. It is interesting to note that the gravitational instanton solution of Atiyah & Hitchin [22] is self-dual but has a negative mass. This is presumably because it has the topology at infinity of $\mathbb{R} \times (S^3/\Gamma)$ where Γ is the binary dihedral group. The crucial point here is whether or not suitable solutions of the Witten equation exist.

The multi Taub-NUT solutions are specified by giving N non-coincident points in $\mathbb{R}^3$. Permutating the points gives the same metric so the moduli space is the well known configuration space $\{(\mathbb{R}^3)^N - \Delta\}/S_N$ where Δ is the points in $(\mathbb{R}^3)^N$ where two or more points coincide and S_N is as before the permutation group on N symbols. The metric on the moduli space is under study. Again the quantum mechanics offers the possibility of multivalued wave functions though whether these monopoles can really be thought of as fermions remains at present unclear.

An important property of the Taub-NUT solutions is that the magnetic charge P satisfies the Dirac quantization condition:

$$eP = 2\pi \tag{10}$$

where e is the basic unit of charge in Kaluza-Klein theory.

This in turn implies that (using the equality in (9)) the mass M is quantized:

$$M = \frac{1}{4\pi\kappa e} \tag{11}$$

Given their stability and the quantization of the mass and magnetic charge it seems reasonable to regard the Taub-NUT solutions as representing solitons though this does require, as in section 2, that some of the topological numbers associated with the object are not conserved. In the present case two such numbers are of interest. The Hirzebruch signature and the Euler number. The multiple monopole has non-vanishing Hirzebruch signature. Roughly it corresponds to magnetic charge. Since this can be read off from the asymptotic boundary conditions one might expect this to be conserved. The Euler number is a different matter however. This cannot be determined from infinity and given the likely occurrence of singularities there seems to be no good reason for it to be conserved. Another argument, due to Hawking, is that the Euclidean action in General Relativity is not scale invariant. This means that it may cost arbitrarily little action to pass from one topological configuration to another. This is unlike the case in Yang-Mills theory in 4-dimensions where the action is scale invariant and typically topologically different configurations differ by an amount $8\pi^2/g^2$ where g is the coupling constant. If one does accept them as solitons one sees a number of striking resemblances with the massive modes of the Kaluza-Klein theory. This is the subject of the next section.

5. Pyrgon-Monopole duality

The physical content of the 5-dimensional Kaluza-Klein theory when viewed from the point of view of 4-dimensions

1) A set of massless states, the graviton, graviphoton and dilaton
2) A tower of massive states of spin 0, 1 and 2 each with mass m and charge e given by

$$m = n \frac{|e|}{2\kappa} \tag{12}$$

where $n = 1,2,3,...$

At the linearized level all the massive states are trivially stable. When one takes into account interactions one might expect the higher mass states to decay into lower mass states but a charged state cannot decay into a neutral state. Thus the lowest mass states, $n=1$, should be absolutely stable except against annihilation with their antiparticle state. These stable lowest mass states have been called Pyrgons [31]. Thus the perturbative physical Hilbert space consists of massless states, pyrgons and antipyrgons. In a supersymmetric theory the pyrgons fit into massive supermultiplets with central charge. In $N=8$ for example the relation (12) corresponds to the maximal central charge allowed. This is necessary to avoid states with spin greater than 2.

Now the G-P-S monopoles possess in the $N=8$ supergravity model of Cremmer [32] the maximum permitted number of Killing spinors and hence supersymmetries. As shown in [30] they fit into supermultiplets when the zero modes are taken into account. There is a rather close analogy, indeed one is tempted to say a duality, between the monopoles of Kaluza-Klein theory and the pyrgons. This suggested duality is analogous to that which has been suggested in Yang-Mills theory [33]. In the present case we suggest that there might exist in the full quantum theory operators which create and annihilate monopole states. In addition there will be operators which create the massless states. If these satisfy an effective field theory it is essentially unique - it must be the original field theory of the pyrgons. This is essentially because of the supermultiplets structure. Thus we have the conjectured dualities:

monopole	$\leftrightarrow$	pyrgon
massless fields	$\leftrightarrow$	massless fields
antimonopole	$\leftrightarrow$	anti-pyrgon.

It is difficult to see with present day techniques how such a conjecture could be verified. In the Yang-Mills case some partial evidence has come from a study of magnetic and electric dipole moments. It has been verified that the gyromagnetic ratio of the ordinary Yang-Mills particles equals the gyroelectric ratio of the monopoles plus fermionic zero-modes [34]. It is known that the gyromagnetic ratios of the pyrgons are anomalous and equal unity, rather than the Dirac value of 2 [35]. It would be interesting to calculate the electric dipole moments of G-P-S monopoles with their fermionic zero-modes.

Further insight into this conjectured duality might come from a study of monopole-pyrgon interactions. A number of authors [36] have pointed out that there is no "Callan-Rubakov" effect [37] which would catalyze the decay of pyrgons. This is most easily seen from the fact that scalar modes on Taub-NUT are well defined and using the covariantly constant spinor fields on Taub-NUT one can obtain all solutions of the Dirac equation.

Thus if ε is a covariantly constant spinor or Taub-NUT and a solution of the wave equation with energy ϕ_ω then:

$$\psi_\pm = [\phi_\omega \pm \frac{i}{\omega} (\not{D}\phi_\omega)]\varepsilon$$

are solutions of the Dirac equation with the same energy.

A striking fact about the scalar modes on the Taub-NUT background is that the massive scalar Pyrgon wave equation separates in 2 different coordinate systems. One system is the standard radial variables in which the metric is

$$ds^2 = (1 + \frac{2N}{\rho})^{-1} 4N^2 (d\psi + \cos\theta d\phi)^2 + (1 + \frac{2N}{\rho})(d\rho^2 + \rho^2 (d\theta^2 + \sin^2\theta d\phi^2)) \qquad (13)$$

where $0 \le \psi \le 4\pi$. Thus $8\pi N = 2\pi R_K$ where R_K is the radius of the Kaluza-Klein circle. The scalar field has the form

$$\phi_\omega = e^{-i\omega t} e^{i\frac{n}{2}\psi} \, {}_{\frac{n}{2}}Y_{\ell m}(\theta) \, e^{im\phi} f_n(r) \qquad (14)$$

where ${}_{\frac{n}{2}}Y_{\ell m}(\theta) e^{im\phi}$ is a spin weighted spherical harmonic and where $f_n(r)$ is a non-relativistic Coulomb wave function with angular momentum ℓ but where the Coulomb potential is energy dependent, i.e. depends upon ω , that is f satisfies

$$\frac{1}{\rho^2} \frac{d}{d\rho}(\rho^2 \frac{df}{d\rho}) - \frac{\ell(\ell+1)f}{\rho^2} + (2N\omega^2 - \frac{n^2}{N})\frac{f}{\rho} + (\omega^2 - \frac{n^2}{4N^2})f = 0 \qquad (15)$$

There are no bound states, just scattering states. Since the radial equation (15) is a Coulomb one one might anticipate that scattering is better described using parabolic coordinates, defined by

$$\xi = (1 + \cos\theta)$$
$$\eta = (1 - \cos\theta) \qquad (16)$$

This is in fact true. The wave equation also separates in the t,ϕ,ξ,η coordinates. Using them one can give a simple description of the scattering. The classical orbits are especially simple being conic sections. They are, when projected into the 3-space spanned by ρ, θ and ϕ, the intersection of a cone centred at $\rho=0$ with a plane, the intersection being a hyperbola in general.

The existence of 2 different coordinate systems in which the wave equation separates is often taken as the indication of hidden symmetries and indeed of a "spectrum generating algebra". The precise nature of this algebra in the present case has not been worked out. It is tempting to speculate that it may be related to the known existence of Kac-Moody algebras in Kaluza-Klein theory [38].

Another tempting speculation is that these ideas will find their full expression in string theory. Mike Green [39] has remarked that if one considers 10-dimensional string theory on where 10-D of the spacelike dimensions form a torus, each of whose radii equals R one obtains string states with masses satisfying

$$(\text{mass})^2 = \sum_{i=0}^{\infty} \left(\frac{M_i^2}{R^2} + \frac{R^2 N_i^2}{\alpha'^2}\right) + \frac{2}{\alpha'}(N_o + \tilde{N}_o) \qquad (17)$$

N_o and $\tilde{N}_o$ are occupation numbers for higher string states. The integers $\{M_i\}$ are Kaluza-Klein charges resulting from the periodicity in the 10-D compact dimensions. The integers $\{N_i\}$ are topological charges associated with the number of times a closed string winds round the i'th compact dimension. Consider the limit

$$R \to 0 \quad \text{and} \quad \frac{\alpha'}{R} \quad \text{constant} = \lambda$$

The resulting D-dimensional field theory has an infinite number of massive spin 2 supermultiplets whose masses are determined by λ. This theory is apparently identical to the theory obtained by starting with 10-dimensional N=2 supergravity and compactifying on a hypertorus with (10-D) dimensions having finite radii $R = \lambda$.

Now set D=5 . The reduction of N=2 d=10 to 5 dimensions gives Cremmer's N=8 D=5 model, with its pyrgon states. On the other hand from (17) we see that the states corresponding to zero Kaluza-Klein charge but non-vanishing topological winding numbers will survive in this limit. These presumably correspond to the magnetic monopole states.

References

[1] G.W. Gibbons in "Supersymmetry, Supergravity and Related Topics" ed. F. del Aguila, J.A. de Azcarraga and L.E. Ibanez, World Scientific 1985.

[2] G.W. Gibbons in "Non Linear Phenomena in Physics" ed. F. Claro, Springer Proceedings in Physics #3. Springer Verlag 1985.

[3] J. Hempel "Topology of 3-Manifolds", Princeton University Press (1976).

[4] R. Schoen and S.T. Yau, Phys. Rev. Lett. 43 1457 (1979).

[5] E. Witten, Commun. Math. Phys. 100 197 (1985).

[6] D. Gannon, J. Math. Phys. 2364 (1975).
D. Gannon, G.R.G. 7 219 (1976).
C.W. Lee, Commun. Math. Phys. 51 157 (1976).

[7] R. Penrose, Ann. N.Y. Acad. Sci. 224 125 (1973).

[8] R. Sorkin, J. Phys. A10 717 (1977).

[9] G.W. Gibbons, unpublished.

[10] C.J. Isham, Phys. Lett. 106B 188 (1981).
C.J. Isham, in "Quantum Structure of Space and Time", eds. C.J. Isham & M.J. Duff. Cambridge University Press (1982).

[11] J. Friedman & R. Sorkin, Phys. Rev. Lett. 44 1100 (1980); see also B. Witt: Milwaukee preprint.

[12] S.W. Hawking, Nature (Lond.) 248 30 (1974).
S.W. Hawking, Commun. Math. Phys. 43 199 (1975).

[13] S. Ferrara & P. van Nieuwenhuizen, Phys. Rev. Lett. 37 1669 (1976).

[14] G.W. Gibbons, in Heisenberg Memorial Symposium, ed. P. Breitenlohner and H.P. Durr, Springer Lecture Notes in Physics #160.

[15] G.W. Gibbons & C.M. Hull, Phys. Letts. 109B 190 (1982).

[16] G.W. Gibbons, in Proc. 4th Silarg Symposium, ed. C. Aragone, World Scientific.

[17] J.B. Hartle and S.W. Hawking, Commun. Math. Phys. 26 87 (1982).

[18] A. Papapetrou, Proc. Roy. Irish. Acad. A51 191 (1947).
S.D. Majumdar, Phys. Rev. 72 390 (1947).

[19] W. Israel and G.A. Wilson, J. Math. Phys. 13 865 (1972).

[20] P. Tod, Phys. Lett. B121 241 (1983).

[21] N. Manton, Phys. Rev.
N. Manton in "Monopoles in Quantum Field Theory" ed. N. Craigie, P. Goddard and W. Nahm, World Scientific (1982).

[22] M.F. Atiyah and N. Hitchin, Phys. Lett. 107A 21 (1985).

[23] G.W. Gibbons, Nucl. Phys. B207 337 (1982).

[24] W. Simon, G.R.G. 17 761 (1985).

[25] R. Schoen and S.T. Yan, Phys. Rev. Lett. 42 547 (1979).

[26] D. Page, Phys. Rev.
B. Allen, Phys. Rev. D30 1153 (1984).

[27] R.E. Young, Phys. Rev. D28 2420 (1983).

[28] G.W. Gibbons and D. Wiltshire, Annals of Phys., in press.

[29] D. Gross and M.J. Perry, Nucl. Phys. B226 29 (1983).
R. Sorkin, Phys. Rev. Lett. 51 87 (1983).

[30] G.W. Gibbons and M.J. Perry, Nucl. Phys. B248 629 (1984).

[31] E.W. Kolb and R. Slansky, Phys. Lett. 135B 378 (1984).

[32] E. Cremmer in "Superspace and Supergravity", ed. S.W. Hawking and S.W. Hawking & M. Rocek, Cambridge University Press 1981.

[33] C. Montonen and D.I. Olive, Phys. Lett. 72B 117 (1977).
[34] H. Osborn, Phys. Lett. 115B 226 (1982).
Bo-Yu. Hou, Phys. Lett. 125B 389 (1983).
[35] A. Hoysoyer et al., Phys. Lett. 134B (1984).
[36] P.C. Nelson, Nucl. Phys. 238B 638 (1984).
H. Ezawa and A. Iwasaki, Phys. Lett. 138B 81 (1984).
M. Kobayashi and A. Sugamoto, Prog. Theor. Phys. 72 122 (1984).
F.A. Bais and P. Batenburg, Nucl. Phys. B245 469 (1984).
[37] V. Rubakov., Pisma Zh. Eksp. Teor. Fiz. 33 658 (1981); Nucl. Phys. 203B 311 (1982).
C.G. Callan, Phys. Rev. D25 2141 (1981).
[38] A. Salam and J. Strathdee, Annals of Phys. 141 316 (1982).
L. Dolan and M.J. Duff, Phys. Rev. Lett. 52 14 (1984).
[39] M. Green "The Status of Superstrings" undated Queen Mary College preprint.

TRUNCATIONS IN KALUZA-KLEIN THEORIES

C.N. Pope

Blackett Laboratory, Imperial College, Prince Consort Road, London SW7 2BZ, UK.

Certain mathematical aspects of Kaluza-Klein theories are discussed, concerned with the ability to truncate the four-dimensional spectrum of states to a finite subsector, including the graviton and Yang-Mills gauge bosons. This yields a criterion by means of which certain exceptional theories are singled out from the generic case.

1. INTRODUCTION

Kaluza-Klein theories provide a natural and geometrical unification of gravity and gauge fields, in which general coordinate invariance and local gauge invariance both arise as subsectors of general coordinate invariance in a higher dimension. However, in a successful unification of all the fundamental forces in nature, one would like the unifying theory to be unique, and in this respect Kaluza-Klein theories seem at first sight to fare rather badly. Not only does one have the usual freedom , as with four-dimensional theories, to pick and choose what fields are to be included in the Lagrangian, but one also has the additional freedom to choose one's favorite dimension!

In order to try to single out the 'right' theory from all the candidate theories, one requires some rather powerful criteria which can be used to restrict the possibilities. Broadly speaking, such criteria tend to divide into two categories; on the one hand there are those based on physical principles derived from phenomenological considerations, while on the other hand there are mathematical principles based on the requirement of self-consistency of the theory.

Examples of physical principles would be the requirement that the theory admit a realistic gauge group with chiral fermions, and a Minkowski space ground state. Superficially such requirements seem very reasonable, but it should be borne in mind that the natural unification scale of any quantum theory of gravity is the Planck scale, 10^{19} GeV or 10^{-33} cm, whilst the physical principles mentioned above are based on observations in particle accelerators at energy scales lower by about 17 orders of magnitude, and cosmological comparisons involve a further extrapolation of about 40 orders of magnitude. Seen in this light, it is perhaps rather premature to be imposing these physical requirements from the outset. Of course the 'correct' theory should ultimately be able to explain the observed phenomena, but it may well be that they emerge in a highly non-trivial way as low-energy collective phenomena which are by no means manifest in the fundamental theory. The phenomenon of superconductivity is possibly a good analogy; it can be understood very satisfactorily within the framework of the non-relativistic Schrödinger equation applied to the theory of electrons in metals, but it is a consequence of highly non-trivial collective effects which were only understood long after the development of quantum mechanics. It would not have been very fruitful at the time when quantum mechanics was being developed to have demanded as a prerequisite of a successful microscopic theory that it be seen at the outset to be able to explain superconductivity.

An illustration of the way in which a quantum theory of gravity might look very different on microscopic and macroscopic scales is provided by the idea of 'spacetime foam', originally suggested by Wheeler[1] and developed by Hawking[2]. In order that spacetime be macroscopically flat, these studies suggested at the Planck scale it should be topologically complex and 'foamlike', with an effective cosmological constant of order -1 in Planck units, compared with the observed Λ at large scales which is smaller by at least 120 orders of magnitude. Thus a huge negative cosmological constant might even be a desirable feature of the fundamental theory! This simple model should probably not be taken too seriously at present, but it does serve to illustrate the point that one should perhaps be wary of imposing our everyday low-energy prejudices at the Planck scale.

We will therefore adopt the view that for now we should be guided more by mathematical principles of self-consistency than by physical considerations. Ultimately, an obvious requirement would be that the theory should be finite, and it is possible that this may be such a powerful criterion that it would lead to a unique unified theory. The recent surge of interest in superstring theories is based in part on such a hope. The issues involved here are far too complicated to be answerable at present, but the same general principle of mathematical consistency is one which can be applied at many more elementary levels. This paper is primarily concerned with certain consistency questions in classical Kaluza-Klein theories, and it is to these that we now turn.

2. INCONSISTENCIES IN GENERIC KALUZA-KLEIN THEORIES

Any (4+k)-dimensional theory which admits spontaneous compactifications to ground state solutions of the form $M_4 \times M_k$, where M_4 is four-dimensional spacetime and M_k is a compact internal space, admits a Kaluza-Klein interpretation in which fluctuations of the

(4+k)-dimensional fields around their ground-state values have a 4-dimensional interpretation as excitations of infinite towers of massless and massive states (see, for example, ref.3). Retaining all these states is necessarily consistent, since it is just a rewriting of the original higher dimensional theory. However, if one tries to truncate to just a finite number of states, then in a generic Kaluza-Klein theory one is liable to run into inconsistencies. These arise because in general the states which are retained act as inhomogeneous source terms for the states which are discarded, and hence setting these states to zero is inconsistent with their equations of motion.

To see this in detail, let us consider the case of pure gravity with a cosmological constant in (4+k) dimensions, with field equation

$$\hat{R}_{MN} - \frac{1}{2} \hat{R}\, \hat{g}_{MN} + \Lambda\, \hat{g}_{MN} = 0, \tag{1}$$

We use the notation that the (4+k)-dimensional fields are 'hatted', and M,N,... are world indices running over 4+k values. These will be decomposed as $M=(\mu,m)$, etc., where μ runs over M_4 and m runs over M_k.

Equation (1) admits ground state solutions of the form $M_4 \times M_k$, where M_k is a compact Einstein space satisfying

$$R_{mn} = \frac{2\Lambda}{k+2}\, g_{mn}, \tag{2}$$

and M_4 is a 4-dimensional spacetime satisfying the Einstein equation with cosmological constant. We will assume Λ is positive, so M_k can be a space with continuous symmetries, such as the k-sphere with SO(k+1) isometry group.

The standard Kaluza-Klein ansatz, which is designed to truncate to just the gravity and gauge boson degrees of freedom in 4 dimensions, corresponds to setting the (4+k)-dimensional metric to be

$$\hat{g}_{MN} = \hat{e}_M{}^A\, \hat{e}_N{}^B\, \eta_{AB}, \tag{3}$$

where

$$\hat{e}^\alpha(x,y) = e^\alpha{}_\mu(x)\, dx^\mu, \tag{4}$$

$$\hat{e}^a(x,y) = e^a{}_m(y)\, dy^m - K^{ia}(y)\, A^i{}_\mu(x)\, dx^\mu, \tag{5}$$

where A,B,... are (4+k)-dimensional local Lorentz indices decomposed into $\alpha,\beta,\ldots$ running over M_4 and a,b,... in M_k, x^μ and y^m are the coordinates on M_4 and M_k respectively, K^i are the Killing vectors on M_k generating its isometry group G (i=1,...dim G), and $A^i{}_\mu(x)$ are the Yang-Mills gauge potentials. The k-bein $e^a{}_m(y)$ is related to the metric on M_k, which is unchanged from the ground state and still satisfies (2), by $e^a{}_m \, e^b{}_n \, \delta_{ab} = g_{mn}$.

A straightforward calculation of the Riemannian curvature for (4) and (5) shows that the Ricci tensor is given by

$$\hat{R}_{\alpha\beta} = R_{\alpha\beta} - \frac{1}{2} K^{ia} K^j{}_a F^i{}_{\alpha\gamma} F^j{}_\beta{}^\gamma, \tag{6}$$

$$\hat{R}_{ab} = R_{ab} + \frac{1}{4} K^i{}_a K^j{}_b F^i{}_{\alpha\beta} F^{j\alpha\beta}, \tag{7}$$

$$\hat{R}_{\alpha b} = - \frac{1}{2} K^i{}_b D_\beta F^i{}_\alpha{}^\beta, \tag{8}$$

where

$$F^i{}_{\alpha\beta} = 2 \nabla_{[\alpha} A^i{}_{\beta]} + c_{ijk} A^j{}_\alpha A^k{}_\beta, \tag{9}$$

the structure constants c_{ijk} are defined by

$$[K^i, K^j] = c_{ijk} K^k, \tag{10}$$

where $K^i = K^{im} \partial_m$, and D_β is the Yang-Mills covariant derivative. $R_{\alpha\beta}$ and R_{ab} are the Ricci tensors for the vielbeins $e^\alpha{}_\mu(x)$ and $e^a{}_m(y)$ on M_4 and M_k respectively.

In the local Lorentz basis $e^A{}_M$, the (4+k)-dimensional field equation (1) reads

$$\hat{R}_{AB} - \frac{1}{2} \hat{R} \eta_{AB} + \Lambda \eta_{AB} = 0. \tag{11}$$

Substituting (6), (7) and (8) into (11) now reveals the inconsistencies. There are three cases to consider, corresponding to AB = $\alpha\beta$, αb and ab in (11). The first of these yields

$$R_{\alpha\beta} - \frac{1}{2} R \eta_{\alpha\beta} + \Lambda \eta_{\alpha\beta} = \frac{1}{2} \left(F^i{}_{\alpha\gamma} F^j{}_\beta{}^\gamma - \frac{1}{4} F^i{}_{\gamma\delta} F^{j\gamma\delta} \eta_{\alpha\beta} \right) K^{ia} K^j{}_a, \tag{12}$$

where $R = R_{\alpha\beta} \eta^{\alpha\beta}$ is the 4-dimensional Ricci scalar. All the terms in (12) depend only on the spacetime coordinates x^μ, except for the Killing vectors K^i, which depend upon the coordinates y^m on M_k. In fact in general one has

$$K^{ia} K^{j}{}_{a} = \delta^{ij} + Y^{ij}(y), \tag{13}$$

where $Y^{ij}(y)$ is symmetric and tracefree in i and j. Thus defining $T^{ij}{}_{\alpha\beta}$ by

$$T^{ij}{}_{\alpha\beta} = F^{(i}{}_{\alpha\gamma} F^{j)}{}_{\beta}{}^{\gamma} - \frac{1}{4} F^{i}{}_{\gamma\delta} F^{j\gamma\delta} \eta_{\alpha\beta}, \tag{14}$$

(12) implies that the tracefree part of $T^{ij}{}_{\alpha\beta}$ must vanish[4],

$$T^{ij}{}_{\alpha\beta} - (\dim G)^{-1} \delta^{ij} T^{kk}{}_{\alpha\beta} = 0, \tag{15}$$

while the y-independent part of (12) gives the standard Einstein equation with Yang-Mills source term. However, the algebraic constraint (15) on the Yang-Mills fields means that the ansatz (4), (5) is inconsistent with ($e^{\alpha}{}_{\mu}(x)$, $A^{i}{}_{\mu}(x)$) being an arbitrary solution of the 4-dimensional Einstein-Yang-Mills equations.

The only way of overcoming this inconsistency of the truncation in (4) and (5) is to restrict the Killing[4] vectors to just those which generate some subgroup G' of G for which

$$K^{i'a} K^{j'}{}_{a} = \delta^{i'j'}. \tag{16}$$

Since (16) implies that each Killing vector has constant (unit) length, the subgroup G' can be non-trivial only if the Euler number χ of M_k vanishes (since if $\chi \neq 0$ then all vector fields must vanish somewhere). Even if $\chi = 0$, then G' is usually a lot smaller than G. For example if M_k is the SO(k+1)-invariant k-sphere, then G' is SU(2) when k=4n+3, and G' can only be U(1) when k=4n+1. On a group manifold H, with $G = H_L \times H_R$, G' can be either H_L or H_R.

If one were prepared to include more 4-dimensional fields in the ansatz (4), (5), consistency could be restored for the entire isometry group G by including massive spin 2 fields, which would introduce a balancing y-dependence on the left-hand-side of (12). But we know on general grounds that massive spin 2 can only be consistently coupled to gravity by coupling infinitely many such fields[5] which would defeat the object of the truncation. If the entire isometry group G is retained, then setting these massive spin 2 fields to zero is inconsistent with their equations of motion[4].

The αb components of (11) present no difficulties, yielding the Yang-Mills equation

$$D_\beta F^{i}{}_{\alpha}{}^{\beta} = 0. \tag{17}$$

However the ab components of (11) yield

$$K^{i}{}_{a} K^{j}{}_{b} F^{i}{}_{\alpha\beta} F^{j\alpha\beta} = 0, \tag{18}$$

another unacceptable constraint. This time there is no subgroup of G for which this inconsistency is avoided. The resolution here is that one must include scalar degrees of freedom $\phi^{i'j'}(x)$ in (4) and (5), corresponding to allowing the metric components g_{mn} to fluctuate as $\phi^{i'j'} K^{i'}{}_{m} K^{j'}{}_{n}$, which means that now (18) is of the general form $\Box\phi^{i'j'} \sim F^{i'\cdot}{}_{\alpha\beta} F^{j'\alpha\beta}$. Of course we must still also restrict to the subgroup G' for which (16) holds, since the introduction of the scalar fields does not resolve the previous inconsistency problem in (12).

The situation described above in the case of the (4+k)-dimensional pure Einstein equation is reasonably representative of any generic Kaluza-Klein theory. In order to write an ansatz which extracts a finite number of 4-dimensional fields including the graviton and gauge bosons, one must restrict the gauge group from the isometry group G to the subgroup G' whose Killing vectors satisfy (16), and also include the Kaluza-Klein scalars $\phi^{i'j'}$, which are in the symmetric product of the adjoint of G' with itself. The need to restrict from G to G' can be understood from a group theoretical point of view: In general the ansatz (4), (5) must be invariant under a transitively acting subgroup K of the isometry group of M_k, and G' is a subgroup of G which is centralized by K in G[6,7]. Of course, in order for G' to be non-trivial, this means that M_k must certainly be an homogeneous space.

The need to restrict from G to G', and include the scalars, arose from insisting that the ansatz should satisfy the higher dimensional field equations. In an alternative approach, the view is sometimes taken that one should substitute the ansatz into the higher dimensional action, and integrate out over y, to obtain an effective 4-dimensional action. In our dicussion, this would be equivalent to averaging (12) over M_k, in which case the y-dependent term in (13) would disappear, and omitting the the scalars $\phi^{i'j'}$ from the ansatz (so the possibility of varying them in the 4-dimensional effective action would of course never arise). Thus one could obtain a 4-dimensional Einstein-Yang-Mills action with gauge group G. However this procedure is a prescription for satisfying certain components of the higher-dimensional field equations and violating others, and such an approach, which would have to be justified on physical grounds, runs counter to the philosophy that it is premature to be imposing low- energy physical prejudices at the Planck scale. One would also lose the correspondence between extrema of the 4-dimensional action and extrema of the higher-dimensional action.

However, perhaps the most compelling reason for adopting the 'consistent' approach is that in certain very exceptional theories, remarkable consistent truncations can be made which cannot be understood within the framework of the general discussion of this Section. Thus we have a mathematical criterion which singles out certain Kaluza-Klein theories as being very special. It is to one such theory that we now turn.

3. ELEVEN DIMENSIONAL SUPERGRAVITY.

We will examine the bosonic sector of d=11 supergravity[8], which comprises the metric tensor $\hat{g}_{MN}$ and a 3-form potential $\hat{A}_{MNP}$, satisfying the field equations

$$\hat{R}_{AB} - \frac{1}{2}\hat{R}\,\eta_{AB} = \frac{1}{3}\left(\hat{F}_{ACDE}\,\hat{F}_B{}^{CDE} - \frac{1}{8}\hat{F}^2\,\eta_{AB}\right), \tag{19}$$

$$\hat{\nabla}_A\,\hat{F}^{ABCD} = -\frac{1}{576}\,\varepsilon^{BCDE_1\ldots E_8}\,\hat{F}_{E_1\ldots E_4}\,\hat{F}_{E_5\ldots E_8}, \tag{20}$$

where $\hat{F}_{ABCD} = 4\,\hat{\nabla}_{[A}\,\hat{A}_{BCD]}$, and we are using local Lorentz indices. These admit ground state Freund-Rubin solutions[9] on $M_4 \times M_7$, in which one sets all components of $\hat{F}_{ABCD}$ to zero except in spacetime, where $\hat{F}_{\alpha\beta\gamma\delta} = 3m\varepsilon_{\alpha\beta\gamma\delta}$; M_7 is an Einstein space satisfying $R_{ab} = 6m^2\,\delta_{ab}$, and in spacetime M_4, $R_{\alpha\beta} = -12m^2\eta_{\alpha\beta}$.

One can easily show that already at the linearized level it is necessary to augment the elfbein ansatz (4) and (5) by an ansatz on $\hat{F}_{ABCD}$ including the gauge bosons, in order to extract the massless spin 1 degrees of freedom in M_4. The correct ansatz for $\hat{F}_{ABCD}$ turns out to be[10,4]

$$\hat{F}_{\alpha\beta\gamma\delta} = 3m\,\varepsilon_{\alpha\beta\gamma\delta}, \quad \hat{F}_{\alpha\beta cd} = \frac{1}{2m}\,\varepsilon_{\alpha\beta\gamma\delta}\,F^{i\gamma\delta}\,\nabla_c\,K^i{}_d. \tag{21}$$

Substituting into (19) and choosing AB = αβ, this yields

$$R_{\alpha\beta} - \frac{1}{2}R\,\eta_{\alpha\beta} - 12m^2\,\eta_{\alpha\beta} = \frac{1}{2}T^{ij}{}_{\alpha\beta}\left(K^{ia}\,K^j{}_a + \frac{1}{2m^2}\nabla_a K^i{}_b\,\nabla^a K^{jb}\right), \tag{22}$$

where $T^{ij}{}_{\alpha\beta}$ is defined by (14). This equation should be compared with

(12). In this case, the equation[4] is consistent for a subgroup G' of G whose Killing vectors satisfy

$$K^{i'a} K^{j'}{}_a + \frac{1}{2m^2} \nabla_a K^{i'}{}_b \nabla^a K^{j'b} = \delta^{i'j'}. \qquad (23)$$

It is easy to see that if $K^{i'a}$ satisfies (16) then $\frac{1}{2} K^{i'a}$ satisfies (23), so the question arises as to whether more Killing vectors can satisfy (23) than (16). For a generic Einstein space M_7 the answer seems to be no, but in just one case the situation is different. On the SO(8)-invariant round 7-sphere, all 28 Killing vectors of SO(8) satisfy (23)[4].

Of course the above calculation has only been concerned with the $\alpha\beta$ components of the Einstein equation (19). Full consistency of (19) and (20) would certainly require the inclusion of Kaluza-Klein scalars in the ansatze for $\hat{g}_{MN}$ and $\hat{A}_{MNP}$. However, the calculation has already exhibited a property of the 7-sphere compactification of d=11 supergravity that seems to be unique; no other known Kaluza-Klein theory could possibly yield all the gauge bosons of a non-abelian isometry group G in 4-dimensions, since all other known theories yield an inconsistency in the $\alpha\beta$ components of the Einstein equation.

In fact there are strong indications that the full d=11 supergravity theory, compactified on S^7, can be consistently truncated to just the massless N=8 supergravity multiplet, whose bosonic sector comprises the graviton, 28 gauge bosons, and 35 each of scalars and pseudo-scalars. The complexities of the spin 0 sector have so far defeated all attempts to perform a complete check of the consistency, but partial results in this direction are encouraging[11,12]. Note that the fact that only a 35 of scalars is needed is another remarkable property of this truncation. Generically, one would have expected to need all the representations occuring in the symmetric product of 28 with itself.

The full N=8 truncation has been intractable to date because of the complications due to the spin 0 fields. A simpler problem, which still exhibits some remarkable properties of the theory, is to first truncate to N=3 supergravity (for which the supermultiplet contains no spin 0 fields) and then discard the fermions. One is then left with just the Einstein-Yang-Mills system, with SU(2) gauge group. The remarkable point here is that consistency should be achievable with no scalar fields atall. The calculations, which is still quite involved, is described in detail in ref. 13. Restricting to the appropriate SU(2) subgroup of SO(8), it turns out that the ansatz (21) for $\hat{F}_{ABCD}$ is correct to all orders, and one finds that (4), (5) and (21) yield an exact solution of (19) and (20), where ($g_{\mu\nu}(x)$, $A^i{}_\mu(x)$) is an arbitrary solution of the 4-dimensional Einstein-Yang-Mills equations with SU(2) gauge group. There is no other theory known to admit this kind of non-trivial embedding of solutions of the Einstein-Yang-Mills equations.

A remarkable feature of this SU(2) truncation is that consistency of the purely bosonic subsector depends crucially upon the presence in the eleven dimensional theory of the $\varepsilon^{M...S}\hat{F}_{M...N}\hat{F}_{P...Q}\hat{A}_{R...S}$ term in the Lagrangian, with precisely the coefficient demanded by supersymmetry. Thus consistency and supersymmetry seem to be intimately related, although the precise way in which this works remains unclear.

4. CONCLUSION

We have seen in section 2 that in a generic Kaluza-Klein theory it is often not possible to make an ansatz which extracts just the massless four-dimensional fields and which satisfies the higher-dimensional equations of motion. In such cases the only way to restore consistency is to reinstate some of the previously truncated fields. In the example of section 2, this would include infinitely many massive spin 2.

Suppose, however, that one were prepared to take the point of view that one should simply take the massless ansatz (even though it is inconsistent) and substitute it into the higher-dimensional action, thereby obtaining an effective four-dimensional action describing just the massless fields. What would go wrong?

As discussed in ref.3, the problem is one of non-uniqueness. Specifying that one should make a massless ansatz is merely a statement that at the linearized level the fluctuations around the ground state are to be expanded as spacetime fields times zero-mode harmonics of the relevant mass-operators on M_k. Provided that one respects the symmetries of the system, one is free to make any non-linear modification of this prescription that one wishes, since it will not affect the criterion of masslessness. For example, one could add a term $\phi^2(x)\,Y^{(1)}(y)$ to the right-hand-side of the massless ansatz $\Phi(x,y)=\phi(x)\,Y^{(0)}(y)$, where $Y^{(0)}$ is a zero-mode and $Y^{(1)}$ is a non-zero mode of the relevant mass operator. Although such a modification will leave the quadratic terms in the effective four-dimensional Lagrangian unchanged, it will of course drastically alter the interaction terms. Which is the correct choice?

In the case of a Kaluza-Klein theory admitting a consistent truncation, the answer is unambiguous: the correct choice is the one that ensures that the ansatz satisfies the equations of motion. (Of course one still has the freedom to make field redefinitions amongst the massless fields.) However if there is no consistent truncation, then there simply is no unique choice of massless ansatz, and so most of the interaction terms in the effective four-dimensional Lagrangian are arbitrary, i.e. their coefficients depend upon which particular massless ansatz one chooses. Thus only for a consistent truncation does it make sense to study the non-linear structure of a Kaluza-Klein theory.

ACKNOWLEDGEMENT

I would like to thank M.J. Duff, G.W. Gibbons, B.E.W. Nilsson and K.S. Stelle for many helpful discussions.

REFERENCES

1) J.A. Wheeler, in: Relativity, groups and topology, eds B.S. deWitt and C.M. deWitt (Gordon and Breach, New York, 1964).

2) S.W. Hawking, Nucl. Phys. B144 (1978) 349.

3) M.J. Duff, B.E.W. Nilsson and C.N. Pope, Phys. Report, in print.

4) M.J. Duff, B.E.W. Nilsson, C.N. Pope and N.P. Warner, Phys. Lett. 149B (1984) 90.

5) D. Boulware and S. Deser, Ann. Phys. 81 (1975) 193.

6) N. Manton, UCSB preprint NSF-ITP-83-04.

7) M.J. Duff and C.N. Pope, Nucl. Phys. B255 (1985) 355.

8) E. Cremmer, B. Julia and J. Scherk, Phys. Lett. 76B (1978) 409.

9) P.G.O. Freund and M.A. Rubin, Phys. Lett. 97B (1980) 233.

10) M.J. Duff and C.N. Pope, in: Supersymmetry and supergravity 82, eds S. Ferrara, J.G. Taylor and P. van Nieuwenhuizen (World Scientific, Singapore, 1983).

11) B.E.W. Nilsson, Phys. Lett. 155B (1985) 54.

12) B. de Wit, H. Nicolai and N.P. Warner, Nucl. Phys. B255 (1985) 29.

13) C.N. Pope, Class. Quantum Grav. 2 (1985) L77.

CANONICAL QUANTIZATION AND COSMIC CENSORSHIP

P. Hajicek
Institute for Theoretical Physics
University of Bern
Sidlerstrasse 5, CH-3012 Bern, Switzerland

1. Introduction

One of the most difficult problems of quantum gravity originates from the well-known feature of the theory that the causal structure of spacetime - the system of light cones - is a function of the dynamical field itself. Moreover, realistic models of matter fields and gravity will be unstable with respect to the gravitational collapse and formation of black holes. This means, in this context, that the deviations of causal structures of possible dynamical developments from each other, or from some standard structure like e.g. that of the Minkowski spacetime, can be drastically large. (We have assumed here that black-hole-like objects exist in nature, but this is very plausible due to cumulating observations - see, e.g. [1] .) By the way, such an assumption could lead to some limits on realistic models for gravity: it seems that the classical limit of such models had to yield more than just the perturbation series for general relativity around the flat background. Then, all such models, even with high derivatives, with a compound graviton, supergravity, Kaluza-Klein, string theory, etc. will suffer from the above problem.

Most of the today investigations are based on the expansion in the number of loops. This means that the true light cones are approximated by the light cones of a given, fixed classical solution (corresponding, say, to the ground state). However, such an approximation is dangerous even in the purely classical theory - the series leading to divergent integrals in higher order contributions [2] .

Another way to avoid the problem seems to be offered by the Euclidean regime: the dynamical equations become elliptic and there is no explicit causal structure. However, the difficulty seems to reappear at a different level: the corresponding quantum theory becomes acausal [3] , non-unitary [4] , not asymptotically complete [5] , and leads to the loss of quantum coherence [4] .

Within the canonical quantization, the above problem takes on the following form. For any canonical formalism to work, we have to foliate the spacetime by Cauchy hypersurfaces. However, the existence of such a foliation, the so-called global hyperbolicity is a very special property of the causal structure. As the causal structure itself is a function of the dynamical field, we do not know whether or not we are in conflict with the quantum dynamics, if we require the global hyperbolicity from the outset.

Let us explain by an example what is meant by a "conflict with the dynamics". First, it is possible to weaken the global hyperbolicity somewhat. Let us consider only asymptotically flat spacetimes. Then, for the scattering problems to be well-defined, it is sufficient that the part $[I^-(\mathcal{J}^+) \cap I^+(\mathcal{J}^-)]$ of each element M' from a "large" class of such spacetimes is globally hyperbolic. This means, roughly speaking, that in each spacetime M' satisfying the requirement, there are no singularities which are visible from $\mathcal{J}^+$ and at the same time influenceable from $\mathcal{J}^-$. Large class means that the exceptions form a set of measure zero. It is convenient to put the measure on the space of regular Cauchy data for asymptotically flat spacetimes. Then, the requirement is clearly closely related to the so-called Weak Cosmic Censorship [6]. In the classical form of this hypothesis, one assumes that the whole system of the classical field equations is satisfied by all spacetimes M'. Thus, a violation of the Weak Cosmic Censorship could be considered as a sort of conflict between the classical dynamics and the weakened global hyperbolicity. Now, it is well known that this form of Cosmic Censorship is very likely to be violated [7].

However, the dynamics we consider is the quantum dynamics. The corresponding Quantum Cosmic Censorship Hypothesis (Q.C.C.H.) is not equivalent to the classical one. In fact, some people believe that the singularities of the classical general relativity will be avoided in the corresponding quantum theory. If this is true, then the Q.C.C.H. will be more likely to be satisfied than the classical one. We shall touch this problem in more detail later on.

Another, more subjective difficulty with the existence of Cauchy hypersurfaces is that the globally hyperbolic spacetimes are "topologically dull" (as Hawking puts it): no change of the space topology is possible (this is, roughly, the content of a classical Geroch theorem [6]). However, as far as I know, there is no proof that the corresponding quantum dynamics will also prohibit any change of space topology and split, in this way, into the corresponding superselection sectors.

Sometimes, one compares the general relativity with the string theory. Any classical (i.e. non-quantized) string is, on one hand, a two-dimensional spacetime with well-defined dynamics. On the other hand, changes in topology of the time-constant folii of the classical string manifold are possible (and even necessary in order that there is any interaction between strings). Why, so one asks, are these two facts compatible in the string theory, and not compatible in the general relativity ? The answer is simple: the strings are able to join their ends to form a regular internal string point or to be torn into pieces with regular end points. No such discontinuities are allowed for classical spacetimes.

2. The method of canonical reduction

Suppose we have some self-consistent field-theoretical model containing gravity in the form of spacetime metric. Then, we can try to quantize it by the so-called canonical reduction method (see, e.g. [8]). This method consists roughly in the following steps. First, one adds some gauge conditions to the dynamical equations. The gauge conditions have the form of equations (mostly differential equations) containing the dynamical variables of the model. Then, the variables are divided into the following classes: the true dynamical variables, the dependent variables, the gauge variables and the Lagrange multipliers. The dynamical equations are divided into gauge conditions, gauge propagating equations, constraints and the true dynamical equations. The gauge conditions, gauge propagating equations and constraints are solved for the dependent variables, gauge variables and Lagrange multipliers, and the true dynamical equations are expressed through the true dynamical variables only. In this way, the constraints and gauge freedom disappear and we obtain only mutually independent dynamical variables. Such a reduction can be performed within a Lagrangian (second order) or Hamiltonian (first order) formulation and the reduced theory can be quantized in the standard way.

In most cases, it is impossible to perform this program explicitly (one has to solve a system of differential equations with arbitrary coefficient functions). However, for our purposes, the abstract existence of solutions to these equations is sufficient. Indeed, one can transform the resulting quantum theory to the form, which is independent of a particular gauge condition, and which enables calculation of relevant physical quantities (like the S-matrix) without an explicit reduction [9, 10] .

For some theories, there is no gauge condition which works for the whole spacetime manifold and for the totality of possible fields (Gribov ambiguity, see, e.g. [11]). However, what we really want to do is the deparametrization of the system, that is, only a partial reduction so that the gauge condition fixes just the spacelike foliation. This should always be possible, or else no reasonable dynamics would exist.

Within the reduction method, the problems with causal structure become even more numerous and involved. First, the family of hypersurfaces defined by the gauge condition can become degenerate (containing, e.g. intersection of the hypersurfaces) or non-spacelike for some values of the dynamical field. Even if the foliation is locally regular and spacelike, it need not represent, globally, a foliation by Cauchy hypersurfaces, irrespectively whether or not the spacetime to be foliated is globally hyperbolic. And, finally, the spacetime need not be globally hyperbolic. However, we have also more conditions on kinematically possible metrics: the gauge conditions, gauge propagating equations and constraints must hold before any dynamics is set up. We call these "predynamical equations". Let us, now, formulate all such predynamical assumptions more carefully.

a) Let the spacetime mandifold (M',g) be 1) smooth (C^2), and 2) asymptotically Minkowskian, 3) causal, orientable and time-orientable.

In many proofs, the requirement a1) can be weakened. However, in this initial state of investigation of these problems, it is very comfortable. The assumption a2) means that (M',g) has a complete $\mathscr{I}$ [12] ; scattering problems can be formulated.

b) Let $\Psi = 0$ be a gauge condition in the form of a (differential) equation for the metric which (possibly supplied with a boundary condition) defines a regular spacelike foliation of some neighbourhood N of i^0 in M' by asymptotically flat hypersurfaces. Let the corresponding time parameter t coincide asymptotically with a proper time and has the bounds $t \in (-\infty, \infty)$.

A foliation t(x) = const is regular and spacelike in N, if the vector field $t_{,i}$ (normal to the hypersurfaces) is well-defined, continuous and timelike everywhere in N.

c) Let there be some $C > 0$ such that the hypersurfaces $t = t_0$ for all $t_0 < -C$ can be extended, as solutions of $\Psi = 0$ in M', to form a regular spacelike foliation of some part N_1 of M' by hypersurfaces, each of which is complete with respect to the positive definite metric induced on it by the metric of (M',g).

Thus, in the "remote past", the folii represent a regular, infinite, asymptotically flat space. It can have a non-trivial topology and contain incoming extremal black holes (Hawking temperature $T = 0$). This is the "assumption of regular initial data".

Let us define M to be the maximal connected neighbourhood of i^0 in M' to which the hypersurfaces t = const can be extended, as solutions of $\Psi = 0$, to form a regular spacelike foliation. Thus, M contains N and N_1.

d) We assume that each hypersurface t = const in M is either complete or has a boundary in M' which coincides with an apparent horizon (AH).

The content of assumption d) is twofold: 1) It states that the kinematically possible fields are regular. Hence, the maximal extension of the foliation $\Psi = 0$ leads either to complete hypersurfaces, or ends at points, which are regular points of M'. Such points can only be singular with respect to the foliation: points of intersection of different t = const hypersurfaces, points, where the hypersurfaces cease to be spacelike, etc. 2) It is a requirement on the gauge condition $\Psi = 0$: the singular points of the corresponding foliation must coincide with the AH of the spacetime (M',g). We shall see in the next section that this can, at least for some models, easily be done. This condition developed during the investigation from the attempt to foliate only the region which was outside of black holes. The AH is chosen because it is locally well-defined.

The assumption of regularity of kinematically possible fields, as contained in d) is, on one hand, a very weak analog of assumptions which are usually done, if one constructs the quantum dynamics of some field. For example, for the linear fields [13] , one considers first some space of Cauchy data which can serve as test fields - they are C^∞ and, say, of compact support. Then, one finds some norm, or scalar product with respect to which the spaces can be completed to Banach, or Hilbert spaces. The norm, or at least the corresponding topology should be preserved by the dynamics in order that the dynamics of the more singular elements be well-defined. Thus, for a construction of quantum dynamics, it seems necessary that the kinematically possible fields are dense in some suitable functional space. On the other hand, the assumptions c) and d) together remind us strongly on the weak cosmic censorship hypothesis. Of course, there are differences: for example, we require that the singularities are hidden beyond AH instead of beyond event horizons. The really important difference is, however, that we do not require the validity of the full system of the classical dynamical equations. We require, first, just the predynamical equations and, second, the possibility to construct a reasonable quantum dynamics.

Let us call the points c) and d) together with the assumption that one can construct a reasonable quantum dynamics with them, a <u>Quantum Weak Cosmic Censorship Hypothesis</u>. It is clear that the quantum censorship could be true even if the classical one is invalid.

e) We assume, finally, that the whole system of the predynamical equations holds in M.

We shall call our field theoretical model for gravity to be <u>completely foliable,</u> if the assumptions a) - e) imply the following properties of M:

(i) M is asymptotically Minkowskian,

(ii) each hypersurface $\Psi = 0$ in M is a Cauchy hypersurface for M.

The gauge condition Ψ can, then, be called a "complete foliation".

Thus, at the end, the singularities beyond the apparent horizon are not visible from $\mathscr{I}^+$.

3. BCMN model

The most simple known field theoretical model in which the dynamics can lead to a truly nontrivial causal structure (namely the formation of black hole horizons) is the Berger-Chitre-Moncrief-Nutku model [14]. We show in this section that the model is completely foliable.

Let us first briefly introduce the model. It results from the Einstein-Maxwell system to which an uncharged scalar field is minimally coupled. All dynamical degrees of freedom are frozen except for the spherically symmetrical ones. We have the following variables: a metric g_{ab} on a two-dimensional manifold M' (t-r-surface of the original four-dimensional spacetime), a real scalar field ϕ on M' (r-coordinate), and the real scalar field ψ on M' (the original scalar). The action has the form:

$$(1) \qquad I = \frac{1}{2} \int d^2x \, |g|^{\frac{1}{2}} \left[\frac{1}{G} + g^{ab} \partial_a \varphi \partial_b \varphi + \frac{1}{2} R \varphi^2 + \right. \left. + G^2 \frac{Q^2 + P^2}{\varphi^2} - \varphi^2 g^{ab} \partial_a \psi \partial_b \psi \right] .$$

(see [15]). Here, G is the Newton constant, Q and P are the electric and the magnetic charges of the possible (incoming) black hole, g is the determinant of g_{ab} and R is the curvature scalar corresponding to g_{ab}.

Let us choose the gauge condition

$$(2) \qquad n^a \partial_a \varphi = 0 ,$$

where n^a is a normal vector to the t = const surfaces, and supplement it by the boundary condition at infinity

$$(3) \qquad \lim_{x \to \infty} g_{00} = -1 .$$

The condition (3) specifies the time parameter t up to an additive constant. In an asymptotically Minkowskian spacetime M', the foliation (2), (3) will be spacelike near i^0 and the assumption b) will be satisfied. The t = const hypersurfaces in the maximal extension M with a regular spacelike foliation can only have AH as boundaries in M'. This is clear from the following considerations.

The foliation is regular at all points, where the direction of n^a is uniquely determined by the condition (2), that is, where $\partial_a \phi$ is a non-zero vector. Any critical point p of ϕ satisfies

$$\ell^a \partial_a \varphi \big|_p = 0 \quad , \quad k^a \partial_a \varphi \big|_p = 0 \, ,$$

where ℓ^a and k^a are the two independent null directions at p. Thus, p is a future and past AH simultaneoulsy, or "double" AH (DAH).

The foliation ceases to be spacelike, if $\partial_a \varphi$ becomes null, that is either at a future AH (FAH) or at a past AH (PAH).

In [16], the following theorem has been shown.

Theorem 1: If the conditions a) - e) are satisfied, then M is asymptotically Minkowskian.

Let us consider the predynamical equations. If we reduce the theory in its Lagrangian form, then the equations read as follows: the Hamiltonian constraint:

$$(4) \qquad \frac{\partial}{\partial x}\left(\frac{x}{\gamma}\right) = -x^2\left[(n^a\partial_a\psi)^2 + (m^a\partial_a\psi)^2\right] + 1 - G^2\,\frac{Q^2+P^2}{x^2} \, ,$$

the momentum constraint:

$$(5) \qquad \frac{\partial\gamma}{\partial t} = 2\alpha\gamma^{\frac{3}{2}}\, x\,(n^a\partial_a\psi)(m^a\partial_a\psi) \, ,$$

and the gauge propagating equations:

$$(6) \qquad \frac{1}{\alpha^2\gamma}\frac{\partial}{\partial x}(x\alpha^2) = x^2\left[(n^a\partial_a\psi)^2 + (m^a\partial_a\psi)^2\right] + 1 - G\,\frac{Q^2+P^2}{x^2} \, ,$$

$$\beta = 0 \, .$$

Here, we have chosen the x coordinate to be

$$x = \sqrt{G}\,\varphi \, .$$

The following relations hold:

$$\gamma = g_{11} \, , \quad \alpha = \sqrt{-g_{00} + \frac{g_{01}^2}{g_{11}}} \, , \quad \beta = g_{01} \, ,$$

$$n^a = \left(\frac{1}{\alpha} \, , \, -\frac{\beta}{\alpha\gamma}\right) \, ,$$

$$m^a = \left(0 \, , \, \frac{1}{\sqrt{\gamma}}\right) \, ,$$

so α is the lapse, β the shift function and m^a is the unit tangential vector to t = const hypersurfaces.

It has been shown in [17] that the three equations (4), (5) and (6) are equivalent to the following tensorial equation:

$$(7) \qquad \frac{\delta I}{\delta g_{ab}} = 0 \ .$$

Thus, it is easily transformable to any coordinate system in M; if written out in a double null coordinates, eq. (7) implies the following

Theorem 2: Let H be the future (past) half of an outgoing (incoming) null hypersurface though a future (past) AH p. Let p lie on the boundary of M and H inside of M. Then, the divergence of the null geodetic generators of H is non-positive (non-negative).

Using the theorems 1 and 2, one can show that the FAH cannot be visible from $\mathcal{J}^+$ the the PAH cannot be influenced from $\mathcal{J}^-$, as well as that the world tube of an apparent horizon at the boundary of M is not timelike.

These properties and the assumptions a) - e) imply the following

Theorem 3: All hypersurfaces t = const in M are complete with respect to the induced metric.

From theorem 3 and the regularity of the foliation in M, it easily follows that the t = const are Cauchy hypersurfaces for M. Hence, the BCMN model is completely foliable and the conditions (2), (3) represent a complete foliation.

4. Hawking effect, positivity problem and unitarity

The picture which the complete foliability of the BCMN model gives concerning the kinematically possible trajectories is quite different from the current ideas about the expectation value of the metric in a spherically symmetric spacetime with a collapse. According to these ideas, a black hole horizon will appear and the most of the information about the collapsing object will be lost inside of it. There will also be a radiation going from the collapse region out to the infinity - the so-called Hawking radiation. The origin of this radiation will be localized to a neighbourhood of the black hole horizon. The energy necessary for the Hawking radiation will be taken directly from the black hole: a locally defined current of negative energy will be pouring through the horizon from outside. Due to this negative current, an apparent horizon can form outside the black hole horizon. Such an AH will be visible from $\mathcal{J}^+$, and its world tube will be timelike.

The crucial question in this respect is which properties of the kinematically possible trajectories can survive the quantization and can, in this way, appear as properties of the expectation value of the metric. Naively, it could seem that all such properties must survive, because the expectation value of the metric can be calculated as a path integral average over the kinematically possible trajectories. However, this is not true in general. For example, the energy density of, say, Klein-Gordon field is everywhere non-negative for any kinematically possible trajectory but its expectation value can be negative at some points.

In [16] , this question has been discussed at some length. One possibility, due to Ashtekar and Horowitz [18] , is the following: if a given general property of all kinematically possible trajectories can be considered as a property of the configuration space of the system, then it will survive the quantization. Such properties are, for example, the absence of AH in M (this is the absence of critical points of φ along t = const surfaces), or the fact that M is asymptotically Minkowskian.

Another, more obvious possibility, is to look at the operators which represent the components of the metric in the quantum theory to see which sort of spacetime they are likely to yield. In our gauge, the component $g^{11} = 1/\gamma$ of the metric gives information whether the t = const surfaces are spacelike ($g^{11} > 0$) or null ($g^{11} = 0$). The latter case would mean an AH. After the reduction, g^{11} is a dependent variable, given by [19] :

$$g^{11}(x) = \frac{1}{x} \int_b^x dy \, F(y) \, e^{T(x)-T(y)} \quad ,$$

where

$$F(y) = 1 - G^2 \frac{Q^2+P^2}{x^2}$$

is a positive function, and

$$T(x) = G \int_b^x dy \left[4 y^{-3} \pi^2(y) + y(\psi'(y))^2 \right]$$

with $\pi(y)$ being the canonical momentum and $\psi(y)$ the canonical coordinate (true dynamical variables). $g^{11}(x)$ is clearly positive classically, but can it be made to an operator with positive spectrum ? Some subtraction procedure is necessary to define $g^{11}(x)$; for example, if this procedure can be applied directly to $T(x) - T(y)$ (so that the exponential of it is already well-defined), then $g^{11}(x)$ will be positive even if $T(x) - T(y)$ itself is not.

It is interesting to notice that the problem with positivity of $g^{11}(x)$ is closely related to the well-known "positivity problem" in the canonical quantization [8] :

A canonical coordinate Q which describes gravity (in two-dimensional theories, this can be $g^{11}(x)$ indeed) has a limited range of its values. If the corresponding operator Q satisfies the canonical commutation rules (CCR) together with its conjugated momentum π_Q, then the spectrum of Q cannot be limited. Two solutions to the problem have been proposed in the literature:

1) Choose other variable (like log Q) and its conjugate [20] , or 2) use the pair $\{Q,(Q\pi_Q)\}$ as the pair of variables whose CCR determine the operator algebra [21] . A third solution, possible only within the reduction method, is to choose such a quantity as a dependent one. (This is analogous to what one does in ordinary quantum mechanics with the canonically conjugated pair {time, energy}).

Another important point to discuss is the unitarity of the resulting quantum theory. We have shown that the canonical qunatization is applicable to the BCMN model, because the relevant part of each kinematically possible spacetime can be foliated by Cauchy hypersurfaces. The dynamical development of the quantum states from one such surface to another will be unitary. However, this relevant part M is not the whole spacetime in general. Thus, the final, $t = \infty$, Cauchy hypersurface can contain a part of the boundary of M in M', e.g. an event horizon, say. It is, however, impossible to perform measurements along the event horizon, and one will be interested to know the state only along that part of the $t = \infty$ hypersurface which does not contain the horizons. Such a state will, in general, be mixed, and we seem to lose the quantum coherence, and unitarity, again.

Here, the Hawking effect could, in fact, help. If all energy of the collapsing object is radiated away again, then we can end up with a "clean" horizon, that is, no horizon at all, or that one which has been present before the collapse (incoming hole). For this to work, it is necessary that the Hawking radiation carries away all information about the collapsing object. Such a transfer of information seems to be impossible according to the current ideas which localize the origin of Hawking radiation to a neighbourhood of the event horizon.

There was another school of thinking about the Hawking effect [22] , let us call it Boulware school, which localized the origin of Hawking radiation to the inside of the collapsing object. The energy of the radiation was taken directly from the object so that the horizon could never form. The calculations of the Boulware school, however, did not reveal any better information transfer than Hawking's. This was due to the assumption that a fixed classical background gravitational field was well-defined everywhere and that it was the only source of the radiation; such a field had certain properties which did not depend on the details of the collapse. The struggle about where the origin of Hawking radiation is to be localized has been won by the Hawking school after the consens has been achieved about the regularization and renormalization of the stress-energy tensor: the expectation value, $<T_{\mu\nu}>$, of this tensor gives a locally well-defined c-number energy current (see, e.g. [23] and the references given there).

Let me very briefly critize these theories. First, such fine localization of the origin of the Hawking radiation that it can distinguish between the inside of the collapsing object and a neighbourhood of the corresponding event horizon need not make any sense at all, in spite of the results about $<T_{\mu\nu}>$. Indeed, the arbitrarily detailed locally defined c-number energy current given by $<T_{\mu\nu}>$ could have some physical reality only if the corresponding mean squared deviation, $<\Delta T^2_{\mu\nu}>$, would be negligible with respect to $<T_{\mu\nu}>$. As far as I know, this was never shown.

Second, it is a non-trivial problem, whether the whole perturbation scheme used to calculate the Hawking effect is applicable. That is the semiclassical aproximation: first, one calculates the classical solution for the collapsing object and the surrounding gravitational field; second, one investigates the behaviour of small quantum disturbances around this fixed classical background; finally, one couples the classical metric to the expectation value of the stress-energy tensor of these disturbances. In our case, the classical background is unstable and the effect it produces, is large.

In this way, my speculations about the information transfer in the Hawking effect need not be completely wrong and unitarity could be saved.

Finally, I would like to stress that my critics concern only the so-called dynamical Hawking effect and not the existence and properties of the thermal quantum states on the static black hole background.

Acknowledgement: I am very indepted to R. Penrose and J. Hartle for important critical remarks.

References

1 D.R. Whitehouse, A.M. Cruise: Nature 315 (1985) 554.

2 D. Christodoulou, B.G. Schmidt: Convergent and Aysmptotic Iteration Method in General Relativity. Preprint MPI-PAE/Astro 177, 1979.

3 S.W. Hawking: in "Qantum Gravity. 2nd Oxford Symposium". Ed. by C.J. Isham, R. Penrose, D.W. Sciama. Oxford, Clarendon Press, 1981.

4 S.W. Hawking: Phys.Rev. D14 (1976) 2460.

5 S.W. Hawking: Commun.Math.Phys. 87 (1982) 395.

6 S.W. Hawking, G.F.R. Ellis: The Large Scale Structure of Spacetime. Cambridge, Cambridge University Press, 1973.

7 D. Christodoulou: Commun.Math.Phys. 93 (1984) 171.

8 C.J. Isham: in "Quantum Gravity. An Oxford Symposium". Ed. by C.J. Isham, R. Penrose, D.W. Sciama. Clarendon Press, Oxford, 1975.

9 J.B. Hartle, K. Kuchar: J.Math.Phys. 25 (1983) 57.

10 E.S. Fradkin, G.A. Vilkovisky: Preprint TH-2332 CERN, 1977; E.S. Fradkin, I.V. Tyutin: Phys.Rev.D2 (1970) 2841.

11 T.P. Killingback: Commun.Math.Phys. 100 (1985) 267.

12 R. Geroch, G.T. Horowith: Phys.Rev.Lett. 40 (1978) 203.

13 B.S. Kay: Commun.Math.Phys. 62 (1978) 55.

14 B.K. Berger, D.M. Chitre, V.E. Moncrief, Y. Nutku: Phys.Rev. D5 (1972) 2467.

15 P. Thomi, B. Isaak, P. Hajicek: Phys.Rev. D30 (1984) 1168.

16 P. Hajicek: Phys.Rev. D31 (1985) 787.

17 P. Hajicek: Phys.Rev. D31 (1985) 2452.

18 A. Ashtekar, G.T. Horowitz: Phys.Rev. D26 (1982) 3342.

19 P. Hajicek: Phys.Rev. D30 (1984) 1178.

20 C.W. Misner: in "Magic without Magic. John Archibald Wheeler. A Collection of Essays in Honor of His Sixtieth Birthday". Ed. by J.R. Klauder. Freeman, San Francsico, 1972.

21 J.R. Klauder: Phys.Rev. D2 (1980) 272; J.R. Klauder: in "Relativity". Ed by M.S. Carmeli, S.I. Flicker, L. Witten. New York, Plenum, 1970; C.J. Isham, A.C. Kakas: Classical Quantum Gravity 1 (1984) 621.

22 D.G. Boulware: Phys.Rev. D13 (1976) 2169.

23 N.D. Birrell, P.C.W. Davies: "Quantum Fields in Curved Space". Cambridge, Cambridge University Pres, 1982.

QUANTUM EFFECTS IN NON INERTIAL FRAMES AND QUANTUM COVARIANCE

Denis BERNARD
Groupe d'Astrophysique Relativiste
C.N.R.S. - Observatoire de Paris-Meudon
92195 Meudon Principal Cedex - France.

Abstract :

We review recent results in non-inertial quantum field theory. By formulating Q.F.T. in a large class of accelerated frames, the classical and the quantum aspects of the theory are unified. We describe the thermal effects, their asymptotic character and the role of the P.C.T. symmetry. A discussion of quantum covariance and detection processes is also given.

0. INTRODUCTION

Quantum field theory in accelerated frames is a possible approach to the understanding of gravitational effects. Most of the known results about quantum field theory in curved space-time, such as the Hawking radiation, can be described by analogies with non-inertial effects in flat space-time. Therefore, non-inertial quantum field theory makes possible the setting up of a "laboratory" for studying field quantization in curved space-time. However, it is also a possible way to discuss non-inertial detection process or a possible way to search for a quantum covariance principle. This paper reviews recent results about non-inertial quantum field theory and presents some new ones, too. In particular, we analyse thermal effects and their asymptotic character and we relate them to proposed quantum covariance laws. A critical discussion of detection processes and their link with quantum field theory in non-inertial frames is presented.

The content of this paper is :

I. Q.F.T. in Rindler frame : the role of the P.C.T. symmetry

II. Q.F.T. in analytic accelerated frames

III. The asymptotic character of thermal effects from a local principle

IV. A hamiltonian formulation

V. Vacuum fluctuation in accelerated frames

VI. Discussion.

1. QUANTUM FIELD THEORY IN RINDLER FRAME : THE ROLE OF THE P.C.T. SYMMETRY.

We shall begin with the Davies-Unruh's |1| result about the quantification of a scalar field ϕ in a uniformly accelerated frame (Rindler frame). There are many ways based on Bogoliubov transformations, Green's functions, and others |2|, to obtain this famous result . But here, we want to present a global description where the role of the P.C.T. symmetry is illustrated |3|. In particular this symmetry becomes crucial for the analysis of the state identification proposed by t'Hooft. First in flat space-time, accelerated trajectories are completely describe by the transport law of the tetrad carrying by the observers. This equation is the Fermi-Walker transport equation |4| which reads :

(1.1) $$\frac{d\vec{e}_m(\tau)}{d\tau} + \vec{e}_n\, \theta^n{}_m(\tau) = 0$$

where $\vec{e}_n(\tau)$ is the tetrad and $\theta^n{}_m(\tau)$ the generator of this transport. $\theta^n{}_m$ is a generator of Lorentz transformation and can be written as

(1.2) $$\hat{\theta} = \frac{1}{2}\, \epsilon^{\alpha\beta} L_{\alpha\beta} \qquad \text{with} \qquad \epsilon^{\alpha\beta} + \epsilon^{\beta\alpha} = 0$$

where $L_{\alpha\beta}$ are the generators of Lorentz transformations. It is useful to introduce the acceleration $\vec{a}$ and the rotation $\vec{\Omega}$ as the "electric" and "magnetic" part of the antisymmetric tensor

(1.3) $$\frac{1}{2}\, \epsilon^{\alpha\beta} \epsilon_{\alpha\beta} = \vec{a}^2 - \vec{\Omega}^2 \quad \text{and} \quad \frac{1}{4}\, {}^*\epsilon^{\alpha\beta} \epsilon_{\alpha\beta} = \vec{a}.\vec{\Omega}$$

We are looking for accelerating trajectories for which ${}^*\epsilon^{\alpha\beta} \epsilon_{\alpha\beta} > 0$ so that the operator $\hat{\theta}$ is a "good" generator of temporel evolution:

$$[\hat{\theta}\, ,\, \xi^\mu \hat{P}_\mu] \neq 0$$

for any time-like vector ξ^μ (P_μ are the translation generators). It is interesting to introduce the non-inertial coordinates (τ, y^k) defined with respect to the inertial ones x^μ by

(1.4) $$x^\mu = \Lambda^\mu{}_k(\tau)\, y^k$$

(1.5) where $$\frac{d}{d\tau} \Lambda^\mu{}_\nu(\tau) = \theta^\mu{}_\sigma(\tau)\, \Lambda^\sigma{}_\nu$$

In terms of these coordinates, the metric take the form

(1.6) $$ds^2 = -\left(\Theta_{0k} y^k d\tau\right)^2 + \delta_{ij}\left(dy^i + \Theta^i{}_k y^k d\tau\right)\left(dy^j + \Theta^j{}_k y^k d\tau\right)$$

Then the hamiltonian

(1.7) $$\hat{H} = \frac{1}{2}\,\epsilon_{\alpha\beta}\,\hat{L}^{\alpha\beta}$$

becomes simply the τ-time evolution generator

(1.8) $$\hat{H} = -i\,\frac{\partial}{\partial\tau}$$

It is generally supposed that these generators represent the observer's hamiltonian. For the sake of simplicity (and because we can always take it as an approximation at least during a small duration $\delta\tau$), we choose Θ τ- independant. For $\vec{a}$ and $\vec{\Omega}$ parallel, the hamiltonian becomes

(1.9) $$H = a\hat{K}_1 + \Omega\hat{J}_1$$

where K_1 and J_1 are respectively the boost-generator and the angular-momentum in the direction 1. Moreover, the previous non-inertial coordinates become the rotating Rindler coordinates

(1.10) $$\begin{cases} t = y^1 \sinh(a\tau) \\ x^1 = y^1 \cosh(a\tau) \\ \varphi = \varphi' + \Omega\tau \\ \rho = \rho' \end{cases}$$

in cylindrical coordinate

The accelerated coordinates $(\tau, y^1, \varphi', \rho')$ cover only a submanifold R_I of the Minkowski space. (see figure)

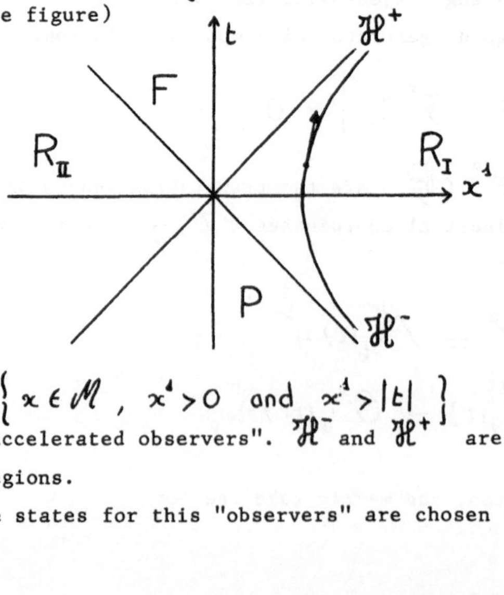

The region $R_I = \{x \in \mathcal{M},\ x^1 > 0 \text{ and } x^1 > |t|\}$, is the field of communication of "Rindler accelerated observers". $\mathcal{H}^-$ and $\mathcal{H}^+$ are the past and futur event-horizons of these regions.

The quantum particle states for this "observers" are chosen to be eigen-functions of the hamiltonian H

$$\hat{H}\,\phi_{\epsilon,\cdot\cdot}(t,\vec{x}) = \epsilon\,\phi_{\epsilon,\cdot\cdot}(t,\vec{x}) \tag{1.11}$$

and we shall require that these wave functions vanish on R_{I} or on R_{II}.
Because $\hat{H}$ is the generator of the Lorentz transformation (1.5), the wave functions $\phi_{\epsilon,\cdot\cdot}$ satisfy the following transformation law

$$\phi_{\epsilon,\cdot\cdot}\left(\Lambda^{\mu}{}_{\nu}(\alpha)x^{\nu}\right) = e^{-i\epsilon\alpha}\,\phi_{\epsilon,\cdot\cdot}(x^{\nu}) \tag{1.12}$$

where $\Lambda(\alpha)$ is the Lorentz transformation (1.5).
That property characterizes the function $\phi_{\epsilon,\cdot\cdot}$, but it is simplest to introduce a plane-wave decomposition

$$\phi_{\epsilon,\cdot\cdot}(t,\vec{x}) = \int d^3\vec{k}\,\frac{e^{i(\vec{k}\vec{x}-E_k t)}}{(2\pi)^3\,2E_k}\cdot G_{\epsilon,\cdot\cdot}(\vec{k})\ ;\ E_k=\sqrt{m^2+k^2} \tag{1.13}$$

Use of (1.12), yields a differential equation for

$$\left[aE_k\frac{\partial}{\partial k^1}+\Omega\frac{\partial}{\partial\varphi_k}\right]G_{\epsilon,\cdot\cdot}(\vec{k}) = -i\epsilon\,G_{\epsilon,\cdot\cdot}(\vec{k}) \tag{1.14}$$

(where φ_k is the cylindrical angle of $\vec{k}$),
whose solutions are

$$\left(\frac{E_k+k^1}{M}\right)^{-i\frac{\epsilon+m\Omega}{a}}\quad e^{im\varphi_k} \tag{1.15}$$

where m is the angular momentum ; $E_k=\sqrt{k^{1\,2}+\vec{q}^{\,2}+m^2}$
and M can be chosen as $M^2=m^2+q^2$.
Therefore, we can built, after normalization, a wave function basis, $\phi_{\epsilon,q,m}$ and $\phi^*_{\epsilon,q,m}$, which can be used to construct the Fock-space of the quantum field:

$$\hat{\phi}(x) = \sum_{\epsilon,q,m}\left(\phi_{\epsilon,q,m}(x)\,\hat{a}_{\epsilon,q,m}+\phi^*_{\epsilon,q,m}(x)\,\hat{a}^{\dagger}_{\epsilon,q,m}\right) \tag{1.16}$$

[in the discrete notation]
The operators of creation-annihilation, $a^{\dagger}_{\epsilon,q,m}$ and $a_{\epsilon,q,m}$, define the vacuum state $|0\rangle$: $a_{\epsilon,q,m}|0\rangle = 0$
Because the $\phi_{\epsilon,q,m}$ have positive minkowskian-energy, this vacuum is the Minkowski one. Now, the region R_{II} is outside the field of communication of the accelerator "observers" inside R_I. Therefore we would like to diagonalize the hamiltonian separatly inside the region R_I and R_{II}. Thanks to the P.C.T. symmetry, we can link

the value of the wave function inside R_I to that inside R_{II}. From (1.13) and (1.15) we get :

$$\phi_{\epsilon,q,m}(-t,-\vec{x}) = e^{-\pi\left(\frac{\epsilon+m\Omega}{a}\right)} \phi_{\epsilon,q,-m}(t,\vec{x}) \quad \text{for } x^1>0 \tag{1.17}$$

(In this region, the logarithm in the equation (1.15) has been defined on the half-upper complex plane). Since,

$$\phi_{\epsilon,q,m}(t,\vec{x}) = \phi_{-\epsilon,q,-m}(-t,-\vec{x}) \tag{1.18}$$

we have

$$\phi^*_{-\epsilon,q,m}(t,\vec{x}) = e^{-\pi\left(\frac{\epsilon+m\Omega}{a}\right)} \phi_{\epsilon,q,-m}(t,\vec{x}) \quad \text{for } x^1>0 \tag{1.19}$$

and a similar relation for $x^1<0$.
Therefore, the states

$$_I\phi_{\epsilon,q,m} = \left|2\sinh\frac{\pi(\epsilon+m\Omega)}{a}\right|^{-1/2} \left\{ e^{\pi\frac{\epsilon+m\Omega}{2a}} \phi_{\epsilon,q,m}(t,\vec{x}) - e^{-\pi\frac{\epsilon+m\Omega}{2a}} \phi^*_{-\epsilon,q,-m}(t,\vec{x}) \right\} \tag{1.20}$$

vanish in the region R_{II} and are eigenfunctions of H.
Similarly, we define

$$_{II}\phi_{\epsilon,q,m}(t,\vec{x}) = {}_I\phi^*_{\epsilon,q,m}(-t,-\vec{x}) \tag{1.21}$$

which is the P.C.T. symmetric image of $_I\phi_{\epsilon,q,m}$. The $_{II}\emptyset$ vanish in the region R_I and are eigenfunctions of H, too. The normalized wave functions $_I\emptyset$ and $_{II}\emptyset$ and their complex conjugates make up a wave function basis which defines the Rindler mode. The quantum field $\emptyset$ reads

$$\phi(x) = \phi_I(x) + \phi_{II}(x) \quad \text{with} \tag{1.22}$$

$$\phi_I(x) = \sum_{\epsilon,q,m} \left\{ {}_I\hat{C}_{\epsilon,q,m}\, \phi_{\epsilon,q,m}(x) + {}_I\hat{C}^{+}_{\epsilon,q,m}\, \phi^*_{\epsilon,q,m} \right\} \tag{1.23}$$

and from (1.32), $\phi_{II} = \Theta^{-1}\phi_I\,\Theta$ where Θ is the antiunitary P.C.T. operator. The creation-annihilation operators $_IC_{\epsilon,q,m}$ and $_{II}C = \Theta^{-1}{}_IC\,\Theta$ define the Rindler vacuum: $|0_R\rangle$; $_IC|0_R\rangle = {}_{II}C|0_R\rangle = 0$.
Because, the definition (1.20) mixes positive and negative frequencies, the Rindler vacuum is not equivalent to the minkowski-one. The different creation-annihilation operators are related by the Bogoliubov transformation

(1.24) $$ {}_{I}\hat{C}_{\epsilon,q,m} = \left|2\sinh \pi\left(\frac{\epsilon+m\Omega}{a}\right)\right|^{-1/2} \left\{ e^{\pi\frac{\epsilon+m\Omega}{2a}} a_{\epsilon,q,m} - e^{-\pi\frac{\epsilon+m\Omega}{2a}} \hat{a}^{+}_{-\epsilon,q,-m} \right\} $$

and similarly for ${}_{II}\hat{C}$.

Therefore, the Minkowski vacuum $|0\rangle$ contains Rindler modes. The density of Rindler modes :

(1.25) $$ \langle 0 | {}_{I}C^{+}_{\epsilon,q,m} \, {}_{I}C_{\epsilon',q',m'} | 0 \rangle = \frac{1}{\exp \frac{2\pi(\epsilon+m\Omega)}{a} - 1} \delta_{\epsilon,\epsilon'} \delta_{q,q'} \delta_{m,m'} $$

describes a Planckian spectrum. The acceleration plays the role of the temperature $T = a/2\pi$ and the rotation velocity appears as a chemical potential.

The unitary transformation linking the Rindler mode to the Minkowski-one can be written as :

$$ |0\rangle = U|0_R\rangle $$

(1.26) $$ U = Z^{-1} \exp\left\{ \sum \left(e^{-\pi\frac{\epsilon+m\Omega}{a}} {}_{I}C^{+}_{\epsilon,q,m} \, {}_{II}C^{+}_{\epsilon,q,m} - h.c. \right) \right\} $$

The pure Minkowski vacuum state contains pairs of Rindler modes. (like the B.C.S. state). Each pair contains one "particle" created in the region R_I and another created outside the horizons $\mathcal{H}$. But, if we restrict ourselves to observable, $\mathcal{O}_I$ say whose support is restricted to the region R_I, it is better to introduce a density matrix $\hat{\rho}$ by

(1.27) $$ \hat{\rho} = \mathrm{Tr}_{II} \, |0\rangle\langle 0| $$

that is, by taking the trace over the states built from ${}_{II}\phi$. Then, the expectation of the observable $\mathcal{O}_I$, in the Minkowski vacuum takes the form :

(1.28) $$ \langle 0 | \mathcal{O}_I | 0 \rangle = \mathrm{Tr}(\hat{\rho}\, \mathcal{O}_I) $$

And the density matrix $\hat{\rho}$, describing a thermal mixed state, is

(1.29) $$ \hat{\rho} = Z^{-1} \sum e^{-\frac{2\pi}{a}(\epsilon+m\Omega)} |n;\epsilon,q,m\rangle\langle n;\epsilon,q,m| $$

where $|n;\epsilon,q,m\rangle = (n!)^{-1/2} \left({}_{I}\hat{C}_{\epsilon,q,m}\right)^{n} |0_R\rangle$ are the n-Rindler mode states.

This thermal character persists in the presence of interactions. By using a path integral approach, W. Unruh and N. Weiss |5| have shown that a thermal quantum field theory in a Rindler frame coïncides, for the Hawking-Unruh temperature, with the euclidean Q.F.T. in an inertial frame.

Remark on electromagnetic analogies.

The description of the accelerated trajectories in terms of Lorentz generators like (1-2) illustrates, once more, the analogie between classical electromagnetic and gravitationals effects| |. The tensor $\epsilon_{\alpha\beta}$ becomes the analog of $(\frac{e}{m})$ times the electromagnetic tensor. In particular all stationary trajectories [such that $\epsilon_{\alpha\beta}$ is τ-independant] can be found directly from the study of trajectories in constant electromagnetic fields. (see ref.(6 bis) and ref. (25) for another derivation of these trajectories).

These analogies persist at the quantum level. Indeed, the Schwinger Lagrangian in presence of an electric field E (B = 0)

$$\mathcal{L}_s = -\frac{m^4}{8\pi^2}\int dt\,\frac{e^{-t}}{t^3}\left(\frac{tE}{\sin tE} - 1 - \frac{t^2E^2}{6}\right)$$

can be written as

$$\mathcal{L}_s = -\frac{m^4}{16\pi^2}\int ds\, f(s)\,\frac{1}{\exp(\beta s) - 1}$$

where $\beta = \pi m^2/eE$

It is quite natural to interpret 2ms as the excitation energy du to virtual particle production and $f(s)$ as the spectral function describing the density of virtual excitations. Then, we may define the temperature describing the average excitation of the ground state in presence of the field E by

$$T = \frac{2m}{\beta} = \frac{eE}{2\pi m}$$

This value is the analog of the Unruh-Hawking temperature

2. Q.F.T. IN ANALYTIC ACCELERATED FRAMES.

The thermal effects analyzed in the previous section are not restricted to the Rindler rotating coordinate (1.10), but persist for a large class of accelerated coordinates. Following on from the work of Sanchez |6| we generalize |7| the formulation of quantum field theories based on analytical mappings in two-dimensional space-times to the massive case in four dimensions by including possible rotation or drifting. We describe the sub-manifold by the curvilinear or accelerated coordinates, $(\vec{x}', t') \equiv (x', \rho', \varphi', t')$, defined by the transformation law :

$$\text{(A.1)}\quad \begin{cases} x \pm t = f(x' \pm t') \\ \rho = \rho' \\ \varphi = \varphi' + \Omega t' \end{cases} \qquad\qquad \text{(B.1)}\quad \begin{cases} x \pm t = f(x' \pm t') \\ z = z' + \alpha t' \\ y = y' + \beta t' \end{cases}$$

where $(\underline{x},t) \equiv (x,\rho,\varphi,t)$ are Minkowskian cylindrical coordinates, f a strictly monotonic function defined on $\mathbb{R}$, Ω an angular velocity and, α and β drifting parameters. Such a transformation ensures that the accelerated coordinates cover completely their domain of communication. The horizons associated with such a region are defined by the singularities of F, the inverse mapping of f :

(A and B.2) $$u_\pm = f(\pm\infty)$$

If $u_\pm = \pm\infty$, there is no horizon, but if $u_\pm$ are finite the accelerated coordinates only cover a limited region of total space-time (Minkowski space-time)

$$u_- < |x \pm t| < u_+$$

In such coordinates, the metric takes the form

(A3) $$ds^2 = -\Lambda(x',t')(dt'^2 - dx'^2) + \rho'^2(d\varphi' + \Omega dt')^2 + d\rho'^2$$

(B3) or $$ds^2 = -\Lambda(x',t')(dt'^2 - dx'^2) + (dz' + \alpha dt')^2 + (dy' + \beta dt')^2$$

with $\Lambda(x',t') = f(x'+t')f(x'-t')$

and the whole cinematic of the "observers" inside the static regions : $\Omega^2\rho'^2 < \Lambda(x',t')$ or $\alpha^2+\beta^2 < \Lambda(x',t')$ is determined by the mapping.
However at the quantum level, the two statements (1) and (3) are not sufficient and it is necessary to make clear the behaviour of $f'(\pm\infty)$ so as to define a Q.F.T. in accelerated coordinates. To illustrate what precedes we limit ourselves to a submanifold with only one horizon : $u_- = 0$ and $u_+ = +\infty$ and to a mapping which has the asymptotic behaviour :

(A and B.4) $$f'(-\infty) = 0 \quad \text{and} \quad f'(+\infty) = +\infty$$

Such a class of mappings includes the Rindler mappings : $f(x'\pm t') = e^{a(x'\pm t')}$ which describe uniformly-accelerated observers. So as to be able to define a complet set of wave funcions for global space-time from the wave functions defined in the submanifold we must also prolong the transformation by (see figure)

(A.5) $$\begin{cases} \mathrm{sgn}(u).u = f(u') \\ \mathrm{sgn}(v).v = f(v') \\ \rho = \rho' \\ \varphi = \varphi' + \Omega t' \end{cases}$$ and (B.5) $$\begin{cases} \mathrm{sgn}(u)\, u = f(u') \\ \mathrm{sgn}(v)\, v = f(v') \\ z = z' + \alpha t' \\ y = y' + \beta t' \end{cases}$$

Figure

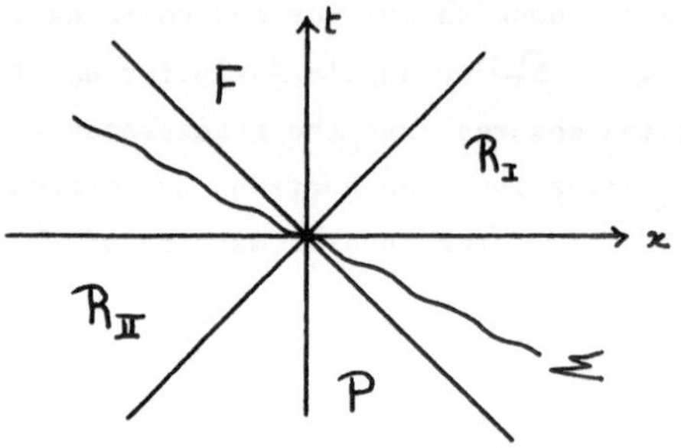

The Fock space associated with the quantization of a massive scalar field ϕ is built up from a basis of wave functions which are solutions of

$$(\Box - m^2)\,\phi = 0$$

and which have a positive "charge" defined by scalar product

$$(\phi, \phi') = i \int_{\Sigma} d\sigma^{\mu} \, (\phi^* \overleftrightarrow{\partial}_{\mu} \phi')$$

Relative to the global coordinates $(\vec{x}, t)$, the Fock space is built upon the cylindrical waves of positive energy :

(A.6) $$\varphi_{k}(\vec{x}, t) = |2(2\pi)^2 E_k|^{-1/2} \, e^{i(k_1 x + \mu \varphi - E_k t)} \sqrt{k}\, J_{\mu}(k\rho)$$

where $k = (k_1, k, \mu)$; $\mu \in \mathbb{Z}$; $E_k = \sqrt{k_1^2 + k^2 + m^2}$

and J_{μ} is a Bessel function.
or upon the plane wave functions

(B.6) $$\varphi_{k}(\vec{x}, t) = |(2\pi)^2 2E_k|^{-1/2} \, e^{i(\vec{k}\vec{x} - E_k t)}$$

The creation-annihilation operators a_k, a_k^+ define the global vacuum $|0\rangle$:

$$\phi(x) = \int d^3k \left(a_k \varphi_k(\vec{x}, t) + a_k^+ \varphi_k^*(\vec{x}, t) \right) \quad ; \quad a_k |0\rangle = 0$$

With accelerated coordinates, we must define the quantum states which can be associated with accelerated "observers" in the region R_I. These wave functions do not make up a complete basis for global space and thus are not sufficient to build a Fock space. In order to form a complete basis from these states we use the PCT symmetry. The wave functions $\{\phi_\lambda\}$ relative to the region R_I are defined by certain Cauchy data on Σ whose support is included in $\Sigma_I = \Sigma \cap R_I$. Under these conditions, ϕ_λ are always null on R_{II} (but not on F and P). Each ϕ_λ is associated with a state Ψ_λ defined as

$$\Psi_\lambda(\vec{x},t) = PCT(\phi_\lambda) = \phi_\lambda^*(-\vec{x},-t)$$

The Ψ_λ are null throughout the region R_I.
Consequently, for $\{\phi_\lambda, \Psi_\lambda\}$ to constitute a complete basis for global space, it is sufficient for ϕ_λ to be a complete basis for the class of wave functions which possess null Cauchy data on $\Sigma_{II} = \Sigma \cap R_{II}$. This can be shown by decomposing $\{\phi_\lambda, \Psi_\lambda\}$ on the basis of the "Rindler states" defined in the previous section of this paper.
The Fock space is thus built upon the creation-annihilition operators c_κ^+, c_κ and d_κ^+, d_κ relative to ϕ_κ and Ψ_κ ; we have

$$\hat{\phi} = \hat{\phi}_I + \hat{\phi}_{II}$$

with

$$\hat{\phi}_I = \int d\kappa \left(c_\kappa \phi_\kappa + c_\kappa^+ \phi_\kappa^* \right)$$

$$\hat{\phi}_{II} = \int d\kappa \left(d_\kappa \Psi_\kappa + d_\kappa^+ \Psi_\kappa^* \right)$$

and

$$[\phi_I, \phi_{II}] = 0$$

The operators c_κ, d_κ define the accelerated vacuum $|0'\rangle$;

$$c_\kappa |0'\rangle = d_\kappa |0'\rangle = 0$$

The PCT construction ensures that the theory in accelerated coordinates is completely determined by its formulation in the region R_I. Indeed, we have

$$d_\kappa = \Theta^{-1} c_\kappa \Theta$$

where Θ is the anti-unitary PCT operator. The Bogoliubov transformation between the two representations of the Fock space is written as

(A and B.7)

$$c_\kappa = \int d^3k \left(A_{\kappa,k}\, a_k + B_{\kappa,k}\, a_k^+ \right)$$

$$d_\kappa = \int d^3k \left(A_{\kappa k}^*\, a_k + B_{\kappa k}^*\, a_k^+ \right)$$

where

$$A_{\kappa k} = (\phi_\kappa, \varphi_k) \quad \text{and} \quad B_{\kappa k} = (\phi_\kappa, \varphi_k)$$

It is desirable to note that the canonical quantization is achieved first of all in the global space-time $\mathcal{M}$. Otherwise the operator PCT could not be built up. The Bogoliubov transformation is simply the unitary transformation linking two choices of possible base states for the Fock space.
In coordinates $(\vec{x}'; t')$ the wave equation takes the form :

$$\left[-\partial_{t'}^2 + \partial_{x'}^2 - \Lambda(x',t')\pi^2 \right] X_{\lambda_1}(x',t') = 0$$

(A.8) with

$$\phi_\kappa(\vec{x}',t') = X_{\lambda_1}(x',t') \cdot \sqrt{q}\, J_\sigma(q\rho') \cdot (2\pi)^{-1/2}\, e^{i\sigma(\varphi' + \Omega t')}$$

and $\Pi^2 = m^2 + q^2$

(B.8)

with $$\phi_{\not\lambda}(x',t') = \chi_{\lambda_1}(x',t')\,(2\pi)^{-1}\, e^{i\lambda_2(y'+\beta t')+i\lambda_3(z'+\alpha t')}$$

and $$\Pi^2 = m^2 + \lambda_2^2 + \lambda_3^2$$

The asymptotic condition (5) implies that the effective mass $\Lambda(x',t')\,\Pi^2$ is null at the horizons and that it is infinite at infinity so that no particle can escape. So we can choose as the base functions, the functions $\phi_\lambda \equiv \phi_\lambda^{in}\,;\,\phi_\lambda^*$ which satisfy :

$$\begin{cases} \lim\limits_{v'\to-\infty} \chi_{\lambda_1}^{in}(u',v') = |2\pi 2\lambda_1|^{-1/2} \exp(i\lambda_1 u') \\ \lim\limits_{u'\to+\infty} \chi_{\lambda_1}^{in}(u',v') = 0 \end{cases}$$

where explicitly :

$$\chi_{\lambda_1}^{in}(u',v') = |2\pi 2\lambda_1|^{-1/2} \int_{-\infty}^{u'} d\xi\,(i\lambda_1)\, e^{i\lambda_1\xi}\, J_0\left(\Pi\sqrt{v(f(\xi)-u)}\right)$$

The functions ϕ_λ are orthonormalized with respect to the scalar product on R_I. Moreover, $\omega = \lambda_1 - \sigma\Omega$ in the first case and $\omega = \lambda_1 - \alpha\lambda_3 - \beta\lambda_2$ in the second one are interpreted as the asymptotic frequencies on the past-horizon.
Another choice of the base states is possible, imposing asymptotic conditions on the future horizon so that :

$$\lim_{u'\to-\infty} \chi_{\lambda_1}^{out}(u',v') = |2\pi 2\lambda_1|^{-1/2} e^{-i\lambda_1 v'} \quad \text{and} \quad \lim_{v'\to+\infty} \chi_{\lambda_1}^{out}(u',v') = 0$$

In the first case, the vacuum is denoted $|0'\ ;\ in\rangle$ whereas in the second the vacuum will be called $|0'\ ;\ out\rangle$. In general, the non-stationary character makes the two vacuums inequivalent (only for the Rindler mapping is $|0';in\rangle = |0';out\rangle$). From here on, we write $|0'\rangle$ for $|0';in\rangle$ unless explicitly stated.
With respect to the region R_I, we note that, by construction, the states defined by $d_{\not\lambda}$ are not observable. The commutator, $[\phi_I, \phi_{II}] = 0$ expresses the absence of a causal relationship between R_I and R_{II}. So, relative to the region R_I, the pure state $|0\rangle$ which corresponds to the global vacuum is described by the density matrix obtained by tracing-out the states $d_{\not\lambda}$:

$$\hat{\rho}_{acc} = \mathrm{Tr}_{II}\, |0\rangle\langle 0|$$

This matrix is completely determined by the population functions :

$$N(\lambda,\lambda') = \mathop{Tr}_{J}(\hat{\rho}_{acc}\, C^+_\lambda C_{\lambda'}) = \int d^3k \; B_{\lambda k} B^*_{\lambda' k}$$

and the correlation functions : $\mathcal{R}(\lambda,\lambda') = \mathop{Tr}_{J}(\rho_{acc} C_\lambda C_{\lambda'}) = \int d^3k \; A_{\lambda k} B^*_{\lambda' k}$

An explicit calculation gives :

$$\begin{cases} A_{\lambda,k} = \delta(q-k)\, \delta_{\mu,\sigma} \times (-1) \dfrac{E_k + k_1}{4\pi\sqrt{\lambda_1 E_k}} \displaystyle\int_0^{+\infty} du \; e^{-i\lambda_1 F(u) + i \frac{E_k + k_1}{2} u} \\ B_{\lambda,k} = \delta(q-k)\, \delta_{\mu,-\sigma} \times \dfrac{E_k + k_1}{4\pi\sqrt{\lambda_1 E_k}} \displaystyle\int_0^{+\infty} du \; e^{-i\lambda_1 F(u) - i \frac{E_k + k_1}{2} u} \end{cases}$$

Thus,

$$\begin{cases} N(\lambda,\lambda') = \delta(q-q')\, \delta_{\sigma,\sigma'} \times \dfrac{-1}{4\pi^2\sqrt{\lambda_1\lambda'_1}} \displaystyle\int_0^{+\infty} du\,du' \; \dfrac{e^{i\lambda_1 F(u) - i\lambda'_1 F(u')}}{(u - u' + i\epsilon)^2} \\ \mathcal{R}(\lambda,\lambda') = \delta(q-q')\, \delta_{\sigma+\sigma'} \times \dfrac{1}{4\pi^2\sqrt{\lambda_1\lambda'_1}} \displaystyle\int_0^{+\infty} du\,du' \; \dfrac{e^{-i\lambda_1 F(u) - i\lambda'_1 F(u')}}{(u - u' + i\epsilon)^2} \end{cases}$$

with

$A:\ \lambda_1 = \omega + \sigma\Omega$ or $B:\ \lambda_1 = \omega + \beta\lambda_2 + \alpha\lambda_3$

So the Bogoliubov coefficients $A_{\lambda k}$ and $B_{\lambda k}$ are not the same as in the non-massive case but $N(\lambda,\lambda')$ and $\mathcal{R}(\lambda,\lambda')$ are not dependent on the mass as the asymptotic condition imposes a total redshift on the past horizon (see dispersion relation). Thus it is the asymptotic behaviour which determines the thermal properties. Indeed the results already obtained by N. Sanchez can be extended.

i) The relation between the mapping f and $N(\lambda,\lambda')$ is reciprocal and we can invert the relation

$$\frac{d}{du'}\left[\operatorname{Ln} f(u')\right] = 4\pi \,\mathcal{R}e \int_0^{+\infty} d\lambda'_1 \; e^{i\lambda'_1 u'} \left[\sqrt{\lambda_1\lambda'_1}\; N_1(\lambda_1,\lambda'_1)\right]_{\lambda_1=\lambda'_1}$$

where N_1 is defined by

(A.9) $$N(\lambda,\lambda') = N_1(\lambda_1,\lambda'_1)\, \delta(q-q')\, \delta_{\sigma,-\sigma'}$$

(B.9) $$N(\lambda,\lambda') = N_1(\lambda_1,\lambda'_1)\, \delta(\lambda_2-\lambda'_2)\, \delta(\lambda_3-\lambda'_3)$$

ii) The above relation makes it possible to show that the Rindler mapping, $f(u') = \exp(au')$, is the only one which satisfies the global thermal balance

$$N(\lambda,\lambda') = N_V(\lambda)\, \delta(\lambda-\lambda') \quad \text{and} \quad \mathcal{R}(\lambda,\lambda') = 0$$

$N_V(\lambda)$ is the population function for a unity of volume and, in the Rindler case, we obtain :

(A. 10) $$N_\nu(\ast) = 1 \Big/ \left[\exp \frac{1}{T}(\omega - \sigma\Omega) - 1\right]$$

and (B.10) $$N_\nu(\ast) = 1 \Big/ \left[\exp \frac{1}{T}(\omega - \alpha\lambda_3 - \beta\lambda_2) - 1\right]$$

where $T = a/2\pi$ appears as the temperature
and Ω; α, β play the role of chemical potentials.

iii) The thermic properties are defined by the asymptotic behaviour of the mapping. For an asymptotic Rindler mapping, $f(u') = \exp(q_{\pm} u')$ when $u' \to \pm\infty$ the population function behaves according to the law

$$N(\ast, \ast') = N_\nu(\ast)\, \delta(\ast - \ast') \quad \text{when } \ast \to \ast'$$

$$\text{with } N_\nu(\ast) = \frac{1}{2}\left\{ \frac{1}{\exp(\omega - \sigma\Omega)/T_+ - 1} + \frac{1}{\exp \omega - \sigma\Omega)/T_- - 1} \right\} \text{ for case A}$$

and there is a simple analogous expression for the case B.
Here, the asymptotic temperature $T_{\pm}$ can be written as

(A and B.11) $$T_{\pm} = \frac{1}{2\pi} \times \frac{d}{du'}\left[\mathrm{Ln}\, f(u')\right]\Big|_{u' = \pm\infty}$$

Contrary to the previous case, there is no global thermal equilibrium but only an asymptotic thermal equilibrium in the region where the coordinates and tend towards infinity.

Moreover, in order to extend the analogy between the examination of the thermal properties linked to these mappings (but in flat space-time) and those that can exist in curved space-time, it is useful to introduce the surface gravity $\mathcal{K}$. $\mathcal{K}$ can be defined by the ratio of the proper acceleration, a', to the temporal compenent, $v^{t'}$ of the speed of the observers that follow the flux lines defined by the normals to the hypersurfaces, t' = constant. Then the asymptotic temperatures are

$$T_{\pm} = \frac{\mathcal{K}}{2\pi}\Big|_{\substack{u' = \pm\infty \\ v' = \pm\infty}} \quad ; \quad \mathcal{K} = \frac{a'}{v^{t'}}$$

This relation can also be interpreted as a generalisation of the Unruh-Hawking temperature $T = a/2\pi$ for uniformly and linearly accelerated observers. The asymptotic character of the thermal effect, and the link between flat space-time and curved space-time effects are clearly shown. In particular, near the horizon of a Kerr black hole the transformation between the Kruskal coordinates (U_K, V_K) and the "tortoise" coordinates $(u' = r^* - t \; ; v' = r^* + t)$

$$r^* = r + \frac{M}{\sqrt{M^2 - a^2}}\left\{ r_+ \mathrm{Ln}\left(\frac{r - r_+}{r_+}\right) + r_- \mathrm{Ln}\left(\frac{r - r_-}{r_-}\right)\right\}$$

$$r_{\pm} = M \pm \sqrt{M^2 - a^2}$$

is basically of type (1) :

$$U_k = f(u') \exp(\kappa_{BH} u')$$

with $\Omega = \Omega_H$ the angular velocity of the horizon of the black hole and κ_{BH} the surface gravity of the Kerr-black-hole:

$$\kappa_{BH} = \frac{r_+ - r_-}{4 M r_+} \quad \text{and} \quad \Omega_H = \frac{a}{2 M r_+} .$$

The Hawking temperature follows from this analogy. But the analogy cannot be pursued further. In particular, the supperradiance effect cannot be reproduced as is shown by the expression (A.10) of $N(\lambda, \lambda')$. If one wished to show schematically such an effect with another mapping, better reflecting the properties of the Kerr metric, the stationary character would be lost ; the vacua $|0';in\rangle$ and $|0';out\rangle$ are then no longer equivalent. In that case, it is no longer possible to distinguish the effects of non-stationarity from the effects of superradiance due to a difference between asymptotic frequencies. The same problems would present themselves if one wished to re-establish the isotropy : the stationary character is destroyed.
This previous study can be extended to mappings with non-constant rotation or drifting unless they becomes constant at the horizons.

Remark 1.
In a thermal equilibrium situation at a temperature T, we typically define the thermal average of an observable $\mathcal{O}$, by computing the expectation of $\mathcal{O}$ at the temperature T and by substracting its value at $T=0$. i.e. :

$$\mathrm{Tr}(\hat{\rho}_T \mathcal{O}) - \langle \mathcal{O} \rangle|_{T=0}$$

In this spirit, the natural definition of the average in an accelerated frame seems to be

$$\langle \mathcal{O} \rangle_{acc} = \mathrm{Tr}\langle \hat{\rho}_{acc} \mathcal{O} \rangle - \langle 0' | \mathcal{O} | 0' \rangle \tag{2-R1}$$

In particular, if $\mathcal{O}$ is the stress tensor in a two dimensional massless case, this definition gives a renormalized stress-tensor which takes into account the energy carried by the "created particles" due to the acceleration. (The meaning of this definition is to give a "physical reality" to the created particles). Namely, for accelerated frames (u',v') :

$$ds^2 = f'(u')\, g'(v')\, du' dv'$$

the stress tensor reads |8|

$$T_{u'u'} = \lambda D_{u'}[f] + \mathcal{U}(u') \quad \text{and} \quad T_{v'v'} = \lambda D_{v'}[g] + \mathcal{V}(v')$$

where $D[f]=f'''/f'-\frac{3}{2}(f''/f')^2$ is the schwarzian derivative.
This stress-tensor definition explicitely breaks covariance by coordinate transformation. Indeed, the choice of the renormalization prescription $(2\text{-}R_1)$ is not a covariant one because the accelerated vacuum $|0'\rangle$ is frame dependent. At this stage, we can either abandon the definition $(2\text{-}R_1)$ and find a covariant one or, find a law which tells us how must transform the vacuum by a frame transformation. The semi-classcial equation of the back reaction problem :

$$G_{\mu\nu} = 8\pi G\langle T_{\mu\nu}\rangle$$

gives us this transformation law. Explicitely, this equation breaks up |9|, in the two dimensional case, into a geometrical equation and into a set of equations linking the accelerated frames to the vacuum states. This relation tell us how to transform the vacuum by frame transformation in order to compensate the non-covariant character of the renormalization scheme.

Remark 2.

It will be observed that our study yields a temperature $T=a/2\pi$ in the Rindler case, and not $T=a/\pi$ as t'Hooft suggested recently |10|. This ambiguity is due to the procedure adopted by t'Hooft for the definition of the associated states in the region R_1. In order to define a quantum covariance principle and to secure a one-to-one correspondance between the global space $\mathcal{M}$ and the region R_I, he identifies the physics of the left region R_I with that of the right region and, he defines a linear relation between a quantum state in $\mathcal{M}$ and a density matrix in R_I. In order to describes his proposal, we introduce the P.C.T. symmetry and we link the Fock-space $\mathcal{H}_I$ associated to the operators $c_{\vec{k}}$ to the Fock space $\mathcal{H}_{II}$ associated to the $d_{\vec{k}}$ by :

$$\mathcal{H}_I \sim \Theta^{-1}\mathcal{H}_{II}\Theta$$

where Θ is the P.C.T. antiunitary operator.

Then, to the state $|\Psi\rangle = \sum_j \alpha_j |\Psi_I^j\rangle|\Psi_{II}^j\rangle$ is associated the density matrix : $\tilde{\rho}_\Psi = \sum_j \alpha_j |\Psi_I^j\rangle\langle\Theta\Psi_{II}^j|$

This relation is independant of the choice of the basis in $\mathcal{H}_I$ and $\mathcal{H}_{II}$ Fock-spaces. It follows that for the Minkowski vacuum state $|0\rangle$, the new density matrix :

$$\tilde{\rho}_0 = \sum_{n;\omega\ldots} e^{-n\pi\left(\frac{\omega-\sigma\Omega}{a}\right)} |n;\omega,\sigma,k\rangle\langle n;\omega,\sigma,k|$$

is now, a thermal state with a temperature twice the standard one. But the hermiticity condition for the density matrix restrict the α_j's by reality conditions. Therefore, we must restrict ourselves to a real quantum mechanics. More accurately, the density matrix $\tilde{\rho}_\Psi$ is a hermitian operator only if the state $|\Psi\rangle$ is P.C.T. invariant. We do not know if such a real formalism has a physical meaning.
Recently, an approach to this problem, based on the construction of symmetric wave functions has been given |11|. Even if after identification, the period in the imagi-

nary time appears to be half the standard one, the resulting Q.F.T. does not have a finite temperature at all.

3. THE ASYMPTOTIC CHARACTER OF THERMAL EFFECTS FROM A LOCAL PRINCIPLE.

The previous study has shown the asymptotic character of the thermal effects. The temperature (A-11) depends only on the behavior of the mapping at the asymptotic regions of the space-time, in particular at the horizon. Recently, Haag and co-workers |12| have deduced the Hawking temperature from a local principle, implemented outside and on the horizon. Contrary to the commutator $G(x,x') = [\phi(x), \phi(x')]$ which is state-independant |13| (in a globally hyperbolic space-time), the anticommutator function $G^{(1)}(x,x') = \{\phi(x), \phi(x')\}$ is state-dependent. This function can be used in order to implement a local criterion satisfied by the "physically allowed" states. Haag and co-workers have chosen to define their local principle on the tangent-space of the space-time $\mathcal{M}$.

Let η_x a map from the tangent-space T_x at x to $\mathcal{M}$.
such that

$$\left| \begin{array}{l} \eta_x(\vec{0}) = x \\ \frac{d}{ds}\eta_x(s\vec{z})\Big|_{s=0} = \vec{z} \in T_x \end{array} \right. \tag{3.1}$$

The local principle becomes :

$$\lim_{s\to 0} s^2 G^{(1)}\left(\eta_x(s\vec{z}_1), \eta_x(s\vec{z}_2)\right) = \frac{1}{2\pi^2(z_1-z_2)^2} \tag{3.2}$$

[It can be shown to be independent of the choice of the map η_x].
This local critirion requires just that the singular part of the $G^{(1)}$ function looks like the Minkowski-one.
The aim of this principle is to show that the only thermal-equilibrium-state (with respect to the Rindler time τ) satisfying this critirion on the horizon has the Hawking temperature $T = a/2\pi$. Indeed, for all observables A and B say, a thermal equilibrium state with temperature $T = 1/\beta$ satisfies K.M.S. condition |14|.

$$\int d\tau \, \langle B A(\tau)\rangle_\beta \, e^{-i\omega\tau} = e^{\beta\omega}\int \langle A(\tau) B\rangle_\beta \, e^{-i\omega\tau} d\tau \tag{3.3}$$

where $A(\tau) = U^{+}(\tau) A \, U(\tau)$; and $U(\tau)$ is the evolution-operator with respect to the τ -time independent hamiltonian.
It is desirable to introduce the commutator |A,B|, and therefore to write the K.M.S. condition as

(3.4) $$\langle BA \rangle_\beta = \frac{1}{2\pi}\int \frac{e^{\beta\omega}}{e^{\beta\omega}-1} \times \langle [A(\tau),B] \rangle_\beta \, e^{i\omega\tau} \, d\omega \, d\tau$$

In particular for A and B being the field operator at different points we get

(3.5) $$\langle \phi(\tau_1,\vec{y}_1)\phi(\tau_2,\vec{y}_2) \rangle_\beta = \frac{1}{2\pi}\int G(\tau+\tau_1,\vec{y}_1;\tau_2,\vec{y}_2)\frac{e^{\beta\omega}}{e^{\beta\omega}-1} e^{i\omega\tau} d\omega d\tau$$

In order to analyse the singular part of the $G^{(1)}$ function, we express the commutator function G in terms of the solutions of the Klein-Gordon equation in the Rindler frame

(3.6) $$K_{\frac{i\omega}{a}}(\Pi y^1)\exp(i\, k_\perp \cdot \vec{y}^\perp) \quad ; \quad \Pi^2 = m^2 + k_\perp^2$$

and we use :

(3.7) $$\int d\tau\, G(\tau+\tau_1,\vec{y}_1;\tau_2,\vec{y}_2) e^{i\omega\tau} = \frac{1}{2\pi^3} e^{-i\omega(\tau_1-\tau_2)} \int K_{\frac{i\omega}{a}}(\Pi y_1^1) K_{\frac{i\omega}{a}}(\Pi y_2^1) e^{i k_\perp(\vec{y}_1^\perp-\vec{y}_2^\perp)} dk_\perp$$

We take :

(3.8) $$\left| \begin{array}{l} \tau_2 = \tau_1 + s z^0 \\ y_2^1 = y_1^1 + s z^1 \\ \vec{y}_2^\perp = \vec{y}_1^\perp + s \vec{z}^\perp \end{array} \right.$$

It is easy to show that the equation (3.5) with the local principle (3.2) imposed outside the horizon allows any value of the temperature, because the singular part of the r.h.s. of (3.5) is independent of β . But <u>on the horizon</u>, $y^1 \to 0$, the behavior of the modified Hankel function becomes

(3.9) $$K_{\frac{i\omega}{a}}(\Pi y^1) \longrightarrow \pi\, \delta(\omega/a)$$

Then, the K.M.S. equation (3.5) yields

(3.10) $$\langle \{ \phi(\tau_1,\vec{y}_1\to 0), \phi(\tau_2,\vec{y}_2) \} \rangle_\beta = \frac{a}{2\pi^2\beta}\int dk^\perp\, K_0(\Pi s z^1)\, e^{-is\, k_\perp \cdot \vec{z}^\perp}$$
$$\xrightarrow[s \to 0]{} \frac{a}{\pi\beta s^2} \times \frac{1}{(z^{12}+z_\perp^2)}$$

Thus the comparison with (3.2) on the horizon shows that the local principle is satisfied only for the Hawking temperature

$$\beta = 2\pi/a = 1/T$$

For quantum fields in accelerated frames, the event horizon plays the role of a thermostat and fixes the temperature.

Remark

This local principle is in fact based on the most singular term of the Hadamard development |15,16| which postulates the expression of the $G^{(1)}$ function as

$$G^{(1)}(x,x') = \frac{1}{4\pi^2}\left[\frac{U(x,x')}{\sigma(x,x')} + V(x,x')\,\mathrm{Ln}\,\sigma + W(x,x')\right] \tag{3.11}$$

at least in a small normal neighbourhood. Where $\sigma(x,x')$ is one-half the square geodesic distance between x and x', and U,V,W some smooth functions. But the previous principle is too local to ensure the unitarity equivalence between differents vacua satisfying this critirion. However, if we impose the Hadamard singularity up to the order one in $\sigma^{;\mu}$ in $U(x,x')$ and $V(x,x')$, the unitarity is ensured |17|. The currently favored methods of renormalisation of the stress-tensor, |18|

$$T_{\mu\nu}(x) = \lim_{x\to x'} \mathcal{D}_{\mu\nu}(x,x')\left[G^{(1)}(x,x') - G^{(1)}_{\substack{\text{local}\\ \text{singular}}}(x,x')\right]$$

with $\mathcal{D}_{\mu\nu}$ a differential operator,
are applicable only if the $G^{(1)}$ function possess a Hadamard development up to the order two. Therefore, it seems quite reasonalbe to extend this locality critirion and to require the asymptotic Hadamard form up to all orders for the anticommutator function. With this asymption, we are able to renormalize |17| all quantities classically defined by an expressions like :

$$\phi(x)_{;\mu_1\cdots\mu_p} \times \cdots \times \phi(x)_{;\nu_1\cdots\nu_q}$$

with an arbitrary number of derivatives and an arbitrary number of products of fields.

4. A HAMILTONIAN FORMULATION

A hamiltonian formulation of these effects is crucial for the study of the Wheeler-De Witt equation for quantum gravitational fields. The equation governing the evolution of scalar field is built on from the action

$$S = -\frac{1}{2}\int d^4x\, |g|^{1/2}\left[g^{\mu\nu}\partial_\mu\phi\,\partial_\nu\phi + m^2\phi^2\right] = \int d^4x\,\mathcal{L} \tag{4.1}$$

by standard procedure. Since, in arbitrary coordinate systems, the metric reads

$$ds^2 = -N^2dt^2 + h_{ij}\left(dx^i + N^i dt\right)\left(dx^j + N^j dt\right) \tag{4.2}$$

the canonical momentum π, defined with respect to the time t is :

(4.3) $$\pi = \frac{\delta \mathcal{L}}{\delta(\partial_t \phi)} = - N |h|^{1/2} g^{00} \partial_0 \phi$$

and the hamiltonian density is

(4.4) $$\mathcal{H} = \pi \, \partial_t \phi - \mathcal{L} = \frac{N h^{1/2}}{2} \left[h^{-1} \pi^2 + h^{ij} \partial_i \phi \partial_j \phi + m^2 \phi^2 \right] + N_i h^{ij} \partial_j \phi . \pi$$

In particular, the Legendre transformation (4-3) becomes singular if N vanishes. The submanifold in which the Hamiltonian formulation remains well defined is bounded by hypersurfaces locally defined by N = 0. They are always null-hypersurfaces in keeping with causal nature of the field propagation. The Schrödinger equation for the wave function Ψ is

(4.5) $$i \frac{d}{dt} \Psi = \int \mathcal{H} . \Psi$$

We shall compare the hamiltonian formulation in Minkowski and Rindler frames and in particular, we analyse the ground states of the differents hamiltonians. For the sake of simplicity, we restrict ourselves to the two dimensional massless case. The other cases are similar and can be found in the reference |19|. The Rindler coordinate system is chosen to be (τ, η) :

(4.6) $$\left| \begin{array}{l} at = e^{a\eta} \sinh(a\tau) \\ ax = e^{a\eta} \cosh(a\tau) \end{array} \right. \qquad ie : \; ay^1 = e^{a\eta}$$

In a Minkowski coordinate system (t,x), with Diricklet boundary conditions, $\phi = 0$ at $\eta \to -\infty$ (the case with Newmann boundary conditions is similar), we expand the wave function Ψ as

(4.7) $$\Psi(t,x) = \int_0^{+\infty} dk \sqrt{\frac{2}{\pi}} \sin(kx) \times q_k(t)$$

Taking q_k as canonical variable, the Schrodinger equation can be written as :

(4.8) $$i \frac{d\Psi}{dt} = \frac{1}{2} \int_0^{+\infty} dk \left[- \frac{\delta^2}{\delta q_k^2} + k^2 q_k^2 \right] \Psi = \int \mathcal{H}_M \Psi$$

The ground state of the Minkowski hamiltonian $\mathcal{H}_M$ is

(4.9) $$\Psi_M^0(q_k) = Z_M^{-1} \exp \left\{ - \frac{1}{2} \int_0^{+\infty} dk \, k \, q_k^2 \right\}$$

For the Rindler case, we expand Ψ as

(4.10) $$\Psi = \int_{-\infty}^{+\infty} dp \, \frac{Q_p}{\sqrt{2\pi}} \, e^{ip\eta} \quad ; \quad Q_p^* = Q_{-p}$$

Taking Q_p as canonical variable, the Schrödinger equation (4.5) is

(4.11) $$i \frac{d\Psi}{d\tau} = \frac{1}{2} \int_{-\infty}^{+\infty} dp \left[-\frac{\delta^2}{\delta Q_p^* \delta Q_p} + p^2 |Q_p|^2 \right] \Psi = \int \mathcal{H}_R \, . \, \Psi$$

The ground state of the Rindler hamiltonian $\mathcal{H}_R$ is

(4.12) $$\Psi_R^o(Q_p) = Z_R^{-1} \exp\left\{ -\frac{1}{2} \int_{-\infty}^{+\infty} dp \, |p| \, Q_p^* Q_p \right\}$$

These two ground states can be compared on the hypersurface $t = \tau = 0$. Inversing the mode decomposition (4.7) and (4.10), yields the Bogoliubov transformaiton between the canonical variable q_k and Q_p :

(4.13) $$q_k = \int_{-\infty}^{+\infty} dp \, \beta_{k,p} \, Q_p$$

(4.14) with $$\beta_{k,p} = \frac{1}{\pi} \int_0^{+\infty} dx \, \sin(kx) \, e^{ip\eta(x)} = \cosh\left(\frac{\pi p}{2a}\right) \times \frac{\Gamma(1 + ip/a)}{a\pi} \left| \frac{k}{a} \right|^{-1-ip/a}$$

We way now substitute this relation into the expression (49) for the Minkowski ground state and we obtain Ψ_M^o in terms of Rindler modes :

(4.15) $$\Psi_M^o = Z_M^{-1} \exp\left\{ -\frac{1}{2} \int_{-\infty}^{+\infty} dp \, p \coth\left(\frac{\pi p}{2a}\right) Q_p^* Q_p \right\}$$

The two ground state appear clearly different. For high p.momentum, the structure of Ψ_M^o approaches that of Ψ_R^o. The difference between Ψ_M^o and Ψ_R^o is significant only for small p-momentum, reflecting the infrared (large distance) nature of the Hawking-Unruh effect. The average of the Rindler number operator for the p-th momentum mode, $N(p)$ say, can be calculated directly by analogy with the simple harmonic oscillator problem :

(4.16) $$\langle \Psi_M^o | N(p) | \Psi_M^o \rangle = \frac{1}{e^{2\pi|p|/a} - 1}$$

which is the familiar result.
We thus see that the Rindler modes in the Minkowski ground state are populated in a thermal distribution. But, by construction, the Minkowski state is described by a coherent state density matrix |20| and not by mixed one.

Remark
In a non-stationary metric, the choice of the vacuum state as the ground state of the hamiltonian becomes delicate. In particular, it does not satisfy the local principle (3.11). Therefore it does not satisfy the unitary condition and it does not possess a well defined renormalized stress-tensor (at least within the standard renormalization schemes |18|).

5. VACUUM FLUCTUATIONS IN ACCELERATED FRAMES.

Hawking radiation is currently interpreted as due to "creation of particles". This interpretation, which appears naturally in the formulation based on a field decomposition on the Rindler modes, is supported by a study of particle detection process. On the other hand, Sciama and co-workers |21,22| have pointed out that this thermal bath has its origin in the zero-point fluctuations of the quantum fields. The spectrum of the zero-point field energy appears to be distorted by the acceleration. Here, following ref.(23), we define the particle density and the energy-density from conserved currents.
For a massless scalar field, the standard density current reads as

(5.1) $$n_\mu = -i\, \phi \overleftrightarrow{\partial_\mu} \phi$$

Moreover, if k^α is a Killing vector for the background space-time, k^α generates a transformation which leaves the action invariant. The conserved Noether current associated to that invariance is

(5.2) $$J_\alpha = -\phi \overleftrightarrow{\partial_\alpha} (k^\beta \partial_\beta \phi)$$

The orbits of the Killing vector can be identified with world lines, $x^\alpha(\tau)$ say, of some observers. The normalized velocity vector is

(5.3) $$u^\alpha = \frac{dx^\alpha}{d\tau} = (k.k)^{-1/2}\, k^\alpha$$

We opt for the following definition of the density of particles , (n) and the energy density (e) seen by those observers

(5.4) $$n = u^\alpha \langle n_\alpha \rangle \qquad \text{and} \qquad e = u^\alpha \langle J_\alpha \rangle$$

where < > stands for the vacuum expectation value.

It is convenient to express the vacuum expectation values in terms of the Wightman functions, $W(x,y)=\langle\phi(x)\phi(y)\rangle$ and to introduce the Fourier transform defined with respect to the proper time along these world lines $x^{\alpha}(\tau)$:

(5.5) $$\tilde{W}^{\pm}(\tau,\omega) = \int ds\, e^{i\omega s}\, W(x(\tau \pm \tfrac{s}{2}), y(\tau \mp s))$$

Then, simple calculations give :

(5.6) $$n = \int_0^{+\infty} d\omega\, \frac{\omega}{\pi} \left[\tilde{W}^{+}(\tau;\omega) - \tilde{W}^{-}(\tau;\omega)\right] = \int 4\pi\omega^2 d\omega\, f(\tau;\omega)$$

(5.7) $$e = (k.k)^{1/2} \int_0^{+\infty} d\omega\, \frac{\omega^2}{\pi} \left[\tilde{W}^{+}(\tau;\omega) + \tilde{W}^{-}(\tau,\omega)\right]$$

Now, interpreting ω as the frequency measured (with respect to the proper time τ) by a detector moving along these trajectories, $f(\tau;\omega)$ and $\frac{de}{d\omega}$ become the particle and energy densities, respectively.

For inertial observers, we get :

(5.8) $$f(\tau;\omega) = \frac{1}{(2\pi)^3} \quad \text{and} \quad \frac{de}{d\omega} = \frac{\omega^3}{2\pi^2}$$

On the otherhand, for Rindler observers, we have :

(5.9) $$f(\tau;\omega) = \frac{1}{(2\pi)^3} \quad \text{and} \quad \frac{de}{d\omega} = (k.k)^{1/2} \frac{\omega^2}{\pi^2}\left[\frac{1}{2} + \frac{1}{e^{2\pi\omega/a_{loc}} - 1}\right]$$

where a_{loc} is the (local) acceleration of these trajectories.

The particles densities are equal in both cases and express that there is "one particle" in each phase-space cell. But the zero-point energy has, in the Rindler case, an additional Planckian term. This is the distortion of the zero-point energy due to the acceleration in agreement with the interpretation of Boyer, Sciama and others.

DISCUSSION

It is quite surprising to note that, despite of its asymptotic character, the Unruh effect admits a local description via the detector models |24|. This special feature has its origin in the high degree of symmetry of the Rindler accelerated frame. The lines, y^1 = constant, coincide with the world-lines of a system of uniformly accelerated observers and furthermore, every where in the region R_I, this coordinate system is locally the Fermi-Walker coordinate system associated with these hyperbolic trajectories. For this reason, the Rindler frame appears as the most adapted one to these trajectories. But in general, for a given flow of trajectories there is no coordinate system which, every where, is locally the Fermi-Walker system associated with them. (The Rindler frame is the unique one which has this property). Therefore, there does not exist a coordinate system "naturally" adapted to the flow of trajectories. This feature has some consequences in non-inertial quantum field theory. In particular, J. Letaw and J. Pfautsch |25,26| have studied the link between the canonical formulation of Q.F.T. in accelerated frames and the models of quantum detection processes by non-inertial observers : A rotating detector plunged into the Minkowskian vacuum responds (the spectrum of the excitations has no simple expression but does not vanish) whereas the rotating vacuum defined by a mode decomposition in a rotating frame is equivalent to the Minkowski one. Here, the choice of the non-inertial rotating frame has been criticized as being highly non-adapted |27|. But, except in the Rindler case, even if it is possible to consider a single point-like detector by the use of the Fermi-Walker coordinate system associated with it, it is impossible to take into account the finite size of the detector. (In particular, we do not know the effect of the acceleration on the internal hamiltonian of this detector). Therefore, up to now, the link between models of detection process and Q.F.T. in accelerated frames is not really etablished. The study of non-inertial quantum field theory seems better adapted to analyse the link between the formulations of quantum theory in a global manifold and in a submanifold respectively, with direct consequences on quantum gravitational effects (section II and vacuum covariance). The problem of quantum detection is an old one. We would like to point-out some new problems which appear from the non-inertial character. First, if we would like to represent the quantum measurement by an observable, without describing the detection process, we must find

a) which element of the observable algebra represents an ideal detection porcess ?

b) A transformation law which tells us how this observable is modified when the same detector is forced to move along some other world-line ?

Up to now, there is no "quantum covariance principle" that answers these questions. Therefore, people looked for models of detection processes. Thus to ensure that the internal hamiltonian of the detector is the same along any trajectories, the currently studied detectors were based on point-like monopoles. By studing the reponse function of the detectors, these models of detectors propose to analyse the "effective particle content" of the quatnum state seen along the detector trajectory. Recently, Hinton |28| and Davies |29| have made important remarks concerning these models.

i) How to normalize the function reponse, in particular in curved space-time

ii) Do different models of detectors give the same "effective particle content" to the same quantum state.

Their studies indicates that the "effective particle content" is detector model dependant. Therefore, these questions are still open : What is really measured by these detection ? or to which measurements do the detection process correspond ? It would be interesting to discuss the problem of anisotropy in the detection of the acceleration radiation in connection with these questions |30|. A way to avoid these problems is to go back to the first point of view, and to search to describe measurement by intrinsic quantities like ϕ^2, $T_{\mu\nu}$ or other vector densities. In the Rindler case, the hyperbolic trajectories are generated by a Killing vector and therefore the Noether current defined in section V is "naturally" adapted to represent the quantum measurement. But in general, for an arbitrary trajectory, such symmetry does not exist and therefore no current is intrinsically defined that answers the questions a) and b). Formulation of a quantum covariance principle as expressed by the previous questions is still an open problem.

ACKNOWLEDGMENTS

I am grateful to Norma Sánchez for numerous discussions, advice and encouragments. I acknowledge Brandon Carter for numerous stimulating discussions and for a critical reading of the draft manuscript.

REFERENCES

1. S.W. Hawking, Com.Math.Phys., 43, 199, (1975).
P.C.W. Davies, J.Phys., A8, 609 (1975).
W.G. Unruh, Phys.Rev. D14, 870 (1976).
S.A. Fulling, PHys.Rev. D7, 2850 (1973).
2. H. Rumpf, Phys.Rev. D28, 2946 (1983).
S.M. Christensen and M.J. Duff, Nucl.Phys. B146, 11 (1978).
P. Candelas and D. Deutsch, ProC.Roy.Soc. A362, 251 (1978) and Proc.Roy.Soc. A354, 79 (1977).
3. R.J. Hughes, Preprint CERN T.H. 3670 (1983).
G.L. Sewell, Ann.Phys.N.Y., 141, 201, (1982).
4. C.W. Misner, K.S. Thorne, J.A. Wheeler, Gravitation,(Freeman, San Francisco 1973).
5. W.G. Unruh and N. Weiss, Phys.Rev. D29, 1656 (1984).
6.b. C. Itzykson and J.B. Zuber, Quantum field Theory, (Mc Graw Hill, N.Y., 1980).
6. N. Sánchez, Phys.Rev. D24, 2100 (1981).
7. D. Bernard and N. Sánchez, preprint in preparation.
8. P.C.W. Davies, Proc.Roy.Soc. Lond, A354, 529 (1977).
9. R. Balbinot and R. Horeanini, Phys.Lett., 151B, 401 (1985).
N. Sánchez, to appear in Nucl. Phys. B.
10. G.t'Hooft, J. Geometry and PHys., 1, 45 (1984).
G. t'Hooft, Utrecht Preprint, December (1984).
11. G. Gibbons, in Cargèse Lectures 1985, to appear.
N. Sánchez and B. Whiting, in preparation.
12. R. Haag, H. Narnhofer and U. Stein, Com.Math.Phys., 94, 219 (1984).
13. Lichnerowitz, in Les Houches, 1963, edited by C. de Witt and B.S. de Witt, Gordon and Beach.
14. R. Kubo, J.Phys.Soc. Japan, 12, 570 (1957).
P.C. Martin and J. Schwinger, Phys.Rev. 115, 1342 (1969).
15. See, for example, F.G. Friedlander, The wave equation on a curved spacetime, (Cambridge University Press, Cambridge, 1975).
16. S.A. Fulling, F.J. Narcowich, R.M. Wald, Ann.Phys. N.Y., 136, 243 (1981).
17. D. Bernard, Meudon Preprint, 1985.
18. N.D. Birrel and P.C.W. Davies, Quantum fields in curved space (Cambridge University Press, Cambridge, 1982).

M.R. Brown and A.C. Ottewill, Proc.Roy.Soc.Lond. A389, 379 (1983).

19. K. Freese, C.T. Hill and R. Mueller, Nucl.Phys. B , (1985).
20. T.D. Lee, Columbia University Preprint (1985).
21. D.W. Sciama, P. Candelas and D. Deutsch, Adv.Phys. 30, 327 (1981).
22. T.H. Boyer, Phys.Rev. D21, 2137 (1980).
23. S. Hacyan et al., Phys.Rev. D32, 914 (1985).
24. W.G. Unruh, Phys.Rev. D14, 870 (1976).
    B.S. de Witt, in General Relativity : an Einstein centenary survey, edited by S.W. Hawking and W. Israel, (Cambridge University Press, Cambridge, 19).
    W.G. Unruh and R.M. Wald, Phys.Rev. D29, 1043 (1984).
25. J.R. Letaw, Phys.Rev. D23, 1709 (1981).
26. J.R. Letaw and J.D. Pfautsch, Phys.Rev. D24, 1491 (1982).
27. N. Myhrwold, Phys. Lett., 100A, 345 (1984).
28. K.J. Hinton, J. Phys. A : Math.Gen., 16, 1937 (1983).
    K.J. Hinton, Class. Quantum Grav., 1, 27, (1984).
29. P.C.W. Davies, in Essays in Honor of the Sixtieth Birthday of B.S. deWitt ; edited by S. Christensen. (Adam Hilger, Bristol, 1984).
30. N. Sánchez, to appear in Phys.Lett.A
    K. Hinton, P.C.W. Davies, J. Pfautsch, Phys. Lett. 120B, 88 (1983).
    W. Israel, J.M. Nester, Phys.Lett., 98A, 329 (1983).

STOCHASTIC DE SITTER (INFLATIONARY) STAGE IN THE EARLY UNIVERSE

A.A. STAROBINSKY
Landau Institute for Theoretical Thysics,
Moscow, 117334, U.S.S.R.

and

ER 176 C.N.R.S. "Département d'Astrophysique Fondamentale"
Observatoire de Meudon
92195 Meudon Principal Cedex
FRANCE

Abstract

The dynamics of a large-scale quasi-homogeneous scalar field producing the de Sitter (inflationary) stage in the early universe is strongly affected by small-scale quantum fluctuations of the same scalar field and, in this way, becomes stochastic. The evolution of the corresponding large-scale space-time metric follows that of the scalar field and is stochastic also. The Fokker-Planck equation for the evolution of the large-scale scalar field is obtained and solved for an arbitrary scalar field potential. The average duration of the de-Sitter stage in the new inflationary scenario is calculated (only partial results on this problem were known earlier). Applications of the developed formalism to the chaotic inflationary scenario and to quantum inflation are considered. In these cases, the main unsolved problem lies in initial pre-inflationary conditions.

1. Introduction

In the models of the early universe with an initial or intermediate metastable de Sitter (inflationary) stage with an effective cosmological constant produced both by quantum gravitational corrections to the Einstein equations |1| and by a scalar field |2-4|, of extreme importance is the exit from this stage that depends on the way of decay of the effective cosmological constant because it determines the spectrum and amplitude of metric perturbations for the subsequent evolution. These perturbations break the homogeneity and isotropy achieved earlier at the inflationary stage and can, in the worst case, destroy all the advantages of inflation. Two ways of decay of the effective cosmological constant are possible : via (quasi) homogeneous classical instability and via inhomogeneous quantum fluctuations. In the first case, the amplitude of perturbations of the de Sitter space-time in the modes which preserve (exactly or approximately) the isotropy and homogeneity of the 3-space in some frame of reference is much more than the amplitude of other, inhomogeneous perturbations. Thus, we have a classical (quasi)-homogeneous perturbation from the very beginning and the subsequent evolution is deterministic ; the duration of the de Sitter stage is totally determined by the initial amplitude of this perturbation. This type of decay takes place, for example, in the author's model |1| for the case of the closed 3-space section if the spatial dimension of this section was of the order of H^{-1} at the beginning of the de Sitter stage (in the paper, we put $\hbar = c = 1$; a(t) is the scale factor of the Friedmann-Robertson-Walker isotropic cosmological model ; $H = \dot{a}/a$).

The existence of a quasi-homogeneous classical scalar field is also assumed in the "chaotic" inflationary scenario |5| (for the inclusion of the R^2 term where R is the Ricci scalar, see |6|). Here, the term "chaotic" simply means the unspecified dependence of the metric and the scalar field on space coordinates though this dependence is weak enough, so that the spatial derivatives of all variables are much less than the temporal ones.

In the second case, we have no large (quasi) homogeneous perturbation at the beginning of the de Sitter stage. This possibility was first pointed in |7| in connection with the model |1|. But, in fact, this situation is more typical for the models where the de Sitter stage arises from the initially radiation-dominated, "hot" universe in the course of a non-equilibrium, close to the II order phase transition (for example, the "new" inflationary scenario). Here, nevertheless, a large quasi-homogeneous "classical" perturbation with characteristic wavelengths $\gg H^{-1}$ can arise during the de Sitter stage from small-scale quantum perturbations. In other words, "classical order" appears from "quantum chaos". In spite of being effectively classical, the evolution of this large-scale perturbation and the space-time metric as a whole is essentially stochastic. The duration of the de Sitter stage also becomes a stochastic quantity in this case.

This is just the process we are interested in. It belongs to the class of the so-called "synenergetic" problems which arise in different branches of science and attract much interest at the present time. We shall consider the new inflationary scenario where the role of the abovementioned perturbation is played by the non-zero large-scale scalar field Φ . It is assumed that $\Phi = 0$ (or sufficiently small) at the beginning of the de Sitter stage. We shall obtain the Fokker-Planck equation for the evolution of the probability distribution of Φ (Sec.2) and calculate the average duration of the de Sitter stage in the new inflationary scenario in Sec.3 (only partial results on this problem or order-of-magnitude estimates were obtained earlier |8-10|). After that, we shall turn to the chaotic inflationary scenario (Sec.4) and discuss the modern state of the problem of the "creation" of the universe briefly (Sec.5).

2. Evolution of a scalar field in the new inflationary scenario.

The de Sitter stage in the new inflationary scenario is assumed to be produced by the vacuum energy of some scalar field with the Lagrangian density

$$L = \frac{1}{2} \Phi_{,i} \Phi^{,k} - V(\Phi) \tag{1}$$

where the vacuum effective potential $V(\Phi)$ has the following properties :

$$V(\Phi_0) = 0$$
$$V(\Phi)_{\Phi \to 0} = V_0 + \frac{1}{2} M^2 \Phi^2 - \frac{1}{3} \nu \Phi^3 - \frac{1}{4} \lambda \Phi^4 , \tag{2}$$

M^2 can have both signes. $\Phi = \Phi_0$ is the flat space-time (true vacuum). $\Phi = 0$ is the false vacuum. We include the term in Φ^3 to describe the case of the so-called "primordial" inflation |11| simultaneously. At the non-zero temperature T, the potential V acquires the additional thermal term which is either small or, with the sufficient accuracy, has the form $\frac{1}{2} B T^2 \Phi^2$, $B \ll 1$, $T \propto a^{-1}$.
At the de Sitter stage, $H = H_0 = \text{const}$, $a = a_0 \exp(H_0 t)$, where $H_0^2 = 8\pi G V_0/3$ (the spatial curvature is negligable). In order to have enough long de Sitter stage and enough small perturbations at the subsequent stages, the following conditions should be fulfilled :

$$|M^2| \lesssim H_0^2/20 \quad ; \quad \nu/H_0 \lesssim 10^{-6} \quad ; \quad \lambda \lesssim 10^{-12} . \tag{3}$$

The Coleman-Weinberg potential does not evidently meet these requirements, so it is usually assumed now that Φ is some weakly interacting scalar field, in particular, it should be the singlet with respect to SU(5) or any other grand unification group. In such a way, the spirit, though not the letter, of the "new" inflationary scenario is maintained.
The de Sitter stage begins when $T^4 \sim V_0$. It can be divided into two successive periods: "hot" and "cold" (vacuum). During the hot period, the temperature $T \gg H_0$ and quantum-gravitational effects caused by the space-time curvature are unimportant. The duration of this period is rather short ; in dimensionless units,

$$H_0 \Delta t_h \sim \ln (V_0^{1/4}/H_0) \sim \ln (G^{-1/2} V_0^{-1/4}) , \tag{4}$$

that is of the order of 10 typically. After that, the cold (vacuum) period begins where $T \ll H_0$ and, in fact, temperature effects can be neglected (except only for the calculation of the initial dispersion of Φ ; see Eq.(13) below). This period is the most interesting because quantum-gravitational effects connected with the space-time curvature play the decisive role here (we denote its beginning by t_0).
To obtain quantitatively (not only qualitatively) correct results one should not use such quantities as $< \Phi >$ or $< \Phi^2 >$ (the approaches based on these quantities have been correctly criticized in [12, 13]). Instead of this, we represent the quantum scalar field Φ (the Heisenberg operator) in the form :

$$\Phi = \overline{\Phi}(t,\vec{r}) + \frac{1}{(2\pi)^{3/2}} \int d^3k .\ \theta(k - \varepsilon a(t) H_0) .$$

$$\cdot [\hat{a}_k \varphi_k(t) \exp(-i\vec{k}\vec{r}) + \hat{a}_k^+ \varphi_k^*(t) \exp(i\vec{k}\vec{r})] + \delta\varphi . \quad (5)$$

$$k = |\vec{k}| , \quad \theta(z) = \begin{cases} 1 , & z > 0 \\ 0 , & z < 0 \end{cases} \qquad \varepsilon = \text{const.} \ll 1 .$$

Here, $\overline{\Phi}(t, \vec{r})$ contains only long wavelength modes with $k \ll H_0 a(t)$, $\delta\varphi$ is the small correction that can be neglected in the leading order in small parameters $|M^2|/H_0^2$, ν/H_0, λ and the second integral term in Eq. (5) satisfies the free massless scalar wave equation in the de Sitter background : $\Box\varphi = 0$. Thus,

$$\varphi_k = H_0 (2k)^{-1/2} \left(\eta - \frac{i}{k}\right) \exp(-ik\eta) \ ; \ \eta = \int \frac{dt}{a(t)} = -[a(t)H_0]^{-1} \quad (6)$$

and $\hat{a}_k^+$ and $\hat{a}_k$ are the usual creation and annihilation Bose-operators. The auxiliary small parameter ε is introduced to refine the derivation, it will not appear in all final equations. In fact, it cannot be arbitrarily small ; the immediate comparison of different terms in Eq. (5) suggests that $\varepsilon \gg |M|/H_0$ but more refined treatment consisting in the substitution of the solution (6) by the solution of the free massive wave equation $\Box\varphi + M^2\varphi = 0$ in the de Sitter background (that does not change Eq. (8) below in the leading approximation in $|M^2|/H_0^2$) shows that the significantly weaker condition $|\ln \varepsilon| \ll \max (H_0^2/M^2 , H_0/\nu , \lambda^{-1})$ is sufficient. It can be also seen immediately that the account of the abovementioned thermal correction to $V(\Phi)$ results in the substitution

$$k \rightarrow (k^2 + k_0^2)^{1/2} , \qquad k_0^2 = BT^2a^2 = \text{const.} \quad (7)$$

in Eq. (6). This gives an effective infrared cut-off that can be important in some problems.
The scalar field Φ satisfies the operator equation of motion $\Box\Phi + dV/d\Phi = 0$

exactly. Using (5, 6) and the conditions of "slow rolling" (3), one obtains the following equation for $\bar{\Phi}$ in the leading order :

$$\dot{\bar{\Phi}}(t,\vec{r}) = -\frac{1}{3H_0}\frac{dV(\bar{\Phi})}{d\bar{\Phi}} + f(t,\vec{r}) \; ;$$

$$f(t,\vec{r}) = \frac{\varepsilon a H_0^2}{(2\pi)^{3/2}} \int d^3k \, . \, \delta(k-\varepsilon a H_0) . \frac{(-i) H_0}{\sqrt{2}\, k^{3/2}} [\hat{a}_k \exp(-i\vec{k}\vec{r}) - - \hat{a}_k^+ \exp(i\vec{k}\vec{r})] . \tag{8}$$

That is the main point : the large-scale scalar field $\bar{\Phi}$ changes not only due to the classical force d $V(\bar{\Phi})$/d $\bar{\Phi}$ but also due to the flow of initially small-scale quantum fluctuations across the de Sitter horizon $k = a(t)H_0$ in the process of expansion. Moreover, the evolution of inhomogeneous fluctuations is linear inside the de Sitter horizon and even in some region outside it ; on the other hand, the evolution of $\bar{\Phi}$ is non-linear but here the spatial and second time derivatives of $\bar{\Phi}$ are small. Below, we shall omit the bar above $\bar{\Phi}$, so Φ will mean the large-scale field only.
Two important consequences follow from Eq. (8). Firstly, there are no spatial derivatives in Eq (8) at all . This means that the evolution of Φ can be studied locally, in the "point" (this "point" has, in fact, spatial dimension $\sim H_0^{-1}$). The temporal evolution of Φ is slow as compared to H_0^{-1} (if the inflation exists at all), so our time "differential" dt can be also chosen $\sim H_0^{-1}$; only the processes with characteristic times $\tau \gg H_0^{-1}$ will be considered. Secondly, though Φ and f have a complicated operator structure, it can be immediately seen that all terms in Eq. (8) commute with each other because $\hat{a}_k$ and $\hat{a}_k^+$ appear only in one combination for each possible $\vec{k}$! Thus, we can consider Φ and f as classical, c-number quantities. But they are certainly stochastic, simply because we can not ascribe any definite numerical value to the combination $[\hat{a}_k \exp(-i\vec{k}\vec{r}) - \hat{a}_k^+ \exp(i\vec{k}\vec{r})]$. As a result, the peculiar properties of the de Sitter space-time — the existence of the horizon and the appearance of the large "friction" term $3H_0\dot{\Phi}$ in the wave equation — simplify the problem of a non-equilibrium phase transition greatly and make its solution possible, in contrast to the case of the flat space-time.
It is clear now that Eq. (8) can be considered as the Langevin equation for $\Phi(t)$ with the stochastic force f(t). The calculation of the correlation function for f(t) is straighforward and gives ($\vec{r}$ is the same throughout) :

$$\langle f(t_1) f(t_2) \rangle = H_0^3 (4\pi^2)^{-1} \delta(t_1 - t_2) \tag{9}$$

Thus f(t) has the properties of white noise. This appears to be the case because different moments of time correspond to different k because of the δ -function in the definition of f, and $\hat{a}_k$ and $\hat{a}_k^+$ with different $\vec{k}$ commute. For spatially separated points,

$$\langle f(t_1,\vec{r}_1)\, f(t_2,\vec{r}_2)\rangle = H_0^3(4\pi^2)^{-1}\delta(t_1-t_2)\cdot\frac{\sin(\varepsilon a H_0|\vec{r}_1-\vec{r}_2|)}{\varepsilon a H_0|\vec{r}_1-\vec{r}_2|} \tag{10}$$

We are interested in the average values $\langle F(\hat{\Phi})\rangle$ where F is an arbitrary function. For that case, one can introduce the normalized probability distribution $\rho(\Phi,t)$ for the classical stochastic quantity Φ

$\left(\int_{-\infty}^{\infty}\rho(\Phi,t)\,d\Phi = 1\right)$, so that

$$\langle F(\hat{\Phi})\rangle = \int_{-\infty}^{\infty}\rho(\Phi,t)\,F(\Phi)\,d\Phi \tag{11}$$

By the standard procedure, the Fokker-Planck (or, better to say, Einstein-Smoluchowski) equation for ρ follows from (8) and (9) :

$$\frac{\partial\rho}{\partial t} = \frac{H_0^3}{8\pi^2}\frac{\partial^2\rho}{\partial\Phi^2} + \frac{1}{3H_0}\frac{\partial}{\partial\Phi}\left(\frac{dV}{d\Phi}\rho\right). \tag{12}$$

This equation has to be supplemented by some initial condition for ρ at $t = t_0$. It should be noted also that Eq. (12) is applicable at the stage of "slow rolling" ($|\dot{\Phi}| \ll H_0\Phi$) only. When this condition ceases to be valid (that takes place at $\Phi \sim \min(H_0\lambda^{-1/2}, H_0^2\nu^{-1})$), the second time derivative of Φ comes into play (though spatial derivatives are still unimportant), the de Sitter stage ends and Φ reaches its flat space-time equilibrium value Φ_0 during the time interval less than H_0^{-1}. After that, a number of oscillations around Φ_0 is possible. Thus, strictly speaking, we can use Eq. (12) if only $|\Phi| \ll \min(H_0\lambda^{-1/2}, H_0^2\nu^{-1})$. But just because we are not interested in time intervals $\Delta t \sim H_0^{-1}$ when calculating such quantities, as e.g., the average duration of the de Sitter stage, we can safely substitute $V(\Phi)$ in Eqs. (8,12) by its expansion for $|\Phi| \ll \Phi_0$ (the second line in Eq. (2)) and use Eqs. (8, 12) for arbitrary Φ. Then the (stochastic) moment of time t_s when the de Sitter stage ends coincides with the sufficient accuracy ($\Delta t \sim H_0^{-1}$) with the moment when $|\Phi|$ reaches infinity according to Eq. (8) (the stochastic force f(t) becomes unimportant at the last stage of evolution).
A note should be added about time reversibility. The microscopic evolution of the total scalar field operator $\hat{\Phi}$ is, certainly, time-reversible, so the apparent, diffusion-like irreversibility of the evolution of $\rho(\Phi,t)$ is due to, as usually, "coarse-graining" that takes place continuously in the process of neglecting more and more information contained in separate modes with different $\vec{k}$.

3. Average duration of the de Sitter stage in the new inflationary scenario.

Now we have to introduce the initial condition for ρ at the beginning of the "cold" part of the inflation : $\rho = \rho_0(\Phi)$ at $t = t_0$. The simplest possible choice would be $\rho_0(\Phi) = \delta(\Phi)$. In fact, the situation is more complicated and depends on the initial conditions at the Planckian moment $t_P = G^{\frac{1}{2}}$. If one assumes thermal equilibrium before the de Sitter stage, then the contribution of thermal quanta of the scalar field Φ with the rest mass $m^2(T) \ll T^2$ $(B \ll 1)$ to $\rho_0(\Phi)$ is gaussian with the dispersion

$$\langle \Phi_T^2 \rangle = \frac{1}{\pi^2} \int_0^\infty k^2 dk \, |\varphi_k|^2 \left[\exp\left(\sqrt{k^2+k_0^2}/aT\right)-1\right]^{-1} \quad (13)$$

At $T \ll H_0$, the main contribution to the integral is due to the region $k \sim k_0 \ll aT$, $|k\eta| \ll 1$. Using (6,7), we obtain [10,14] :

$$\langle \Phi_T^2 \rangle = \frac{H_0^2 a T}{2\pi^2} \int_0^\infty \frac{k^2 dk}{(k^2+k_0^2)^2} = H_0^2/8\pi\sqrt{B} \gg H_0^2 \quad (14)$$

This expression is valid if the modes with $k \sim k_0$ are inside the horizon at the beginning of the de Sitter stage that requires $B \gg GV_0^{\frac{1}{2}} \sim H_0/M_P$, where $M_P = G^{-\frac{1}{2}}$ is the Planck mass. In the opposite case, the modes with $k < k_1 = H_0\, a(t=H_0^{-1})$ are never inside the horizon. For these modes, $\varphi_k \approx$ const. and, in fact, nothing definite can be said about their occupation numbers. The probability distribution needs not be gaussian either, but it is independent of time (we do not include the term $R\Phi^2/12$ into the Lagrangian (1) because then the fine-tuning between M^2 and H_0^2 is necessary for the inflation to occur). In this case, the reasonable lower limit on the initial dispersion can be obtained by integrating from k_1 to ∞ in Eqs. (13,14) that gives

$$\langle \Phi^2(t=t_0) \rangle \gtrsim H_0 V_0^{1/4} \sim H_0^{3/2} M_P^{1/2} \gg H_0^2 \quad , (15)$$

if thermal equilibrium is assumed in the whole region inside the horizon at the beginning of the de Sitter stage.

Thus, the initial dispersion of Φ , in general, exceeds H_0^2 significantly. Nevertheless, it appears (see below) that if

$$\langle \Phi^2(t=t_0) \rangle \ll \min.\left(H_0^2/\lambda^{1/2} \, , \, H_0^{8/3}/\nu^{2/3} \right), \quad (16)$$

then the initial dispersion can be neglected because its effect on the average duration of the de Sitter stage proves to be small. Therefore, there exists a set of possible (though not necessary) initial conditions at $t = t_P$ for which we can use the initial condition $\rho_0(\Phi) = \delta(\Phi)$ at $t = t_0$.

Note that, if the last term in Eq.(12) can be neglected (that takes place in the bening of the "cold" period of inflation), then Eq. (12)is the usual diffusion equation. Thus, the initially gaussian distribution $\rho(\Phi)$ remains gaussian in the course of time evolution and its dispersion changes as

$$\langle \Phi^2 \rangle = \langle \Phi^2 (t=t_0) \rangle + \frac{H_0^3}{4\pi^2}(t-t_0) \; . \tag{17}$$

This is just the result obtained in [9,10,15]. In the presence of the quadratic potential $V = M^2\Phi^2/2$, the distribution remains gaussian and the dispersion can be obtained from the "one-loop" equation |10|

$$\frac{d}{dt}\langle \Phi^2 \rangle = -\frac{2M^2}{3H_0}\langle \Phi^2 \rangle + \frac{H_0^3}{4\pi^2} \; . \tag{18}$$

In this case, Eq.(20) below reduces to that of the harmonic oscillator and can be solved analytically.

In the general case, the solution of Eq.(12) is :

$$\rho(\Phi, t) = \exp\left(-\frac{4\pi^2 V(\Phi)}{3H_0^4}\right) \sum_n c_n \psi_n(\Phi) \exp\left(-E_n \frac{H_0^3}{4\pi^2}(t-t_0)\right), \tag{19}$$

where $\psi_n(\Phi)$is the complete orthonormal set of eigenfunctions of the Schrodinger equation

$$\frac{1}{2}\frac{d^2\psi_n}{d\Phi^2} + (E_n - W(\Phi))\psi_n = 0 \; ;$$

$$|\psi_n(\pm\infty)| = 0 \; ; \tag{20}$$

$$W(\Phi) = \frac{8\pi^4}{9H_0^8}\left(\frac{dV}{d\Phi}\right)^2 - \frac{2\pi^2}{3H_0^4}\frac{d^2V}{d\Phi^2} = \frac{1}{2}(v'^2 - v'') \; ;$$

$$v(\Phi) = 4\pi^2 V(\Phi)/3H_0^4 \; .$$

It was explained at the end of Sec.2 that we may set $V(\infty) = -|V(-\infty)| = -\infty$. Therefore, $W(\pm\infty) = \infty$ and Eq.(20) has the discrete spectrum of eigenvalues only. For $V(\Phi)$ given in Eq.(2), it is the equation of the anharmonic (or doubly anharmonic) oscillator. The coefficients c_n are obtained from the initial condition for $\rho(\Phi, t)$ at $t = t_0$:

$$c_n = \int_{-\infty}^{\infty} d\Phi \, \rho_0(\Phi) \exp(v(\Phi)) \, \psi_n(\Phi) \; . \tag{21}$$

The behaviour of $\rho(\Phi,t)$ at large times is, as usually, determined by the lowest energy level E_0. E_0 is strictly positive that follows from the "supersymmetric" form of the potential $W(\Phi)$.
In practice, we are more interested not in $\rho(\phi,t)$ itself but in $w(t_s)$ - the probability distribution for the stochastic moment t_s when the de Sitter stage ends. $w(t_s)$ can be obtained from $\rho(\phi,t)$ by the following way. Let the rolling of the scalar field to both sides is possible : $V(\pm\infty) = -\infty$. The integral $\int d\phi(\frac{dV}{d\phi})^{-1}$ converges at $|\phi|\to\infty$ that means that $|\phi|$ approaches infinity in finite time. For $|\phi|\to\infty$, the evolution of Φ becomes deterministic ; both the stochastic force in Eq.(8) and the second derivative with respect to Φ in Eq.(12) can be neglected. Then the solution of Eq.(12) for $\Phi\to\pm\infty$ is, correspondingly,

$$\rho = \left(\frac{dV}{d\Phi}\right)^{-1} g\left(t + 3H_0\int_{\pm\infty}^{\Phi} d\Phi\left(\frac{dV}{d\Phi}\right)^{-1}\right) \quad , (22)$$

where g is some unknown function that has to be determined from the previous evolution. The form of the solution represents the fact that the probability is transported without changing along the classical paths

$$t + 3H_0\int_{\pm\infty}^{\Phi} d\Phi\left(\frac{dV}{d\Phi}\right)^{-1} = \text{const.} = t_s \quad . \qquad (23)$$

Therefore, one can introduce $w(t_s)\propto g(t_s)$. The exact coefficient of proportionality is determined by the condition of probability conservation

$$w(t_s) = \rho(\Phi,t)\;\left|(\partial\Phi/\partial t_s)_t\right| \quad , \qquad (24)$$

along the path (23). If we do not make difference between rolling down to the left and to the right sides, then the resulting expression for $w(t_s)$ is

$$w(t_s) = \frac{1}{3H_0}\left(\lim_{\Phi\to+\infty} + \lim_{\Phi\to-\infty}\right)\left|\frac{dV(\Phi)}{d\Phi}\right|\rho(\Phi,t_s). \qquad (25)$$

If the rolling of the scalar field is possible to the right side only ($V(-\infty) = \infty$, $V(\infty) = -\infty$; e.g., when $\lambda = 0$ in Eq.(2)), the second limit in Eq.(25) has to be omitted. The distribution $w(t_s)$ is certainly non-gaussian. Its behaviour for large t_s is exponential and is determined by the lowest energy level E_0. Though $w(t_s)$ cannot be computed analitically, it is remarkable that the closed explicit expressions for all moments $\langle(H_0(t_s-t_0))^n\rangle$ with integer n can be obtained in the form of successive integrals. The approach used here is similar to the Stratonovich's "first time passage" method.
Let us consider a set of the functions

$$Q_n(\Phi) = \int_{t_0}^{\infty} (t-t_0)^n \rho(\Phi,t)\, dt \; ; \quad n = 0, 1, 2 \ldots \tag{26}$$

Then

$$\frac{1}{3H_0}\left(\lim_{\Phi\to+\infty} + \lim_{\Phi\to-\infty}\right)\left|\frac{dV}{d\Phi}\right| Q_n(\Phi) = \int_{t_0}^{\infty}(t-t_0)^n w(t)\,dt = \langle (t_s - t_0)^n \rangle \tag{27}$$

Integrating both sides of Eq.(12) over t from $t = t_0$ to $t = \infty$, we obtain the ordinary differential equation

$$\frac{H_0^3}{8\pi^2} Q_0'' + \frac{1}{3H_0}\left(\frac{dV}{d\Phi} Q_0\right)' = -\rho_0(\Phi). \tag{28}$$

Its solution, subjected to the boundary conditions $Q_0(\pm\infty) = 0$ (because $\rho(\pm\infty, t)=0$), is

$$Q_0 = \frac{8\pi^2}{H_0^3} e^{-2v(\Phi)} \int_{\Phi}^{\infty} e^{2v(\Phi_1)} d\Phi_1 \left(\int_{-\infty}^{\Phi_1} \rho_0(\Phi_2)\, d\Phi_2 - C \right);$$

$$C = \frac{\int_{-\infty}^{\infty} e^{2v(\Phi)} d\Phi \int_{-\infty}^{\Phi} \rho_0(\Phi_1)\, d\Phi_1}{\int_{-\infty}^{\infty} e^{2v(\Phi)}\, d\Phi} = \text{const.}\, ; \; 0 < C < 1 \tag{29}$$

If the rolling is possible to the right (left) side only, then C=0 (C=1). For the symmetric case $V(-\Phi) = V(\Phi)$ and $\rho_0(-\Phi) = \rho_0(\Phi)$, $C = \frac{1}{2}$. Now,

$$\frac{1}{3H_0}\left(\lim_{\Phi\to+\infty} + \lim_{\Phi\to-\infty}\right)\left|\frac{dV}{d\Phi}\right| Q_0(\Phi) = (1-C) + C = 1 = \int_{t_0}^{\infty} w(t)\,dt \tag{30}$$

Thus, the probability $w(t_s)$ introduced according to Eq.(25) is properly normalized. By multiplying both sides of Eq.(12) by $(t-t_0)^n$ and integrating over t from t_0 to $t = \infty$, the recurrence relation between Q_n can be found. It has the form $(n \geq 1)$:

$$\frac{H_0^3}{8\pi^2} Q_n'' + \frac{1}{3H_0}\left(\frac{dV}{d\Phi} Q_n\right)' = -n\, Q_{n-1} \, . \tag{31}$$

The boundary conditions are $Q_n(\pm\infty) = 0$ for all n. Then

$$Q_n = \frac{8\pi^2 n}{H_0^3} e^{-2v(\Phi)} \int_{\Phi}^{\infty} e^{2v(\Phi_1)} d\Phi_1 \left(\int_{-\infty}^{\Phi_1} Q_{n-1}(\Phi_2)\, d\Phi_2 - C_n \right);$$

$$C_m = \frac{\int_{-\infty}^{\infty} e^{2v(\Phi)}\,d\Phi \int_{-\infty}^{\Phi} Q_{m-1}(\Phi_1)\,d\Phi_1}{\int_{-\infty}^{\infty} e^{2v(\Phi)}\,d\Phi} = \text{const.} \tag{32}$$

Using Eq.(27), we obtain

$$\langle H_0^n (t_s - t_0)^n \rangle = n H_0^n \int_{-\infty}^{\infty} Q_{n-1}(\Phi)\,d\Phi \tag{33}$$

In particular, the average dimensionless duration of the de Sitter stage is equal to

$$\langle H_0 \Delta t \rangle = H_0 \Delta t_h + \langle H_0 (t_s - t_0) \rangle = H_0 \Delta t_h + H_0 \int_{-\infty}^{\infty} Q_0(\Phi)\,d\Phi \,, \tag{34}$$

where Δt_h is given in Eq.(4) and Q_0 is presented in Eq.(29).
Let us now consider several particular cases. Let $\gamma = 0$ in Eq.(2) (that corresponds to the original picture of the "new" inflation) and $\rho_0(\Phi) = \delta(\Phi)$. Then Eq.(34) simplifies ($C = \frac{1}{2}$) :

$$\langle H_0 \Delta t \rangle = H_0 \Delta t_h + \frac{8\pi^2}{H_0^2} \int_0^{\infty} d\Phi\, e^{-2v(\Phi)} \int_{\Phi}^{\infty} e^{2v(\Phi_1)}\,d\Phi_1 \,; \tag{35}$$

$$v(\Phi) = \frac{4\pi^2 V(\Phi)}{3H_0^4} = \frac{4\pi^2}{3H_0^4}\left(\frac{M^2\Phi^2}{2} - \frac{\lambda\Phi^4}{4}\right)$$

(the constant term in the potential may be omitted because it cancels in Eq.(35)).
After some manipulation, the expression (35) can be represented in the form containing only one integration :

$$\langle H_0 \Delta t \rangle = H_0 \Delta t_h + \frac{\pi\sqrt{3}}{\sqrt{\lambda}}\left[\sqrt{\frac{\pi}{2}} \int_0^1 \frac{dx}{\sqrt{x(1-x^2)}} \exp\left(\frac{\alpha^2 x}{2}\right) + \right.$$

$$\left. + \alpha \int_0^1 \frac{dx}{\sqrt{1-x^2}}\, \Phi\left(1, \frac{3}{2}, \frac{\alpha^2 x}{2}\right)\right], \quad \alpha = \frac{2\pi M^2}{\sqrt{3\lambda}\, H_0^2}, \tag{36}$$

where Φ is the confluent hypergeometric function.
Three more particular cases are of special interest.
1) $M^2 < 0$; $\lambda^{\frac{1}{2}} H_0^2 \ll |M^2| \ll H_0^2$; $|\alpha| \gg 1$.
Then

$$\langle H_0 \Delta t \rangle = H_0 \Delta t_h + \frac{3H_0^2}{2|M^2|}\left(\ln \frac{16\pi^2 M^4}{3\lambda H_0^4} + \gamma\right), \tag{37}$$

where $\gamma = 0.577$ ... is the Euler constant. In this case, one-loop approximation which consists in the substitution of $\langle\Phi^4\rangle$ by $3(\langle\Phi^2\rangle)^2$ in the equation for $\langle\Phi^2\rangle$ gives the result which is correct with the logarithmic accuracy :

$$\langle H_0 \Delta t \rangle_{\text{one-Loop}} = H_0 \Delta t_k + \frac{3 H^2}{2|M^2|} \ln \frac{8\pi^2 M^4}{9\lambda H_0^4} \cdot \tag{38}$$

However, more accurate approach was developed in |10| for this case which gave the right answer. It consists in the observation that in this case the stochastic force f(t) in Eq.(8) is important then and only then when the classical force $(-dV(\Phi)/d\Phi)$ can be neglected and vice versa. Thus, Eq.(8) can be integrated directly that gives the following result for the stochastic quantity t_s itself |10| :

$$H_0(t_s - t_0) = \frac{3H_0^2}{2|M^2|} \ln \frac{|M^2|}{\lambda \Phi_1^2} , \tag{39}$$

where Φ_1 is a gaussian stochastic quantity with zero average and the dispersion

$$\langle \Phi_1^2 \rangle = 3 H_0^4 / 8\pi^2 |M^2| \tag{40}$$

(the thermal contribution to $\langle\Phi_1^2\rangle$ is neglected here for simplicity). After averaging $\ln \Phi_1$ in Eq.(39) over the gaussian distribution, just the correct result (37) appears.

2) $|M^2| \ll \lambda^{1/2} H_0^2$; $|\alpha| \ll 1$.

For this case, only one-loop |10| or order-of-magnitude |9| estimates were known earlier. It follows from Eq.(36) that

$$\langle H_0 \Delta t \rangle = H_0 \Delta t_k + \frac{\pi\sqrt{3}}{4} \Gamma^2(1/4) \lambda^{-1/2} \simeq H_0 \Delta t_k + 17.88 \lambda^{-1/2} \tag{41}$$

One-loop approximation gives the numerical coefficient in the second term equal to $\pi^2/\sqrt{2} \simeq 6.98$ that is 2.56 times less.

It is intructive to consider the case of a many-component scalar field Φ_a with the symmetry group O(N) and see how the one-loop approximation becomes exact in the limit $N \to \infty$. Let $\Phi = (\Phi_a \Phi^a)^{1/2}$. The strightforward application of the developed approach shows that the corresponding generalization of Eq.(12) to the $N \neq 1$ case is :

$$\frac{\partial \rho}{\partial t} = \frac{H_0^3}{8\pi^2 \Phi^{N-1}} \frac{\partial}{\partial \Phi}\left(\Phi^{N-1} \frac{\partial \rho}{\partial \Phi}\right) + \frac{1}{3 H_0 \Phi^{N-1}} \frac{\partial}{\partial \Phi}\left(\frac{dV}{d\Phi} \Phi^{N-1} \rho\right) \tag{42}$$

$$S_N \int_0^\infty \Phi^{N-1} \rho(\Phi, t)\, d\Phi = 1 \; ; \quad S_N = N \pi^{N/2} / \Gamma\left(1 + \frac{N}{2}\right),$$

where S_N is the area of the N-dimensional sphere (O(N)-symmetrical initial condition for ρ is also assumed). If $\rho(\Phi,t) = \delta(\Phi)$ at $t = t_0$, then, instead of Eq.(35), the following expression for the average duration of the de Sitter stage results :

$$\langle H_0 \Delta t \rangle = H_0 \Delta t_h + \frac{8\pi^2}{H_0^2}\int_0^\infty d\Phi \, \Phi^{N-1} e^{-2v(\Phi)} \int_\Phi^\infty d\Phi_1 \, \Phi_1^{1-N} e^{2v(\Phi_1)} \tag{43}$$

For $V(\Phi) = V_0 - \lambda\Phi^4/4N$,

$$\langle H_0(t_s - t_0)\rangle = \frac{\pi^2\sqrt{3}}{\sqrt{2\lambda}} \cdot \frac{\sqrt{N}\,\Gamma(N/4)}{2\,\Gamma(1/2 + N/4)} ;$$
$$\langle H_0(t_s - t_0)\rangle_{\text{one-Loop}} = \frac{\pi^2\sqrt{3}}{\sqrt{2\lambda}} \cdot \sqrt{\frac{N}{N+2}} . \tag{44}$$

Thus, both expressions tend to the same limit $\pi^2\sqrt{3}/\sqrt{2\lambda} \simeq 12.09\,\lambda^{-1/2}$ at $N \to \infty$ (but from different sides).

Now we return to the N = 1 case and calculate the dispersion of the quantity $H_0(t_s - t_0)$. By the use of Eqs.(32,33), we obtain

$$\langle H_0^2(t_s - t_0)^2\rangle = \frac{96\pi^2}{\lambda}\int_0^\infty e^{u^4/2}\,du \int_u^\infty e^{-x^4/2}\,dx \int_0^x e^{y^4/2}\,dy \int_y^\infty e^{-z^4/2}\,dz ;$$

$$\delta^2 = \langle H_0^2(t_s - t_0)^2\rangle - (\langle H_0(t_s - t_0)\rangle)^2 = \tag{45}$$

$$= \frac{12\pi^2}{\lambda}\left(\frac{\Gamma^2(1/4)}{64} - \int_0^1 \frac{F^2(\arccos t, 1/\sqrt{2})}{1 - t^4}\,dt\right);$$

$$\delta \simeq 0.6408\,\langle H_0(t_s - t_0)\rangle ,$$

where $F(\varphi,k)$ is the elliptic integral of the first kind. Also interesting is to calculate the change in the result (41) due to the spreading of the initial condition at $t = t_0$ (the "thermal" correction). If $\rho(\Phi)$ is the gaussian distribution with the zero average and the dispersion Φ_T^2 (see Eqs.(14,15), then by applying Eq.(29) with $C = \frac{1}{2}$ the following result can be found :

$$\langle H_0(t_s - t_0)\rangle = 17.88\,\lambda^{-1/2} - 4\pi^2\,\Phi_T^2\,H_0^{-2} ;$$
$$\sqrt{\lambda}\,\Phi_T^2 \ll H_0^2 . \tag{46}$$

Thus, if the condition (16) is satisfied, then the thermal correction is small ; in the opposite case, the inflationary stage is very short.

3) $\lambda^{1/2} H_0^2 \ll M^2 \ll H_0^2$; $\alpha \gg 1$.

In this case, the result (36) simplifies to the form :

$$\langle H_0 \Delta t\rangle = H_0 \Delta t_h + \frac{3\pi H_0^2}{\sqrt{2}\,M^2} \exp\left(\frac{2\pi^2 M^4}{3\lambda H_0^4}\right) . \tag{47}$$

The exponent just coincides with the result obtained by Hawking and Moss |8| with the help of the de Sitter instanton. Thus, our approach reproduces the instanton results without using instantons at all. Moreover, we have obtained a little more - the coefficient of the exponential, that corresponds to the summation of all one-loop diagrams on the instanton background in the standard functional integral approach. The corresponding probability distribution $w(t_s)$ is determined by the lowest energy level E_0 of Eq.(20) with the excellent accuracy and, thus, is purely exponential :

$$w(t_s) = \frac{E_0 H_0^3}{4\pi^2} \exp\left(-E_0 . \frac{H_0^3 (t_s - t_0)}{4\pi^2}\right) ;$$
$$\frac{E_0 H_0^2}{4\pi^2} = \frac{\sqrt{2}\, M^2}{3\pi\, H_0^2} \exp\left(-\frac{2\pi^2 M^4}{3\lambda H_0^4}\right) = \langle H_0 (t_s - t_0) \rangle^{-1} , \tag{48}$$

It is clear in our approach that the transition of the scalar field through the potential barrier takes place only locally, that is, in the volume $\sim H_0^{-3}$ (in fact, somewhat larger), but not in the whole 3-space. This fact can be also understood in the functional integral approach if one rewrites the de Sitter instanton in the static, "thermal" form :

$$dS^2 = (1 - H_0^2 r^2)\, d\tau^2 + (1 - H_0^2 r^2)^{-1} dr^2 + r^2 (d\theta^2 + \sin^2\theta\, d\varphi^2) \tag{49}$$
$$\Phi = \Phi_{max} = M \lambda^{-1/2} ,$$

where τ is periodic with the period $2\pi H_0^{-1}$. Then the instanton tells us that Φ has reached the top of the potential barrier inside the horizon ($r < H_0^{-1}$) but gives us no information about the behaviour of Φ outside the horizon.

That is enough for the case of the "new" inflation. Now we shall turn to the so-called "primordial" inflation |11| where it is assumed that $\nu \neq 0$, $\lambda = 0$ and present the most interesting results briefly. In this case, the average duration of the de Sitter stage is given by Eqs.(34,29) with C = 0. Two limiting cases are the most important and representative.

1) $|M^2| \ll H_0^{4/3} \nu^{2/3}$.

Then

$$\langle H_0 (t_s - t_0) \rangle = 4 \left(\frac{\pi}{3}\right)^{2/3} \Gamma^2(1/3) \left(\frac{H_0}{\nu}\right)^{2/3} \simeq 29.60\, (H_0/\nu)^{2/3} . \tag{50}$$

2) $H_0^{4/3} \nu^{2/3} \ll M^2 \ll H_0^2$.

In this case,

$$\langle H_0 (t_s - t_0) \rangle = \frac{6\pi H_0^2}{M^2} \exp\left(\frac{4\pi^2 M^6}{9 H_0^4 \nu^2}\right) \tag{51}$$

Again, the exponent is just the action for the Hawking-Moss instanton which is equal to the difference between the actions for the de Sitter instantons (49) with $\Phi = \Phi_{max} = M^2/\gamma$ and $\Phi = \Phi_{min} = 0$. The third case $M^2<0$, $H_0^{4/3}\gamma^{2/3} \ll |M^2| \ll H_0^2$ reduces, in fact, to the second one after shifting the scalar field : $\Phi = \Phi_1 - |M^2|/\gamma$.
The quantitative results presented in the Sec. 2,3 were first published by the author in the shorter form in Russian in |16,17|. Two points should be emphasized, however.
Firstly, though the quantity $\ln(a(t_s)/a(t_0)) = H_0(t_s-t_0)$ has the well-defined probability distribution $w(t_s)$, the quantity $a(t_s)/a(t_0)$ does not, because $E_0H_0^2 \ll 1$ in all cases. Thus, it seems that the quantity $\ln a(t)$ is more suitable for the description of the stochastic inflation than the scale factor $a(t)$ itself.
Secondly, the calculated duration of the de Sitter stage gives us the typical size of causally connected regions. However, only a minor last part of this inflation produces regions those remain approximately homogeneous and isotropic in the course of subsequent evolution. This follows from the fact that after the inflation, the space-time metric at scales much larger than the cosmological post-inflationary particle horizon has the following simple structure in the proper ("ultra-synchronous") gauge |10|

$$ds^2 = dt^2 - \exp(h(\vec{r}))\, a^2(t)(dx^2 + dy^2 + dz^2) \;; \qquad (52)$$
$$h(\vec{r}) = 2 \ln(a(t_s(\vec{r}))/a(t_0)),$$

where $h(\vec{r})$ is not assumed to be small and $a(t)$ is the scale factor for the strictly isotropic and homogeneous solution. The quantity $h(\vec{r})$ is essentially stochastic, its rms value is of the order of its average (see, e.g., Eq.(45)). Thus, the metric (52) becomes anisotropic and inhomogeneous in the course of the after-inflationary expansion when spatial gradients of $h(\vec{r})$ (omitted in Eq.(52) in the leading approximation) come into play. This situation illustrates the well-known fact that "general" inflation produces neither isotropy nor homogeneity of the present-day universe and, therefore, cannot "explain" them without further assumptions. Nevertheless, if the conditions (3) are fulfilled, then the last, "useful" part of inflation does produce sufficiently large regions with the degree of homogeneity and isotropy that matches the observations. It is important that during this part of inflation the stochastic force $f(t)$ in Eq.(8) becomes small as compared to the classical force $(-dV(\Phi)/(d\Phi)$. Then, for regions those are not too large, $h(\vec{r})$ can be represented in the form which was used in |10,18-21| :

$$h(\vec{r}) = \text{const} + \delta h(\vec{r}) \;;\; \delta h(\vec{r}) = -2H_0\, \delta\Phi(t,\vec{r})/\dot{\Phi} \,, \qquad (53)$$

where $\delta\Phi$ is the small fluctuation of $\Phi(t)$ produced by $f(t,\vec{r})$. Here $\delta h(\vec{r})$ is really small.
The duration Δt_1 of this "useful" part of inflation (when $|\delta h|<1$) is easily estimated using the expression for perturbations (53) :

$$H_0 \Delta t_1 \sim \lambda^{-1/3} \quad ; \quad |M^2| \ll \lambda^{1/3} H_0^2 \, , \; \nu = 0 \, ;$$

$$\sim \frac{H_0^2}{|M^2|} \ln \frac{|M|^6}{\lambda H_0^6} \quad ; \quad |M^2| \gg \lambda^{1/3} H_0^2 \, , \; \nu = 0 \, ;$$

$$\sim \left(\frac{H_0}{\nu}\right)^{1/2} \quad ; \quad |M^2| \ll \nu^{1/2} H_0^{3/2} \, , \; \lambda = 0 \, ; \tag{54}$$

$$\sim \frac{H_0^2}{|M^2|} \ln \frac{M^4}{\nu H_0^3} \quad ; \quad |M^2| \gg \nu^{1/2} H_0^{3/2} \, , \; \lambda = 0 \, .$$

$H_0 \Delta t_1$ contains no exponentially large multipliers. If λ or ν are fixed, then Δt_1 is maximal and the amplitude of perturbations at the given present-day scale is minimal when $|M^2| \ll \lambda^{1/3} H_0^2$ or $|M^2| \ll \nu^{1/2} H_0^{3/2}$; the upper limits on λ and ν presented in Eq.(3), strictly speaking, refer just to these cases. If M does not satisfy these conditions, the duration of the "useful" part of inflation diminishes ; however, the numerical restrictions on λ and ν remain practically unchanged due to the first condition in Eq.(3). It should be pointed also that the case $M^2>0$ presents no more advantages than the case $M^2<0$.

4. Evolution of the scalar field in the chaotic inflationary scenario.

In the chaotic inflationary scenario, it is assumed that the initial value of the quasi-homogeneous scalar field Φ is non-zero and, in fact, large ; typically, $|\Phi|>M_P$ at $t=t_P$. The potential $V(\Phi)$ can be a rather arbitrary function ; the only condition is that it should grow less faster than exp(const.$|\Phi|$) for $|\Phi|\to\infty$. Typical examples are $V(\Phi) = \lambda\Phi^4/4$ |5| and even $V(\Phi)=M^2\Phi^2/2$ with $M^2>0$ (the dynamics of the latter model was studied in |22-26|). Here, the quantity $H = \dot{a}/a$ cannot be constant in general, but if $|\dot{H}| \ll H^2$, then the expansion of the universe is quasi-exponential. Thus, the notion of the quasi-de Sitter stage with the slow varying H arises. The scalar field should also change slowly during this stage : $|\dot{\Phi}| \ll H\Phi$. Then, $H^2 = 8\pi GV(\Phi)$.

We can now repeat the derivation of Eqs.(8,12) (Sec.2) for this case. Because of the dependence of H on t, the quantity $\ln a(t) = \int H(t)dt$ appears to be more proper and fundamental independent variable than the time t. Eq.(6) retains its form with the change : $H_0 \to H$. It is straightforward to obtain the following equation for the large-scale scalar field :

$$\frac{\partial \Phi}{\partial \ln a} = -\frac{1}{3H^2}\frac{dV}{d\Phi} + \frac{f}{H} \; ; \tag{55}$$

$$< f(\ln a)\, f(\ln a_1) > = H^4 (4\pi^2)^{-1} \delta(\ln a - \ln a_1)$$

Then the corresponding Fokker-Planck equation takes the form (H^2 can be expressed through $V(\Phi)$) :

$$\frac{\partial \rho}{\partial \ln a} = \frac{G}{3\pi} \frac{\partial^2}{\partial \Phi^2} (V\rho) + \frac{1}{8\pi G} \frac{\partial}{\partial \Phi} \left(\frac{d \ln V}{d\Phi} \rho \right) . \tag{56}$$

It is worthwhile to note that this equation has just the form one would expect to follow from quantum cosmology because it is no longer depends on such classical quantities as t or H, but contains only fundamental variables $\ln a$ and Φ which remain in quantum case.

Now, the problem of the initial condition for $\rho(\Phi, \ln a)$ arises. In the studies of classical chaotic inflation, it is usually assumed that $\Phi = \Phi_0$ at $t=t_p$ that corresponds to $\rho_0(\Phi) \propto \delta(\Phi - \Phi_0)$ for some $\ln a_0$. But such a condition contradicts the whole spirit of quantum cosmology. A natural idea is to consider stationary solutions (e.g., independent of $\ln a$) of Eq.(56). They can be thought of as being in "equilibrium with space-time foam" which may arise at planckian curvatures. At first, we introduce the notion of the probability flux $j(\Phi, \ln a)$ by rewriting Eq.(56) in the form

$$\frac{\partial \rho}{\partial \ln a} = - \frac{\partial j}{\partial \Phi} ;$$

$$-j = \frac{G}{3\pi} \frac{\partial}{\partial \Phi} (V\rho) + \frac{1}{8\pi G} \frac{d \ln V}{d\Phi} \rho . \tag{57}$$

Then, two types of stationary solutions arise : with no flux and with a constant flux j_0 :

$$\rho = \text{const.}\, V^{-1} \exp(3/8\, G^2 V) - 3\pi j_0 (GV)^{-1} \exp(3/8\, G^2 V) \int^{\Phi} d\Phi_1 \exp(-3/8\, G^2 V(\Phi_1)). \tag{58}$$

The first solution (with $j = 0$) is just the envelope of the Hartle-Hawking time-symmetric wave function |27| in the classically permitted region ($a^2 \gg (8\pi GV)^{-1}$) ; the exponent is the action for the de Sitter instanton with Φ = const (with the correct sign). Moreover, we have obtained the coefficient of the exponent, so the solution appears to be normalizable. It is easy to verify that the average value of Φ calculated with the use of this solution practically coincides with Φ_s —the value of Φ for which $|\dot{H}| \sim H^2$ and the de Sitter stage ends ($\Phi_s \sim M_p$ if $V(\Phi) = \lambda_n \Phi^n/n$). This does not mean that the dimension of the universe after inflation is small (because all $\ln a$ are equally probable for stationary solutions) but suggests that the "useful" part of inflation is typically very small (if exists at all) in this case. It is possible to obtain the "useful" part of inflation that is long enough, but with the very small probability $\sim \exp(-3/8 G^2 V(\Phi_s)) \sim \exp(-10^{10})$.

It is interesting that the second solution with $j \neq 0$ does not, in fact, contain any exponential at all. For $G^2 V(\Phi) \ll 1$ that corresponds to curvatures much less than the planckian one, its form for $j_0<0$ is :

$$\rho_2 \simeq |j_0| \cdot 8\pi G V \Big/ \frac{dV}{d\Phi} . \tag{59}$$

In this case, the stochastic force is unimportant. Thus, we have only two possibilities :either the stationary solution contains the instanton contribution exp(-S) (where S is the action for the instanton, S<0) or the solution is non-exponential. We have not obtained the solution proportional to exp(S) = exp(-|S|) which was advocated by several authors (including the author of this paper) some time ago |28-30|. It seems that the latter solution describing the process of "quantum creation" of the universe via quantum tunneling to the de Sitter stage, though possible formally, has a very small probability also (with the same order of magnitude as above). This conclusion is similar to that obtained by Rubakov |31| though we suppose that his terminology of "catastrophic particle creation in the process of quantum tunneling" is inadequate ; in fact, no real particle creation takes place at the de Sitter instanton solution.

5. Conclusions and discussion.

We introduced and elaborated the approach consisting in taking into account the change in a large-scale scalar field due to the continuous flow of small-scale quantum perturbations of the same scalar field across the de Sitter horizon during the de Sitter (inflationary) stage. That gave us the possibility to find the explicit expressions for the average duration of the de Sitter stage (and for any higher moment if necessary) in the case when the initial probability distribution of the scalar field before the beginning of de Sitter stage was known. Certainly, the method used in the paper (as any other mathematical method) cannot solve the problem of initial pre-inflationary conditions ; new physical hypothesises (or "principles") are necessary for this purpose.

What can be said now about the possibility of "spontaneous quantum creation of the universe" which was so extensively discussed in |32-35| ? To make the terminology more precise, the author proposed some time ago |36| (see also |30|) to call the "quantum creation of the universe" the situation when we have a solution for the wave function of the universe with a non-zero probability flux emerging from the region of small values of a (or, equivalently, large values of space-time curvature). This proposal can be used in our stochastic approach also. Then the first stationary solution of Eq.(56) (the first term in Eq.(58)) corresponds to the time-symmetric universe which has no beginning and was not created. This coincides with the Hawking's interpretation of the Hartle-Hawking wave function in quantum cosmology. In the case of our first solution, we encounter the serious difficulty connected (as was explained in Sec. 4) with the very small probability of having the large duration of the "useful" part of inflation.

The second stationary solution with the non-zero probability flux does correspond to the "creation" of the universe but this creation has very little in common with the picture that was introduced in |32-35|. In particular, no quantum tunneling takes place, and the evolution of the metric and the scalar field remains classical up to the planckian curvatures. This type of creation was called the "classical creation" in |28| but it should be clear that the "classical creation" is not a new concept but simply the paraphrase of the standard classical picture of a singularity as a boundary of the space-time through which the space-time cannot be continued ; the only difference is that now this boundary is assumed to have a finite thickness .

The difficulties with the second solution are connected with our impossibility at the present time to prove the very existence of such a solution (in other words, to prove the possibility of the quantum change of topology) and to say something definite about the value of j_o, if it is non-zero. Thus, the problem of the possibility of the quantum creation of the universe remains open.

The author would like to thank Prof. Norma Sanchez for the hospitality in the Groupe d'Astrophysique Relativiste de l'Observatoire de Paris-Meudon where this paper was completed and the Centre National de la Recherche Scientifique for financial support.

References

1. A.A. Starobinsky, Phys.Lett. 91B (1980) 99.
2. A.H. Guth, Phys.Rev. D23 (1981) 347.
3. A.D. Linde, Pys.Lett. 108B (1982) 389.
4. A. Albrecht, P.J. Steinhardt, Phys.Rev.Lett. 48 (1982) 1220.
5. A.D. Linde, Pisma v ZhETF 38(1983) 149 ; Phys.Lett. 129B (1983) 177.
6. L.A. Kofman, A.D. Linde, A.A. Starobinsky, Phys.Lett. 157B (1985) 361.
7. V.F. Mukhanov, G.V. Chibisov, Pisma v ZhETF 33 (1981) 549.
8. S.W. Hawking, I.G. Moss, Phys.Lett. 110B (1982) 35.
9. A.D. Linde, Phys. Lett. 116B (1982) 335.
10. A.A. Starobinsky, Phys.Lett. 117B (1982) 175.
11. J. Ellis, D.V. Nanopoulos, K.A. Olive, K.Tamvakis, Phys.Lett. 120B (1983) 331.
12. G.F. Mazenko, W.G. Unruh, R.M. Wald, Phys.Rev. D31 (1985) 273.
13. M. Evans, J.G. Mc Carthy, Phys.Rev. D31 (1985) 1799.
14. A. Vilenkin, Phys.Lett. 115B (1982) 91.
15. A. Vilenkin, L.H. ford, Phys.Rev. D26 (1982) 1231.
16. A. Starobinsky, in : Proc. 6th Sov.Gravit.conf., Moscow 3-5 july 1984 (MGPI Press, Moscow, 1984), vol. 2, p. 39.
17. A. Starobinsky, in : Fundamental Interactions, ed. V.N. Ponomarev (MGPI Press, Moscow, 1984) p. 54.

18. S.W. Hawking, Phys.Lett. 115B (1982) 295.
19. A.H. Guth, S.-Y. Pi, Phys.Rev.Lett. 49 (1982) 1110.
20. J. Bardeen, P.I. Steinhardt, M.S. Turner, Phys.Rev. D28 (1983) 679.
21. A.A. Starobinsky, Pis'ma Astron.Zh. 9 (1983) 579 |Sov.Astron.Lett. 9 (1983) 302|.
22. L. Parker, S.A. Fulling, Phys.Rev. D7 (1973) 2357.
23. A.A. Starobinsky. Pis'ma Astron.Zh. 4 (1978) 155 |Sov.Astron.Lett 4 (1978) 82|.
24. S.W. Hawking, Nucl.Phys. B239 (1984) 257.
25. D.N. Page, Class.Quantum Grav. 1 (1984) 417.
26. V.A. Belinsky, L.P. Grishchuk, I.M. Khalatnikov, Ya.B. Zeldovich, Phys.Lett. 155B (1985) 232.
27. J.B. Hartle, S.W. Hawking, Phys.Rev. D28 (1983) 2960.
28. Ya.B. Zeldovich, A.A. Starobinsky, Pis'ma Astron.Zh. 10 (1984) 323 |Sov.Astron. Lett. 10 (1984) 135|.
29. A.D. Linde, Zh.Eksp.Teor.Fiz. 87 (1984) 369 ; Lett. Nuovo Cimento 39 (1984) 401.
30. A. Vilenkin, Phys.Rev. D30 (1984) 509.
31. V.A. Rubakov, Pis'ma ZhETF, 39 (1984) 89 ; Phys.Lett. 148 (1984) 280.
32. E.P. Tryon, Nature 246 (1973) 396.
33. P.I. Fomin, Dokl.Akad.Nauk Ukrain. SSR (1975) 831.
34. L.P. Grishchuk, Ya.B. Zeldovich, in : Quantum structure of space-time, eds. M. Duff and C. Isham (Cambridge U.P., Cambridge, 1982) p. 409.
35. A. Vilenkin, Phys.Lett. 117B (1982) 25 ; Phys.Rev. D27 (1983) 2848.
36. A.A. Starobinsky, talk at the seminar in the P.K. Sternberg State Astronomical Institute (Moscow, November 1983), unpublished.

SOME MATHEMATICAL ASPECTS OF STOCHASTIC QUANTIZATION

G.Jona-Lasinio
Dipartimento di Fisica - Università "La Sapienza",
GNSM and INFN - Roma

In this report I would like to briefly outline some of the mathematical problems encountered in a rigorous implementation of the program of stochastic quantization first proposed by Parisi and Wu Yong-Shi /1/ and then developed in its formal aspects by several authors /2/.

We recall that the basic idea of stochastic quantization consists in considering the Euclidean measure associated to a quantum mechanical system with finite or infinite degrees of freedom, as the stationary state of some stochastic process.

The standard proposal for the construction of such a process is to solve the following stochastic differential equation

$$\frac{\partial\phi(\underline{x},t)}{\partial t} = -\frac{\delta S(\phi)}{\delta\phi(\underline{x},t)} + \frac{\partial W(\underline{x},t)}{\partial t} \qquad (1)$$

where $S(\phi)$ is the Euclidean action describing the system and $W(\underline{x},t)$ in the Wiener process characterized by the covariance

$$E\,(W\,(\underline{x},t)\,W\,(\underline{x}',t')) = \min\,(t,t')\,\delta(\underline{x}-\underline{x}') \qquad (1.2)$$

The typical form of the functional $S(\phi)$ is

$$S(\phi) = \int d^{\nu}x \left(\frac{1}{2}\left[(\nabla\phi)^2 + \phi^2\right] + V(\phi)\right) \tag{1.3}$$

where $V(\phi)$ is a local polynomial in ϕ of even degree and ν is the dimension of the space. Introducing (1.3) in (1.1) we obtain

$$\frac{\partial\phi}{\partial t} = \Delta\phi - \phi - V'(\phi) + \frac{\partial W}{\partial t} \tag{1.4}$$

Δ is the Laplacian.

As it is eq.(1.4) is only formal because the Wiener process is non differentiable. This is a well known difficulty already in the case of ordinary stochastic differential equations where it is overcome by integrating with respect to the time and transforming the equation into an integral equation/3/.

The natural thing to do in the case of (1.4) is to obtain an integral equation by using the Green function of the linear part, that is

$$\phi(t,\underline{x}) = -\int d^{\nu}x' \int_0^t dt'\ G(t,t',\underline{x},\underline{x}')\ V'(\phi(t',\underline{x}')) + Z(t,\underline{x}) \tag{1.5}$$

where G satisfies

$$\frac{\partial G}{\partial t} - \Delta G + G = \delta(t'-t)\ \delta(\underline{x}'-\underline{x}) \tag{1.6}$$

and Z is the Gaussian process

$$Z(t,\underline{x}) = \int d^{\nu}x' \int_0^t dt'\ G(t,t',\underline{x},\underline{x}')\ \frac{\partial W(t',\underline{x}')}{\partial t'} + \phi_0(t,\underline{x}) \tag{1.7}$$

where ϕ_o is a solution of the linear homogeneous part of (1.4). The next step depends crucially on the dimensionality ν.

$\nu = 1$. - In this case (1.5) is a meaningful equation. This depends on the circumstance that the typical trajectories of the process Z(t,x) are continuous functions in both variables. (1.5) can then be solved for each continuous realization of the input Z(t,x). A rather complete treatment of this case can be found in ref./4/. The ν =1 theory covers the stochastic quantization of systems with a finite number of degrees of freedom, that is the case of Quantum Mechanics. It is interesting to note however that its relevance, goes beyond quantum mechanical applications. In recent years in fact the theory developed in /4/ has been useful in entirely different domains /5/-/6/.

$\nu = 2$. - At ν =2 the typical difficulties of Quantum Field Theory appear. The free field $Z(t,\underline{x})$ is not anymore a continuous function but a distribution. If one evaluates for example the expectation value $E_W\ (Z^2(t,\underline{x}))$ this diverges logarithmically. E_W means expectation with respect to the Wiener process (1.2).
Eq.(1.5) has therefore to be modified by introducing counterterms. In this way however, since in the end the counterterms become infinite, the equation itself does not have a mathematical meaning. The way out to this problem taken in the physical literature consists in developing the solution in perturbation theory and then adjusting the renormalization terms in such a way that expectations of the form

$$E_W\ (\phi(t_1, \underline{x}_1)\ \phi(t_2, \underline{x}_2)\ - -\ \phi(t_n, \underline{x}_n))$$

make sense. For a rigorous mathematical treatment this is not sufficient and one has to resort to a non perturbative approach. Therefore the problem facing us consists in defining a "solution" of (1.5) in spite of the fact that the equation does not have a meaning. To solve this problem we appeal to what probabilists call a weak solution of a stochastic differential equation. The basic idea is as follows. Let us first regularize (1.5) by introducing the counter terms. In the present situation this means taking the Wick product of the nonlinear term $V'(\phi)$ and then introducing a cut-off κ in evaluating it. That is the nonlinear term will be: $V'(\phi_\kappa)$: where ϕ_κ is the cut-off field. The Wick product can be taken with respect to the covariance of the free field $C(\underline{x},\underline{y}) = (-\Delta+1)^{-1}$. If for example $V(\phi) = \frac{\lambda}{4}\phi^4$, $:V'(\phi): = \lambda(\phi^3 - 3C(\underline{x},\underline{x})\phi)$. With these modifications (1.5) becomes

$$\phi = -G * :V'(\phi_\kappa): + Z \tag{1.8}$$

and as long as $\kappa < \infty$ this is a meaningful equation which can be solved for each imput Z. The process Z will now induce a measure μ_ϕ^κ on the solutions of (1.8). If now μ_ϕ^κ converges to a limiting measure (in the sense of weak convergence of measures) as $\kappa \to \infty$ we shall say that (1.8) has a weak solution when the cut-off in removed. Our goal therefore consists in implementing non perturbatively this idea. In stochastic calculus there is a well known formula, the Girsanov-Cameron-Martin formula, which provides the Radon-Nikodym derivative of the measure μ_ϕ^κ with respect to μ_Z. This is[3]

$$\frac{d\mu_\phi^\kappa}{d\mu_Z} = \exp\left\{ - \int_0^T (:V'(Z_\kappa):, dW) - \frac{1}{2}\int_0^T dt' \, \| :V'(Z_\kappa): \|^2 \right\} \tag{1.9}$$

where

$$(:V'(Z_\kappa):, dW) = \int_\Lambda d^2x: V'(Z_\kappa(t,\underline{x})): dW(t,\underline{x})$$

is a scalar product in the space variables and $\|\cdot\|$ is the norm induced by it. We work in the finite volume Λ. The stochastic integral appearing in (1.9) is a Ito integral, that is it must be considered as a limit of sums

$$\lim_{n\to\infty} \sum_{i=0}^{n} (:V'(Z_\kappa(t_i)):, W(t_{i+1}) - W(t_i)) \tag{1.10}$$

where the t_i represent a partition of $[o, T]$. Notice that with this definition the increment is uncorrelated with the integrand due to the Markov property of the Wiener process. This integral, very naturally probabilistically, does not obey as it is well known to the usual rules of differential calculus/3/.

Our basic problem now consists in showing that when $\kappa \to \infty$ (1.9) is a good stochastic variable and in particular

$$E_{Z_o} \left(\frac{d\mu_\phi}{d\mu_Z}\right) = 1 \tag{1.11}$$

that is, the measure μ_ϕ is normalized.

Z_o means that the expectation is taken with $Z\ (o, \underline{x}) = Z_o(\underline{x})$.

The problem now reminds of constructive field theory, only a rather special lagrangian is involved. To make the connection even more explicit we notice that by the rules of the Ito calculus it follows

$$\int_0^T (:V'(Z\ ):,\ dW) = \frac{1}{4} \int_\Lambda :\ Z^4\ (T,\underline{x}):\ d^2x - \frac{1}{4} \int_\Lambda :Z^4(o,\underline{x}):\ d^2x + \tag{1.12}$$

$$+ \frac{1}{2} \int_0^T dt\ :(Z^3,\ (-\Delta +1)\,Z):$$

Using (1.12), the exponential in (1.9) takes a less exotic form.

At this point everything seems to be ready to apply the methods of constructive field theory, in particular the rather straight forward methods by which $P(\phi)_2$ was constructed/7/. One realizes immediately however that our problem is more difficult. In fact neither term in the exponent of (1.9) is a well defined stochastic variable. For example the expectation of the second term diverges. The remarkable thing is that this type of divergence does not show up in perturbation theory! The divergence we just mentioned is cancelled by the square of the first term in the expansion of the exponential and a similar mechanism operates with higher order contributions. The reason is the special structure of (1.9) which is such as to insure in any case the normalization condition $E\left(\frac{d\mu_\phi}{d\mu_\infty}\right) = 1$. This reminds of the cancellation mechanisms in supersymmetric theories(*). These divergences however constitute a difficulty in a non perturbative approach.

At this stage there are two possibilities. If we insist on the specific form of eq.(1.8) as the basis for stochastic quantization we must conclude that the methods devised for $P(\phi)_2$ are not sufficient to treat its stochastic counterpart. We must in such a case look for more powerful methods like for example the phase space cell expansion or more generally the renormalization group methods which led to the construction of ϕ_3^4. In fact the above difficulties of stochastic $P(\phi)_2$ seem of a similar nature as those encountered in ϕ_3^4.

The other possibility consists in modifying eq.(1.8) in such a way that the usual $P(\phi)_2$ Euclidean theory still represents its equilibrium state. This was the way followed recently by Jona-Lasinio and Mitter/9/. There is in fact a whole family of stochastic differential equations which admit the same equilibrium measure. The one considered in/9/ is

$$d\phi(t,\underline{x}) = dW(t,\underline{x}) - \frac{1}{2}\left(C^{-\varepsilon}\phi(t,\underline{x}) + C^{1-\varepsilon} :V'(\phi(t,\underline{x}))\right) dt \qquad (1.13)$$

with $0<\varepsilon<1$ and

$$E\left(W(t,\underline{x})\, W(t', \underline{x}')\right) = C^{1-\varepsilon}(\underline{x},\underline{x}') \min(t,t') \qquad (1.14)$$

In/9/ it was shown that for $\varepsilon < \frac{1}{10}$ the methods used for $P(\phi)_2$ are sufficient to prove the existence of an ergodic weak solution of (1.13). The previous equation (1.8) corresponds to $\varepsilon = 1$. The approach to equilibrium is slower for (1.13).

We conclude with some comments. It would certainly be worth to push the analysis to treat the case $\varepsilon = 1$ i.e. eq.(1.8) with the more powerful methods mentioned before.

(*) The connection between stochastic calculus and supersymmetry has been considered by many authors/8/.

In[9] only the ultraviolet problem was studied. It would be interesting to take the limit $\Lambda \to \infty$. In this connection we remark that the formalism of the cluster expansion applies also to the study of (1.13).

In conclusion I would like to express my gratitude to P.K. Mitter. My understanding of the subject discussed here owes much to our pleasant and fruitful collaboration.

References

1) G.Parisi, Wu Yong-Shi, Sci.Sin. 24, 483 (1981).

2) For a review see for example B.Sakita, 7th Johns Hopkins Workshop, ed. G.Domokos, S.Kovesi-Domokos (World Scientific, Singapore 1983).

3) See e.g. I.I. Gihman, A.V.Skorohod, "Stochastic Differential Equations", Springer 1972.

4) W.Faris, G.Jona-Lasinio, J.Phys.A, 15, 3025 (1982).

5) R.Benzi, A.Sutera, J.Phys.A, 18, 2239 (1985).

6) M.Cassandro, E.Olivieri, P.Picco, Ann.Inst. H.Poincaré, in Press.

7) E.Nelson, in "Constructive Quantum Field Theory" Lecture Notes in Phys. Vol. 25, Springer 1973;
B.Simon, "The P $(\phi)_2$ Euclidean (Quantum) Field Theory" Princeton NJ, Princeton University Press 1974;
J.Glimm, A.Jaffe, "Quantum Physics" Springer 1981.

8) S.Cecotti, L.Girardello, Phys.Lett. 110B, 39 (1982);
G.Parisi, N.Sourlas, Nucl.Phys. 206B, 321 (1982);
E.Gozzi, Phys.Lett. 129Bn 432 (1983);
V.de Alfaro, S.Fubini, G.Furlan, G.Veneziano, Phys.Lett. 142B, 399 (1984).

9) G.Jona-Lasinio, P.K.Mitter, Comm.Math.Phys. 101, 409 (1985).

SUPERSTRINGS AND THE UNIFICATION OF FORCES AND PARTICLES

Michael B. Green,
Physics Department, Queen Mary College, University of London, U.K.

The question of how to reconcile the classical description of the gravitational force embodied in Einstein's general theory of relativity with the principles of quantum theory is a central issue in theoretical physics. A simple application of the uncertainty principle shows that at a distance , Δx, around the Planck scale, i.e.

$$\Delta x \approx \sqrt{G\hbar/c^3} \approx 10^{-35} \text{ meters} \tag{1}$$

(where G is the gravitational constant) quantum fluctuations become so large that space-time must be considered to contain a sea of virtual black holes. Since perturbative calculations in quantum gravity assume that the curvature of space-time is small on all length scales they are invalid and lead to non-renormalizable infinities. Non-perturbative methods have not led to calculable consequences.

It appears likely that superstring theories unite gravity and quantum mechanics in a consistent manner. This is acheived by a modification of general relativity at short distances so that Einstein's theory emerges as a long distance approximation. Furthermore, the quantum consistency of superstring theories provides very stringent restrictions on the possible unifying Yang-Mills gauge groups. As a result gravity is unified with the other forces and particles in an almost unique manner. The only possible unifying groups are

$$SO(32) \text{ or } E_8 \times E_8 \tag{2}$$

[SO(32) is a large orthogonal group while E_8 is the largest exceptional Lie group.] The dimensionality of space-time is also required to take a special (or "critical") value

$$D = 10 \tag{3}$$

in order to obtain a consistent superstring quantum theory. Clearly, in order to have any chance of describing the observed physics of our (approximately) four-dimensional world, six dimensions must turn out to be curled-up (or "compactified") to a very small size.

The idea of higher dimensions arose in modern physics in the proposal by Kaluza and Klein[1] in the 1920's to unify electromagnetism with gravity by assuming the existence of a fifth dimension which forms a very small circle. This idea has been revived in the context of supergravity theories which have tried to unify all the interactions in this manner. [A theory in eleven dimensions has been particularly popular.] In these Kaluza-Klein schemes the gauge symmetries of the effectively four-dimensional theory arise from the symmetries of the compactified space. In this respect superstring theories are very different. Already in the ten-dimensional theory there is more gauge symmetry than anyone could wish for since the possible gauge groups (in (2)) are so large. The compactification of the extra dimensions is here expected to reduce the gauge symmetry down to a smaller symmetry group. This should lead to something like a "Grand Unified" symmetry in the effective four-dimensional theory at high energies. Furthermore, to explain observed accelerator physics, this symmetry must in turn break down to the standard model with symmetry groups SU(3) (for colour) and SU(2) × U(1) (for the electro-weak forces).

Although a completely realistic way in which this might happen is not yet understood it is already clear that superstring theories have a good chance of making contact with observed physics. The programme is very ambitious since these theories contain no free input parameters (although the space of solutions may logically have free parameters). The techniques required for analysing both the phenomenological predictions and the theoretical structure of these theories involve the use of many ideas in modern mathematics that have not been used by particle physicists until now. Conversely, many aspects of superstring theory raise issues of interest in pure mathematics.

CHIRALITY

A key constraint on any theory is that it must give rise to the observed chirality (i.e. parity violation) of the four-dimensional world due to the weak interactions. Over the last few years the study of the Kaluza-Klein mechanism has indicated that chiral physics can probably only emerge from a higher-dimensional theory if two conditions are satisfied[2]:

(a) the higher-dimensional theory is chiral (which excludes odd-dimensional theories, since chirality only exists in even dimensions) and

(b) there is a gauge group, G, in the higher-dimensional theory. This seems to be necessary to avoid losing the chirality property in the process of compactification. The gauge fields can twist up into a topologically non-trivial configuration (such as a magnetic monopole) in the internal space - this then distinguishes the different four-dimensional chiralities.

CHIRAL ANOMALIES

Any chiral theory is likely to be plagued by inconsistencies known as chiral gauge "anomalies". These represent the breakdown in the quantum theory of sacrosanct conservation laws that were built into the classical theory. Anomalies may in general arise in the conservation of Yang-Mills currents and in the conservation of gravitational currents i.e. the energy-momentum tensor (as well as in the supersymmetry current). In four dimensions a theory with Weyl fermions also contains anti-fermions of the opposite chirality. Only if the fermions lie in a complex representation of a gauge group (so that the anti-fermions lie in the complex conjugate representation) is the theory chiral. In that case there may be Yang-Mills anomalies but no gravitational anomalies since gravity is insensitive to the gauge group quantum numbers. However, in ten dimensions (and generally in 4n+2 dimensions) a fermion and its anti-particle have the same chirality and there can be both Yang-Mills anomalies (with fermions in any representation) and gravitational anomalies.

The existence of anomalies renders a theory inconsistent because they lead to a violation of unitarity due to the coupling of unphysical longitudinal modes of gauge particles to the physical transverse modes. Up to last summer it had been thought that there were no anomaly-free chiral theories with gauge groups in ten dimensions. It was then discovered[3] that anomalies may be absent from theories with the gauge groups mentioned earlier.

Superstring theories with these gauge groups are both free from anomalies as well as the infinities that plague quantum theories of gravity (as far as has been checked). These successes are unprecedented in any quantum theory of gravity.

WHAT ARE SUPERSTRINGS?

In contrast to usual relativistic field theories, in which the fundamental constituents are structurless point particles, the constituents of any string field theory have extension in one dimension. This leads to significant differences between string field theory and conventional "point" field theories such as Yang-Mills or general relativity. A single classical relativistic string can vibrate in an infinite set of normal modes with unlimited frequencies. The separation between the frequencies of these modes is determined by the rest tension of the string, T. The modes can be quantized so that the quantum mechanics of a single string describes an infinite set of states with masses which increase without bound, their separation given by

$$\Delta(\text{mass})^2 = 2\pi T \qquad (4)$$

These states also have spins which increase without bound since they lie on straight-line Regge trajectories (with slope $\alpha' = 1/2\pi T$). This is not an accident - string theory originated in the late 1960's with the dual resonance model[4] which was developed to explain hadronic phenomena. The earliest string theory (the bosonic string theory[5]) had a critical dimension D = 26 while the spinning[6] string theory which also incorporated fermions had D = 10. It was noticed that the spectrum of the spinning string theory could be truncated to give a supersymmetric spectrum[7] i.e. at every mass level there are an equal number of boson and fermion states. This gave rise to the explicit construction of theories with space-time supersymmetry over the last five years[8]. I shall refer to these theories as superstring theories. The ground states of superstring theories are massless (in contrast to the earlier string theories which were plagued by having tachyonic ground states i.e. states with negative $(\text{mass})^2$). These massless states form supersymmetry multiplets corresponding to the familiar massless states in ten-dimensional super-Yang-Mills and supergravity.

The mass scale set by the string tension is supposed to be the Planck scale (in ten dimensions). This means that, for many purposes, when considering <u>momentum scales much less than the Planck scale</u> the higher mass states are effectively infinitely massive and they decouple leaving an effective "low-energy" theory of the massless ground states. This is just a conventional point field theory such as supergravity and super-Yang-mills. The fundamental particles observed in nature (the quarks, leptons, gauge particles,....) should occur among the massless ground states since their masses are negligible compared to the Planck mass. However, the fact that the low energy theory has arisen from an almost unique superstring theory suggests that the parameters measured in experiments (such as the masses and coupling strengths) should be determined with little ambiguity from the theory.

At <u>momentum scales around the Planck scale</u> the massive states of the string can be excited so that superstring theory then differs radically from any point field theory. This scale is just where the problems with quantum gravity arise. It is because they differ from Einstein's theory (or any supergravity field theory) at these scales that certain superstring theories avoid quantum inconsistencies. In a space-time picture the strings have an average size of the Planck length so they appear as points when looked at coarsely but their non-zero extension is crucial when calculating quantum fluctuations at small scales.

SUPERSTRING DYNAMICS

I will give a very sketchy outline of the way in which the dynamics of a free superstring is formulated.

As a string moves through space-time it sweeps out a (two-dimensional) world-sheet just as a point particle traces out a world-line. The space-time coordinate of any point on the string at a given time, $X^{\mu}(\sigma,\tau)$, is a function of the two parameters of the world-sheet, σ and τ, and μ (= 0,1,...9) is a space-time vector index. In superstring theories there are additionally one or two <u>anticommuting</u> coordinates $\Theta^a(\sigma,\tau)$ which are Weyl spinors (which have 16 components in ten dimensions) labelled by the index a = 1,2,...16. These spinor coordinates embody the supersymmetry of the theory (X^{μ} and Θ^a are superspace coordinates).

The classical dynamics of a relativistic string is obtained from an action principle that generalizes that of a relativistic point particle. Just as the action for a relativistic point particle is the length of its world-line, the action for a string is taken to be proportional to the area of the world-sheet[9]. This is a geometrical quantity which does not depend on the way in which the world-sheet is parametrized. In the case of the superstring theories the notion of the area is generalized so that, roughly speaking, the action is proportional to the area of the world-sheet in superspace. The fact that the action, S, is independent of the parametrization of the two-dimensional world-sheet makes it like a theory of gravity in the two-dimensional σ-τ space[10]

$$S = \int d\sigma\, d\tau\, \sqrt{g}\, \eta_{\mu\nu}\, g^{\alpha\beta}\, \partial_\alpha X^\mu\, \partial_\beta X^\nu + \Theta \text{ terms} \tag{5}$$

where $g_{\alpha\beta}$ is a two-dimensional metric ($\alpha,\beta = \sigma,\tau$) and g is its determinant. This metric is a non-dynamical auxiliary field in two dimensions which can be eliminated by substituting the solution of its equation of motion back into the action. The Θ terms in (5) are designed to ensure the supersymmetry of the action. The above action describes a string moving in flat ten-dimensional Minkowski space where $\eta_{\mu\nu}$ is the flat space-time metric. Much work of recent months has concerned generalizations to background spaces with six compactified dimensions. Requiring the compactified string theory to be consistent puts severe restrictions on the possible background space-times as I will describe later.

An important feature of the action, in addition to the manifest reparametrization invariance, is invariance under rescalings $g^{\alpha\beta} \to \Lambda g^{\alpha\beta}$ where $\Lambda(\sigma,\tau)$ is an arbitrary function. These symmetries allow the choice of a class of gauges in which $g_{\alpha\beta} = 1$ (the conformal gauges) and in which the theory is invariant under (pseudo) conformal transformations. This conformal invariance plays a crucial role in the consistency of the quantum mechanics of a single free string in ensuring that the states created by the time-like ocillations of the string decouple from the physical space of states. This is important since the time-like modes have negative norm and are therefore ghost states. The choice of such a gauge is only possible in the quantum theory in the critical dimension which is ten for superstring theories. It is conceivable that superstring theories could be obtained in lower dimensions (D = 3, 4 or 6) by techniques advocated by Polyakov[11] but that issue is somewhat murky at present. Only in ten dimensions are the physical modes of the string purely transverse just as gauge fields are transversely polarized in Yang-Mills theories.

The solutions of the classical equations, derived from the action, can be expanded in an infinite set of normal modes which can then be quantized. The spectrum depends on the boundary conditions.

A string with free endpoints can carry internal quantum numbers associated with a classical group[12] (SO(n), U(n) or USp(n)). The charges are attached to the ends of the string (rather like the old picture of a meson as a string with a quark at one end and an anti-quark at the other). It turns out that a string with free endpoints has a massless vector particle among its massless states. This apparently accidental feature is the reason why string theories reduce to Yang-Mills theories in the low-energy limit[13] (when all the massive states effectively decouple).

A closed string contains a massless spin-2 particle which is the graviton associated with the fact that the low energy effective theory contains general relativity[14].

When the interactions between strings are included there is a remarkable unification between gravity and Yang-Mills. In the theories containing open strings (known as type I theories) two open strings interact by joining at their endpoints to form a single open string or an open string splits into two strings.

Fig.(1)

This is a local interaction and consistency requires the same interaction to couple the two ends of a single open string to form a closed string as illustrated by

Fig.(2)

so that the existence of open strings (and hence the Yang-Mills sector) requires the existence of closed strings (and hence the gravity sector). The gravitational constant, κ, and the Yang-Mills coupling, g, are related by $\kappa \propto g^2T$.

There are also theories with only closed strings. For example, type II theories describe closed strings which have an orientation i.e. they have excitations corresponding to waves running around the string independently in either direction (the II refers to the fact that these theories have twice as much supersymmetry). These theories may have no net chirality (type IIa) or may be chiral (type IIb). The latter theory is striking since its low energy limit yields a point field theory[15] which is free from all gravitational anomalies[16]. However, type II theories do not have an internal symmetry group and so do not reduce in any obvious way to a chiral four-dimensional theory.

The most interesting kind of superstring theory is the heterotic superstring[17]. This describes closed strings which carry internal symmetry (unlike the other closed superstring theories) with charges which are smeared out as densities along the string. These theories are built from modes of the ten-dimensional superstring running around the string in one sense with modes of the 26-dimensional bosonic string theory running around in the other sense! This apparently bizarre mixture of dimensionalities is reconciled by the identity[18] 26 = 10+16 where the first ten dimensions of the right polarized modes are taken to be the space-time dimensions. The other sixteen dimensions become internal coordinates forming a hypertorus associated with a sixteen dimensional lattice. The consistency of the theory requires this lattice to be even and self-dual. There are known to be only two such lattices[19] which are related to the root lattices of the groups $E_8 \times E_8$ and SO(32) (or, more accurately, the group (Spin 32)/Z_2 which has the same algebra as SO(32)). Therefore the heterotic string theory is only consistent for the two groups that were already known to be selected by requiring the absence of anomalies. In the heterotic string theory $\kappa \propto g/\sqrt{T}$.

SUPERSTRING INTERACTIONS

Superstring scattering amplitudes can be calculated in perturbation theory by constructing a series of diagrams that generalize the Feynman diagrams of familiar point field theories.

Tree Diagrams

The tree approximation to the scattering amplitude of, for example, four closed strings is represented by a continuous world-sheet that joins the two incoming and two outgoing strings.

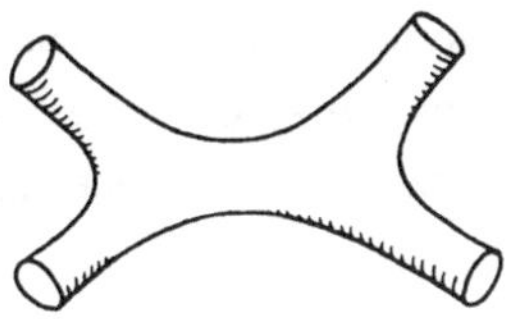

Fig.(3)

This diagram describes two incoming closed strings which join together by touching at a point to form one intermediate closed string which subsequently splits into the two final strings (time is taken to be increasing from left to right). It is possible to derive the amplitude for diagrams like fig.(3) either by a string generalization of Feynman's path integral approach to quantum mechanics or from a second-quantized formalism expressed in terms of string fields which create and destroy complete strings. Unfortunately, for the moment the only complete formulation of the field theory of strings[20] is in a special gauge[21] (the light-cone gauge) which is not satisfactory for understanding the geometric structure of the theory but suffices for perturbative calculations. A string field, $\Psi[X(\sigma),\Theta(\sigma)]$, is a functional of the string configuration. A particularly important aspect of closed string theories is that the interactions only involve terms which are cubic in closed-string fields (as can be seen by slicing through the world-sheet of fig.(3) at the place where one of the interactions takes place) corresponding to the local joining or splitting of the strings

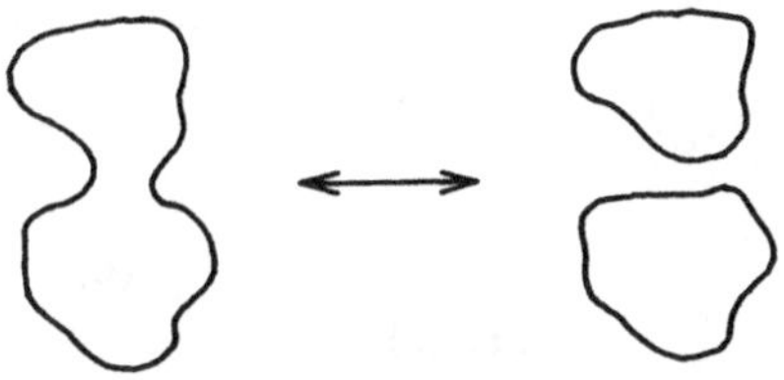

Fig.(4)

(which is the closed-string analogue of the open-string interaction of fig.(1). There are no higher order contact interactions whereas in the perturbative treatment of gravity based on the Einstein-Hilbert action there are an infinite number of interaction terms involving contact interactions between arbitrary numbers of gravitons. All these contact terms emerge as low energy effective interactions arising from the exchange of the massive string states between cubic string vertices. This is analogous to the way in which the four-Fermi model of the weak interactions is now known to emerge as an effective theory from the Weinberg-Salam theory at energies much less than the W or Z boson mass.

Fig.(3) generalizes the four-graviton tree diagram contribution to the scattering amplitude of Einstein's theory (since the graviton is one of the massless string states). Taking the external states to be gravitons this amplitude is given by

$$T(s,t,u) = \begin{bmatrix}\text{Point field theory} \\ \text{result}\end{bmatrix} \times \frac{\Gamma(1-\frac{s}{8\pi T})\Gamma(1-\frac{t}{8\pi T})\Gamma(1-\frac{u}{8\pi T})}{\Gamma(1+\frac{s}{8\pi T})\Gamma(1+\frac{t}{8\pi T})\Gamma(1+\frac{u}{8\pi T})} \tag{6}$$

(where s,t,u are the Mandelstam invariants defined by $s = (p_1+p_2)^2$, $t = (p_1+p_4)^2$, and $u = (p_1+p_3)^2$ where p_1,p_2,p_3,p_4, are the external momenta). In this expression all the string features are contained in the Γ functions. In the low energy limit $s,t,u \ll T$ the expression manifestly reduces to the familiar result based on general relativity.

Loop Diagrams

Higher order diagrams in the perturbation expansion are given by a series of loop diagrams analogous to those of point field theories. For example, the world-surface of the one-loop contribution to the scattering of four closed strings is represented by either of the diagrams.

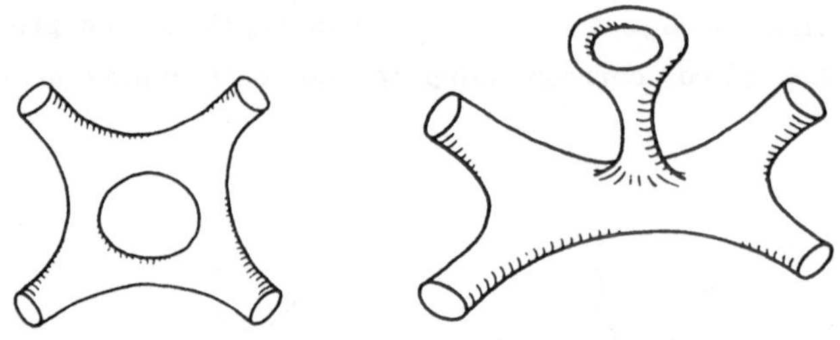

(a) (b)

Fig.(5)

The rules of string theory make these two diagrams equivalent because they can be distorted into each other. This equivalence between different distortions of the world-sheet is a striking property of string theories ("duality") which has no analogue in point field theory. Fig. 5(a) looks like a string box diagram while fig. 5(b) looks like a tadpole diagram. In the original bosonic and spinning string theories there is an infinity[22] in this amplitude. This infinity has a very simple interpretation in terms of the tadpole diagram configuration (fig. 5(b)). The propagator in the leg of the tadpole is singular since it includes (among an infinite set of states) the contribution from the massless scalar partner of the graviton which has the form $1/k^2$ where the momentum in the leg, k, is zero by momentum conservation. This description of the divergence of a loop diagram as an infra-red effect is unique to string theories. The discovery that the type II superstring theories are finite at one loop[23] was the first indication that superstring theories might be consistent quantum field theories (remember that the loop integral is evaluated in 10 dimensions where ordinary point field theories have terribly ultra-violet divergences). This result has also recently been established for the heterotic superstring theory[24].

It has also now been established that the open-string one-loop amplitudes with gauge group SO(32) are finite with four[25] (or more[26]) external states and infinite for any other gauge group.

It is probable that any possible divergences at higher loops can also be associated with the emission of massless scalar particles at zero momentum via generalized tadpoles. For example, at two loops the divergent tadpole contribution to a closed-string amplitude is represented by the ("E.T.") diagram.

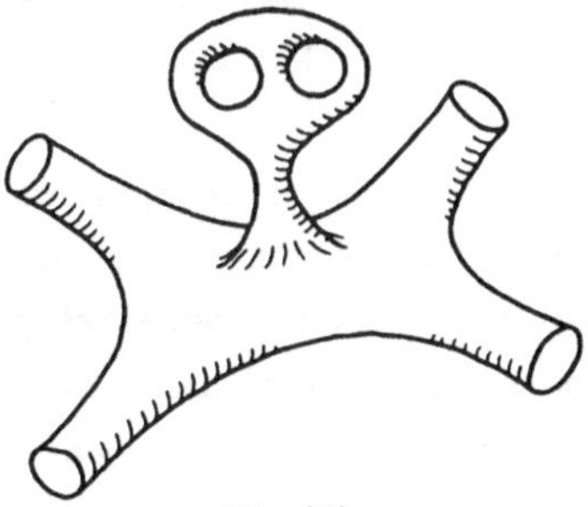

Fig.(6)

From this it follows that the condition for an amplitude to be finite at any number of loops is that  = 0 where  is the on-shell coupling of the massless scalar particle to the general tadpole. But this condition is precisely the requirement that there be an unbroken supersymmetry. Since supersymmetry can only be broken in perturbation theory in ten dimensions if there are anomalies it follows that <u>freedom from anomalies ↔ finiteness</u>[27].

It is important to establish by explicit calculation whether the theories are finite to all orders. This is being intensively studied at the present time[28,29] and a complete proof of finiteness (at least for the type II and heterotic superstring theories) should be forthcoming in the near future.

ANOMALIES AND THEIR CANCELLATION

The signal for an anomaly is the presence of a non-zero coupling between an unphysical longitudinal mode of a gauge particle and any physical transverse modes. Just as in four dimensions an anomaly can arise from a triangle diagram, in ten dimensions an anomaly can arise from a hexagon diagram with circulating chiral fermions and external gauge particles. For example, the evaluation of the anomaly in the Yang-Mills current in a type I theory requires the calculation of

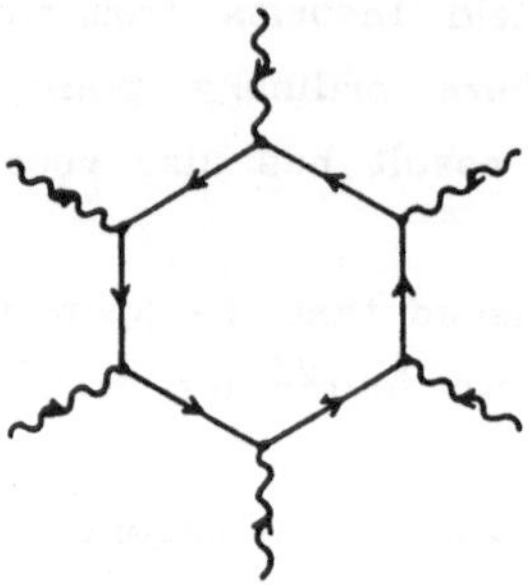

Fig.(7)

where the internal lines are open string propagators and one of the external states is a longitudinal mode of a Yang-Mills particle while the others are transverse modes. The result of this calculation is that the anomaly vanishes[3] when the gauge group is SO(32).

This may appear puzzling because the hexagon diagrams of the corresponding low energy massless point field theory do not give a vanishing anomaly. The explanation is based on the fact that certain open-string hexagon diagrams contain closed-string bound states in various channels. These bound states include the massless states of the supergravity sector which means that the low-energy theory (in which all the massive states decouple) gets contributions from these states in addition to the expected anomalous contribution from the massless hexagon diagrams. These extra terms have the form of *tree* diagrams in which the supergravity particles are exchanged and which have anomalies which exactly cancel the usual anomaly of the massless hexagon diagrams. This explains the absence of an anomaly in the language of the low-energy point field theory as being due to a cancellation between the expected quantum anomaly (due to the usual hexagon diagrams) and a new, anomalous, term in the classical theory (associated with these tree diagrams). The new term can be thought of as an additional (anomalous) term in the effective point field theory action which is a *local* polynomial in the fields. This mechanism depends on a delicate interplay between gravitational and Yang-Mills effects.

The anomalies associated with gravitational currents (due to hexagon diagrams with external gravitons) and the mixed anomalies (due to hexagon diagrams with external Yang-Mills particles together with gravitons) have not yet been calculated in the superstring theories. However, the analysis of the low energy point field theory was carried out for these anomalies also[3]. It turned out that all the anomalies can be cancelled by adding local anomalous terms to the action if the Yang-Mills group, G, is such that :

(a) The dimension of the adjoint representation of G = 496 (which guarantees the absence of the gravitational anomalies).

(b) An arbitrary matrix, F, in the adjoint representation of G satisfies

$$\mathrm{Tr}F^6 = \frac{1}{48}\,\mathrm{Tr}F^2\,(\mathrm{Tr}F^4 - \frac{1}{300}\,(\mathrm{Tr}F^2)^2) \qquad (7)$$

(which ensures the absence of the Yang-Mills and mixed anomalies). The only groups for which these conditions are satisfied are SO(32) and $E_8 \times E_8$ (apart from the presumably uninteresting cases ($U(1)^{496}$ and $E_8 \times U(1)^{248}$). The type I theories do not admit exceptional groups but the heterotic string incorporates both of them. For completeness it would be desirable for the analysis of possible supersymmetry anomalies to be carried out[30].

The preceding discussion referred to anomalies associated with infinitessimal gauge transformations. There is still the possibility of anomalies in the "large" transformations which are not continuously connected to the identity. There are, for example, known to be 991 types of large general coordinate transformations in ten dimensional spherical space-time. These have been shown not to be anomalous for the cases in which the gauge group is one of those free of infinitessimal anomalies[31]. This result has also been generalized[32] with some assumptions to more general spaces than the ten-sphere in the case of the $E_8 \times E_8$.

COMPACTIFICATION OF EXTRA DIMENSIONS

The structure of space-time and hence the possibility of compactification to four dimensions should be determined by the solution of the equations of the superstring field theory (which has so far only been formulated in the light-cone gauge). This approach has not yet been productive. However, there are many other observations that suggest that realistic four-dimensional physics may well emerge. Many of these are based on topological features of the theory. For example, one of the crucial features of the anomaly cancellation argument discussed above is the presence of the massless second-rank antisymmetric tensor field $B_{\mu\nu}$ which has a field strength $H_{\mu\nu\rho}$ defined by

$$H_{\mu\nu\rho} = \partial_{[\mu}B_{\nu\rho]} - \frac{1}{30}\,\omega^{Y}_{\mu\nu\rho} + \omega^{L}_{\mu\nu\rho} \tag{8}$$

where [ ] denotes antisymmetrization of the indices and the Chern-Simons terms are defined by $\partial_{[\sigma}\omega^{Y}_{\mu\nu\rho]} = \mathrm{Tr}F_{[\sigma\mu}F_{\nu\rho]}$ (where the Yang-Mills field strength $F^{ab}_{\mu\nu}$ is a matrix in the adjoint representation of the gauge group) and $\partial_{[\sigma}\omega^{L}_{\mu\nu\rho]} = \mathrm{tr}R_{[\sigma\mu}R_{\nu\rho]}$ (where the Riemann curvature $R^{mn}_{\mu\nu}$ is a matrix in the tangent space with $m,n=0,1,\ldots9$). The condition that $H_{\mu\nu\rho}$ should be single-valued is that the integral of its curl over any four-dimensional subspace should vanish[33], i.e.

$$\int(\mathrm{tr}R_{[\sigma\mu}R_{\nu\rho]} - \frac{1}{30}\,\mathrm{Tr}F_{[\sigma\mu}F_{\nu\rho]}) = 0 \tag{9}$$

This cohomology constraint has two immediate and important consequences :

(a) It is the condition that ensures that the anomalies which were previously shown to vanish in flat ten-dimensional space continue to be absent when some dimensions are curved.

(b) It indicates that in general when there is non-zero curvature, so that $R_{\mu\nu} \neq 0$, the Yang-Mills field strength is also non-zero, i.e. $F_{\mu\nu} \neq 0$. This is just what is needed, since a non-zero field strength will lead to a breaking of the (very large) gauge group of the ten-dimensional theory to a smaller group in the compactified theory (which will hopefully be of more direct interest for physics).

TOWARDS FOUR-DIMENSIONAL PHYSICS

A particularly interesting class of possible compactified solutions has been proposed in ref. 34 in which the ten-dimensional space is the product of four-dimensional flat Minkowski space and a curved six-dimensional compact space with vanishing Ricci curvature. The holonomy group of this compact six-dimensional space is SU(3) i.e. the curvature, $R^{\overline{\mu\nu}}_{\mu\nu}$, is an SU(3) matrix in the tangent space - a condition that leads to a four-dimensional theory possessing an unbroken supersymmetry. Such compact Ricci-flat spaces have come to be called "Calabi-Yau" spaces (they were conjectured to exist by Calabi and shown to exist by Yau). Although the original motivation for suggesting that Ricci-flat spaces may be of interest was based on analysing the effective point field theory that approximates the superstring at low energy, it is possible to argue convincingly that they are also solutions of the full string theory.

The equations of the low energy effective point field theory also suggest that the curvature and the Yang-Mills field strengths should be proportional (which is yet another example of the unification of the gravitational and Yang-Mills aspects of the theory), i.e.

$$R_{\mu\nu} \propto F_{\mu\nu} \tag{10}$$

(suppressing the matrix indices) so that the field strength is also non-zero in an SU(3) subgroup of $E_8 \times E_8$ (the SO(32) case does not appear to hold much prospect of describing physics). As a result the symmetry is broken down to a subgroup that commutes with that SU(3), namely

$$E_6 \times E_8 \tag{11}$$

The E_6 factor plays the rôle of a Grand Unified group for the effectively four-dimensional theory (familiar from the phenomenology of the mid-1970's[35]) but with certain novel features. The E_8 factor describes another sector of matter (that has been dubbed 'shadow matter') consisting of particles which are neutral under the E_6 forces and therefore undetectable except via their gravitational interactions with the matter that we observe. It is highly non-trivial that it is possible for the symmetry to break in this manner and yet be consistent with the topological conditions implied by eq.(9).

Consequences

There isn't yet any systematic procedure for classifying all possible Calabi-Yau spaces but it has been conjectured by Yau that there are a discrete but large number of such spaces (approximately 10000!) and so it is to be hoped that further theoretical constraints will restrict the choice. For the moment the particular Calabi-Yau space used for the six compact dimensions is a discrete parameter that can be chosen arbitrarily. However, the phenomenological considerations described below provide such severe restrictions that no single space so far constructed encompasses all desirable features. Nevertheless, the way in which some of the known spaces come close to explaining various aspects of "low energy" physics makes use of such remarkable features that it seems likely that the scheme has the potential for making contact with physics. Some of the major features of this scheme follow.

(a) Unlike with conventional kinds of unified theories, there are few adjustable parameters once the Calabi-Yau space has been selected. For example, the number of species of massless particles (and hence the number of fermion generations) is determined by a topological property of the space, namely, it is equal to ½ × (Euler characteristic). An aspect of this scheme that is crucial for describing E_6 phenomenology is that the massless chiral fermions in the effectively four-dimensional theory lie in the complex E_6 representation 27 (which is again consistent with eq.(9)). There is no freedom to adjust the particle content so this is already a notable success for the scheme.

(b) Those spaces which give rise to a small number of fermion families (the E_6 phenomenology restricts the number to be three or four) turn out to have holes in them (they are not simply connected). This in turn plays a key rôle in allowing the E_6 symmetry to break down to a realistic low energy symmetry. In conventional unified point field theories symmetry breaking can be introduced by arbitrarily adding Higgs fields. This cannot be done in superstring theories since they allow no such adjustable parameters. However, loops of E_6 flux can become trapped in the holes in the compact space, which has the same effect as having an effective Higgs field in the adjoint representation. With this topological mechanism the values of the effective Higgs fields are determined, up to a discrete choice, which in turn determines, up to a discrete choice, the way in which E_6 breaks to a low energy symmetry group. Among possible low energy symmetry groups are[36] $SU(3) \times SU(2) \times U(1) \times U(1) \times U(1)$, $SU(3) \times SU(2) \times SU(2)$, ... A generic feature of this mechanism is that there is always extra low energy symmetry in addition to the standard model (the residual group has to have at least rank 5).

The mechanism of breaking the symmetry by flux loops cannot be continuously switched off. This means that the E_6 symmetry is never an exact symmetry of the four-dimensional theory, even at high energy (in contrast to its previous rôle in so-called Grand Unified schemes). Although there is unification of the gauge couplings the Yukawa couplings do not satisfy the E_6 relations which avoids some of the bad predictions of E_6 mass relations.

(c) All the couplings of the massless particles are determined by topological considerations[37] and do not depend on detailed knowledge of the metric of the Calabi-Yau space (which is just as well since none of the metrics of these spaces has ever been constructed!). A possible problem arises with proton stability since there is no apparent reason for the Yukawa couplings responsible for proton decay at the tree level to vanish. Nevertheless, explicit calculation of these couplings in many of the known spaces shows that for most of them the proton is stable[38] (up to the usual considerations about decay caused by radiative corrections).

(d) There are possible states associated with strings winding through holes in the compact space which would lead to stable, unconfined particles with fractional electric charges with masses around the Planck mass[39] (the value of the fundamental charge depending on which Calabi-Yau space is used). Correspondingly, magnetic monopoles are predicted by the theory which have charges which are a multiple of the usual Dirac value.

(e) The point of view adopted in this whole scenario is that much physics can be obtained from treating the low energy effective point field theory in lowest order in perturbation theory. However, certain aspects clearly require a much deeper understanding of the theory. The effective four-dimensional low-energy classical theory has the form of a no-scale supergravity theory[40] and it is important to understand how supersymmetry gets broken. One suggestion[41] invokes a condensate of the gluinos in the "shadow" E_8 sector which triggers a breaking of the supersymmetry without leading to the usual problem of generating a cosmological constant. Despite this virtue, the suggested mechanism requires a non-perturbative effect that goes outside of the approximations on which the scheme is based.

(f) (Mildly) pessimistic note. The consideration of Calabi-Yau spaces can be well-motivated provided the radius of the compact dimension (which is a free parameter that should be approximately the inverse of the unification mass) is much more than the Planck scale. The calculations also assume that the string coupling constant (which should ultimately be determined by the theory) is weak. There are convincing arguments[42] that neither of these approximations can be valid and that superstring theory is intrinsically a strongly coupled theory. It is important to establish, in that case whether the results based on topological considerations might still have some validity.

Since superstring theories are so very different from point field theories it would be more satisfying to find a qualitatively new kind of experimental prediction rather than trying to predict details of presently measured accelerator data. [Examples of such predictions are the existence of extra low energy symmetries and their associated gauge particles, the occurrence of unconfined fractionally charged particles and the existence of shadow matter.]

THEORETICAL DEVELOPMENTS

(a) String Theories in Curved Space-Time.

The string actions of the form of eq. (5), which describe the motion of a string in a flat Minkowski space background (with metric $\eta_{\mu\nu}$) can be thought of as two-dimensional field theories of gravity (in which the "fields" are the superspace coordinates). The tree diagrams of the closed-string theory are associated with a two-dimensional world-sheet (as in fig. 3) which is a closed surface that is topologically equivalent to a sphere. The n-loop corrections (illustrsted in fig. 5) correspond to two-dimensional surfaces with n handles. Therefore to any order in string perturbation theory a string theory is equivalent to a two-dimensional gravitational theory evaluated on a manifold of particular genus. This viewpoint is a theme in many interesting developments.

(i) The heterotic superstring[17] possesses not only two-dimensional (world-sheet) coordinate invariance and two-dimensional supersymmetry in common with the other superstring theories but is also chiral in the two-dimensional sense (due to the asymmetric treatment of the right and left polarized modes). It is noteworthy that all these properties are also properties of the theory in the ten-dimensional sense. Consistency of string theory requires it to be free of anomalies in two- dimensional coordinate transformations that cannot be continuously connected to the identity (these "large" coordinate transformations are modular transformations). This is the ingredient in the heterotic superstring theory that restricted the possible gauge groups to just those previously obtained by requiring the absence of the infinitessimal ten-dimensional chiral anomalies. Furthermore, this generalizes to the situation in which the ten-dimensional space is curved provided the identification in eq.(10) is made[43]. This identification is also required in order to ensure that two-dimensional chiral anomalies vanish in the compactified theory[44].

(ii) In order to describe a string propagating in a curved background it is necessary to replace the flat metric in eq.(5) by a curved metric, $G_{\mu\nu}(X)$, which is a function of X. This gives the action of a non-linear sigma model mapping the world-sheet into the curved space-time. There are, however, strong constraints on formulating consistent string theories in a curved background due to the requirement that the theory can be formulated in a parametrization in which it is conformally invariant (recall that this was necessary to provide the gauge conditions required to decouple the negative-normed states). In general this will require the presence of other terms in the two-dimensional action involving, in addition to $G_{\mu\nu}(X)$, an antisymmetric tensor background, $B_{\mu\nu}(X)$, a scalar background, $\Phi(X)$, and fermionic terms[45]. [These background fields correspond to the massless field content of the string field theory.] In addition, in order for one of these generalized sigma models to correspond to a compactification of the ten-dimensional superstring the algebra satisfied by the generators of two-dimensional conformal transformations (the "Virasoro" algebra) must have a central extension with a particular coefficient. The condition that such a theory be conformally invariant is that the renormalization group β functions should all vanish (there is one β function for each kind of background field). It is known that a non-linear sigma model on a group manifold can be made conformally invariant by the addition of the term involving $B_{\mu\nu}$ if this is normalized to a special value[46] (this corresponds to adding a torsion term that parallelizes the curvature of the group manifold). However, it seems unlikely that such string theories[47] can be supersymmetric in space-time and they do not correspond to a compactification of a ten-dimensional theory (they have the wrong value for the central extension term in the Virasoro algebra).

(iii) The equations implied by the vanishing of the β functions, and hence by the requirement of conformal invariance, have been studied for a wide class of sigma models in perturbation theory (where the expansion parameter is the inverse string tension). As expected by general arguments[48] these equations are just the field equations for the massless components of the superstring fields expanded in a power series in the inverse string tension i.e. in a low energy expansion. This analysis provides more evidence, at least in low orders in this expansion, for the fact that Ricci-flat spaces of SU(3) holonomy (Calabi-Yau spaces) are solutions of the string theory together with the identification of the curvature with the Yang-Mills field strength (eq.(10)).

(iv) There is now a proof[49] that supersymmetric non-linear sigma models in which $G_{\mu\nu}$ is the metric on a Calabi-Yau space are finite to all orders in this expansion and so have vanishing β function (and are therefore conformally invariant) and are suitable candidates for compactified superstring theories. There is also evidence that the Ricci-flat condition is not by itself sufficient but it is crucial for the space to be Kahler[50] (which is equivalent to demanding SU(3) holonomy), thus lending further support to the scheme of ref.(34).

(v) Since no metric on a Calabi-Yau space has ever been explicitly constructed it is not possible to give an explicit solution for a superstring theory in which six dimensions are compactified on such a space. However, certain Calabi-Yau spaces reduce, in a singular limit, to six-dimensional tori with discrete isometries divided out. These singular spaces are called "orbifolds". Superstring theories defined with orbifold backgrounds can be analyzed explicitly and behave as if the background were a Calabi-Yau space (the singularities of the orbifold are irrelevant in the string theory)[51].

(vi) The covariant formulation of superstring theories can either be deduced by starting from the spinning string theory and then truncating to a supersymmetric subset of states (as mentioned earlier) or from a manifestly supersymmetric action in superspace as implied by eq.(5)[8]. The latter formulation has a much more geometrical interpretation[52]. It has also been generalized to curved gravitational backgrounds for the type I theories[53] (and, recently, for the type II theories[54]).

It seems unlikely that the physics of two dimensions will determine all the constraints on the theories although it is remarkable how restricted the possibilities are for constructing a suitable conformally invariant sigma model on a two-dimensional manifold of arbitrary genus. A suggestion that non-perturbative string effects must play an important rôle is that flat ten-dimensional Minkowski space satisfies all the restrictions we know of that follow from the two-dimensional viewpoint and yet we hope to prove it is not a possible solution of the theory.

(b) **Towards a gauge-invariant field theory of superstrings.**

In order to arrive at a more geometrical understanding of string field theory it is probably necessary to formulate it in a gauge-invariant manner. A preliminary step was the understanding of the free bosonic string field theory in a Lorentz-covariant gauge using the BRS technique[55]. This has led to a gauge invariant formulation of the free string field theory[56] (as well as some aspects of the interactions[57]).

(c) **Other Topics**

(i) Throughout the development of string theories there has been a parallel development of Kac-Moody algebras in mathematics. The representation theory of Kac-Moody algebras has deep connections with the dynamics of string theories[58]. These infinite-dimensional algebras express the algebra of local currents in the string world-sheet. The connection between Kac-Moody algebras and string theories has been actively studied and was crucial in developing the heterotic superstring[59].

(ii) Although appealing from a geometrical point of view, the formulation of superstring theories in terms of a manifestly supersymmetric Lorentz-covariant action like eq.(5) has not been quantized in a covariant manner. This seems likely to be solved by extending the symmetries of the action[60]. Furthermore, there has been progress towards formulating a manifestly supersymmetric first-quantized theory by directly constructing the quantum operators of the theory[61]. This may lead to a proof of the absence of supersymmetry anomalies at n loops for the type II and heterotic superstring theories, and hence to their finiteness.

(iii) It is now plausible that the requirements that a chiral ten-dimensional theory with a Yang-Mills gauge group be supersymmetric and also free of anomalies leads inexorably to superstring theory. The "minimal" ten-dimensional field theory of super-Yang-Mills coupled to supergravity has anomalies. These can be cancelled by adding extra terms to the theory[3] (motivated by the low energy limit of superstring theory) which spoil its supersymmetry. The process of restoring the supersymmetry by adding yet more terms should eventually reconstruct the infinite number of terms that constitute the exact expansion of the superstring theory in terms of the massless fields. At the lowest non-trivial order in this expansion this has been shown[62] to imply the existence of terms in the low energy action which are quadratic in the Riemann curvature and which have the structure initially conjectured in ref. 63.

CONCLUSION

Superstring theories have passed all the tests of consistency with quantum mechanics that cause problems with conventional theories of gravity, based on Einstein's theory. Furthermore, this consistency restricts the possible Yang-Mills groups almost uniquely and therefore holds the exciting prospect of a unified and consistent quantum theory of all the interactions.

These are early days, however, and there are many questions to be answered about how superstring theories may make contact with observed physics. Up until now, quite apart from the phenomenological issues described earlier, these theories have not provided a natural explanation for several of the most accurately known numbers in physics:

- At the very least we must understand how it is that these theories, formulated initially in D=10 dimensions predict that, to a very good approximation,

$$D = 4$$

in the world that we see at accessible energies which is an assumption in the scheme outlined above. [It would at least be satisfying to discover that ten-dimensional Minkowski space is not a solution of the theory.]

- The fact that the cosmological constant is zero to an amazing accuracy has not yet been explained in a natural way in superstring theory. [Although the scheme of ref.(34) outlined earlier does not generate a cosmological constant in the process of compactification there is no obvious mechanism that would prevent one being generated in the subsequent symmetry breaking transitions.]

- Another outstanding question is where the mass scale associated with weak symmetry breaking comes from.

The present theoretical understanding of superstring theories is somewhat primitive. The theories are formulated in terms of invariance principles related to the two-dimensional world-sheet. They contain both the massless Yang-Mills particle and the massless graviton which is why they reduce, at low energies, to (super) Yang-Mills coupled to (super) gravity. However, this appears to be a fortuitous accident since the theories were not explicitly based on any geometrical principle in space-time. The discovery of such a priciple, which would be a generalization of the principle of general relativity, would lead to a much more profound understanding of the basis of the theory and therefore of its predictions.

REFERENCES

(1) Th. Kaluza, Sitzungsber. Preuss. Akad. Wiss. Berlin, Math. Phys. Kl (1921) 966;
O. Klein, Z. Phys. 37 (1926) 895; Nature 118 (1926) 516.
(2) C. Wetterich, Nucl. Phys. B233 (1983) 109;
S. Randjbar-Daemi, A. Salam, E. Sezgin and J. Strathdee, Nucl. Phys. B214 (1983) 491;
E. Witten, Proceedings of the 1983 Shelter Island Conference, ed. N. Khuri et. al. (MIT press, 1985).
(3) M.B. Green and J.H. Schwarz, Phys. Lett., 149B (1984) 117;
(4) G. Veneziano, Nuovo Cim. 57A (1968) 190;
M.A. Virasoro, Phys. Rev. 177 (1969) 2309;
J.A. Shapiro, Phys. Lett. 33B (1970) 361.
(5) Y. Nambu, Int. Conf. on Symmetries and Quark Models, Wayne State University 1969 (Gordon and Breach,1970) p.269;
H. B. Nielsen, Proceedings of the 15th Int. Conf. on High Energy Physics, Kiev 1970;
L. Susskind, Nuovo Cim. 69A (1970) 457; Phys. Rev. D1 (1970) 1182.
(6) P.M.Ramond, Phys. Rev. D3 (1971) 2415;
A. Neveu and J.H. Schwarz, Nucl. Phys. B31 (1971) 86; Phys. Rev. D4 (1971) 1109.
(7) F. Gliozzi, J. Scherk and D.I. Olive, Nucl. Phys. B122 (1977) 253.
(8) M.B. Green and J.H. Schwarz, Nucl. Phys. B181 (1981) 502; Phys. Lett. 109B (1982) 444; Nucl. Phys. B198 (1982) 252; B198 (1982) 441; Phys. Lett. 136B (1984) 367; Nucl. Phys. B243 (1984) 285;
M.B. Green, J.H. Schwarz, L. Brink, Nucl. Phys. B198 (1982) 474.
(9) Y. Nambu, Lectures at the Copenhagen Symposium, 1970;
T. Goto, Progr. Theor. Phys. 46 (1971) 1560.
(10) S. Deser and B. Zumino, Phys. Lett. 65B (1976) 369;
L. Brink, P. Di Vecchia and P.S. Howe, Phys. Lett. 65B (1976) 471.
(11) A.M. Polyakov, Phys. Lett. 103B (1981) 207; Phys. Lett., 103B (1981) 211.
(12) J. H. Schwarz, Proc. Johns Hopkins Workshop on Current Problems in Particle Theory 6, (Florence, 1982) 233;
N. Marcus and A. Sagnotti, Phys. Lett. 119B (1982) 97.
(13) A. Neveu and J. Scherk, Nucl. Phys. B36 (1972) 155.
(14) T. Yoneya, Nuovo Cim. Lett. 8 (1973) 951; Progr. Theor. Phys. 51 (1974) 951.
J. Scherk and J.H. Schwarz, Phys. Lett. 52B (1974) 347.
(15) M.B. Green and J.H. Schwarz, Phys. Lett. 122B (1983) 143;
J.H. Schwarz and P.C. West, Phys. Lett. 126B (1983) 301;
J.H. Schwarz, Nucl. Phys. B226 (1983) 269;
P. Howe and P.C. West, Nucl. Phys. B238 (1984) 181
(16) L. Alvarez-Gaume and E. Witten, Nucl. Phys. B234 (1983) 269.
(17) D.J. Gross, J.A. Harvey, E. Martinec, and R. Rohm, Phys. Rev. Lett. 54 (1985) 502; Nucl. Phys. B256 (1985) 625.
(18) P.G.O. Freund, Phys. Lett., 151B (1985) 387.
(19) J.-P. Serre, A Course in Arithmetic (Springer-Verlag, 1973);
P. Goddard and D.I. Olive, in Vertex Operators in Maths and Phys., ed. J. Lepowsky, S. Mandelstam and I.M. Singer, Math. Sci. Research Inst. Publication #3 (Springer-Verlag 1985).

(20) S. Mandelstam, Nucl. Phys. B64 (1973) 205; B69 (1974) 77;
M. Kaku and K. Kikkawa, Phys. Rev. D10 (1974) 1110, 1823.
E. Cremmer and J.-L. Gervais, Nucl. Phys. B76 (1974) 209.
J.F.L. Hopkinson R.W. Tucker and P.A. Collins, Phys. Rev. D12 (1975) 1653.
M.B. Green and J.H. Schwarz, Nucl. Phys. B218 (1983) 43.
M.B. Green, J.H. Schwarz and L. Brink, Nucl. Phys. B219 (1983) 437.
M.B. Green and J.H. Schwarz, Phys. Lett. 140B (1984) 33; Nucl. Phys. B243 (1984) 475.
(21) P. Goddard, J. Goldstone, C. Rebbi and C.B. Thorn, Nucl. Phys.B56 (1973) 109.
(22) J. A. Shapiro, Phys. Rev. D5 (1972) 1945.
(23) M.B. Green and J.H. Schwarz, Phys. Lett. 109B (1982) 444.
(24) D.J. Gross, J.A. Harvey, E. Martinec, and R. Rohm, "Heterotic String Theory II. The Interacting Heterotic String", Princeton preprint (June 1985).
(25) M.B. Green and J.H. Schwarz, Phys. Lett. 151B (1985) 21.
(26) P.H. Frampton P. Moxhay and Y.J. Ng, Harvard preprint HUTP-85/A059 (1985);
L. Clavelli, Alabama preprint (1985).
(27) M.B. Green, Caltech preprint CALT-68-1219 (1984), to be published in the volume in honour of the 60 th birthday of E.S. Fradkin.
(28) S. Mandelstam, Proceedings of the Niels Bohr Centennial Conference, Copenhagen, Denmark (May, 1985);
A. Restuccia and J. G. Taylor, King's College preprint (1985).
(29) S. Mandelstam, Proceedings of the Workshop on Unified String Theories, ITP Santa Barbara, 1985.
(30) R. Kallosh, Phys. Lett. 159B (1985) 111.
(31) E. Witten, "Global Gravitational Anomalies", Princeton preprint (1985).
(32) E. Witten, Proceedings of the Cambridge Nuffield Workshop (CUP 1985)
(33) E. Witten, Phys. Lett., 149B (1984) 351.
(34) P. Candelas, L. Horowitz, A. Strominger and E. Witten, Nucl. Phys. B258 (1985) 46.
(35) F. Gursey, P. Ramond and P. Sikivie, Phys. Lett. 60B (1976) 177.
(36) E. Witten, Nucl. Phys. B258 (1985) 75;
J.D. Breit, B.A. Ovrut, G. Segrè, Phys. Lett. 158B (1985) 33;
M. Dine, V. Kaplunovsky, M. Mangano, C. Nappi and N. Seiberg, Princeton preprint (1985).
(37) A. Strominger and E. Witten, Princeton preprint (1985).
(38) A. Strominger, Princeton preprint (1985).
(39) X.G. Wen and E. Witten, Princeton preprint (1985).
(40) E. Witten, Phys. Lett. 155B (1985) 151.
(41) M. Dine, R. Rohm, N. Seiberg and E. Witten, Phys. Lett. 156B (1985) 55;
J.-P. Derendinger, L.E. Ibañez and H.-P. Nilles, Phys. Lett. 155B (1985) 65.
(42) M. Dine and N. Seiberg, Phys. Rev. Lett. 55 (1985) 366; Princeton preprint (1985);
V. Kaplunovsky, Phys. Rev. Lett. 55 (1985) 1033.
(43) E. Witten, Proceedings of the Argonne-Chicago Symposium on Anomalies, Geometry and Topology (World Scientific, 1985).

CONFORMALLY INVARIANT FIELD THEORIES IN TWO DIMENSIONS CRITICAL SYSTEMS AND STRINGS

J.-L. GERVAIS
Physique Théorique, Ecole Normale Supérieure
24 rue Lhomond 75231 Paris cedex 05

At the present time it is hardly necessary to emphasize the fundamental importance of string models since super string theories are the most promising candidates for a completely unified theory of all interactions. An other key point is that string concepts have plaid an important role in the recent developments of theoretical physics and mathematics by suggesting many new important ideas, such as in particular supersymmetry (1), and have led to very interesting progress in the related critical models in two dimensions. In these notes I shall mostly concentrate on this latter aspect which is not, presently, so directly aimed at a unified theory of all interactions but is quite interesting in its own right .

The unifying feature of string theories and critical systems is that they are both associated with conformally invariant field theories. We shall first review the essential features of this connection considering only, for simplicity, bosonic strings. At the level of the present discussion, supersymmetric strings are not basically different. One essentially replaces the conformal group by its superconformal generalization.

The position of a string at time τ is specified by a field $X_\mu(\sigma,\tau)$ where τ distinguishes the various points along the line. Hence one has a two dimensionnal field theory in parameter space. When σ varies the string sweeps out a world sheet and one sets up the dynamics in such a way that it be invariant under reparametrization of the corresponding geometrical surface. One can rigourously show that it is always possible to choose the parametrization in such a way that the curves σ =cste and the curves τ=cste intersect at right angle. This choice is not unique since this orthogonality condition is left invariant by all conformal transformations of σ and τ . In string theories the conformal group is thus the residual symmetry of the system with an orthogonal choice of σ,τ parameters. Basically it is the group of all transformations of the form

$$\sigma'+\tau'=f(\sigma+\tau) \qquad \sigma'-\tau'=g(\sigma-\tau) \tag{1}$$

where f and g are two arbitrary real functions of one

(44) Ref.(34);
C. Hull and E. Witten, "Supersymmetric Sigma Models and the Heterotic String", MIT preprint (1985).
(45) E.S. Fradkin and A.A. Tseytlin, Lebedev preprint #261 (October,1984).
(46) E. Witten, Comm. Math. Phys. 92 (1984) 455;
T. Curtright and C. Zakhos, Phys. Rev. Lett. 53 (1984) 1799.
(47) D. Nemeschansky and S. Yankielowicz, Phys. Rev. Lett. 54 (1985) 620
R.S. Jain, R. Shankar and S. Wadia, Tata Institute preprint (January, 1985);
E. Bergshoeff, S. Ranjbar-Daemi, A. Salam, H. Sarmadi and E. Szesgin ICTP, Trieste, preprint (1985).
(48) D. Friedan, Ph.D. Thesis, LBL preprint LBL-11517 (August, 1980); ref.45;
A. Sen, Fermilab preprints FERMILAB-PUB-85/60-T (April, 1985) FERMILAB-PUB-85/77-T (May, 1985);
C.G. Callan, E. Martinec M.J. Perry and D. Friedan, "Strings in Background Fields", Princeton preprint (1985).
(49) L.Alvarez-Gaumé, S. Coleman and P.Ginsparg, Harvard preprint (1985)
(50) D. Gross and E. Witten, to appear.
(51) L. Dixon, J.A. Harvey, C. Vafa and E. Witten, "Strings on Orbifolds" Princeton preprint (1985).
(52) M. Henneaux and L. Mezincescu, Phys. Lett. 152B (1985) 340;
T.L. Curtright, L. Mezincescu, C.K. Zachos, Argonne preprint ANL-HEP-PR-85-28 (1985).
(53) E. Witten, "Twistor-Like Transform in Ten Dimensions", Princeton preprint (May, 1985).
(54) M.T. Grisaru, P. Howe, L. Mezincescu, B. Nilsson, P.K. Townsend, D.A.M.T.P. Cambridge preprint (1985).
(55) W. Siegel, Phys. Lett. 149B (1984) 157, 162.
(56) T. Banks and M. Peskin, Proceedings of the Argonne-Chicago Symposium on Anomalies, Geometry and Topology (World Scientific, 1985); SLAC preprint SLAC-PUB-3740 (July,1985);
D. Friedan, E.F.I. preprint EFI 85-27 (1985);
A. Neveu and P.C. West, CERN preprint CERN-TH 4000/85;
K. Itoh, T. Kugo, H. Kunitomo and H. Ooguri, Kyoto preprint (1985);
W. Siegel and B. Zweibach, Berkeley preprint UCB-PTH-85/30 (1985);
M. Kaku and Lykken, CUNY preprint (1985);
M. Kaku, Osaka University preprints OU-HET 79 and 80 (July, 1985);
S. Raby, R. Slansky and G. West, Los Alamos preprint (1985).
(57) A. Neveu and P. West, ref.56.
(58) M. Halpern, Phys. Rev. D12 (1975) 1684 (appendix B);
J. Lepowsky and R.L. Wilson, Comm. Math. Phys. 62 (1978) 43;
I.B. Frenkel and V. Kac, Inv. Math. 62 (1980) 23;
G. Segal, Comm. Math. Phys. 80 (1981) 301.
(59) P. Goddard and D.I. Olive, ref.20.
(60) W. Siegel, Berkeley preprint UCB-PTH-85/23 (May, 1985).
(61) D. Friedan, S. Shenker and E. Martinec, E.F.I. preprint EFI 85-32 (April 1985).
(62) L.J. Romans and N.P. Warner, Caltech preprint CALT-68-1291 (1985)
(63) B. Zweibach, Phys. Lett. 156B (1985) 315.

variable. In the present dicussion we stick to the Lorentz covariant string quantization where conformal invariance is not explicitely broken.
Since a physical string has a finite length, σ varies over a finite range. It is always possible to redefine the parameters in such a way that the dynamics is periodic in σ with period 2π. It is often very convenient to go to Euclidean time by letting

$$\nu = i\tau \tag{2}$$

For real ν the conformal group can best be described as the group of analytic transformations of

$$z = e^{\nu + i\sigma} \tag{3}$$

Indeed with this variable, the strip $0 \leq \sigma \leq 2\pi$ is represented by the whole complex z plane. In this picture, a conformal transformation in given by

$$z' = F(z) \tag{4}$$

where F is an arbitrary complex function of one variable.
In general,[2] a quantity $\mathcal{O}(z, z^*)$ is called conformally covariant if it transforms according to

$$\mathcal{O}'(z', z'^*) = \left(\frac{dF}{dz}\right)^{-\delta} \left(\frac{dF^*}{dz^*}\right)^{-\bar{\delta}} \mathcal{O}(z, z^*) \tag{5}$$

where δ and $\bar{\delta}$ are parameters depending on the quantity considered which are called conformal weights. This notion was rediscovered recently[3] and the corresponding fields were called primary. If we separate the real and imaginary parts according to

$$z = x_1 + i x_2 \tag{6}$$

the differential transforms as

$$dx'_\ell = \mu R_{\ell m}\, dx_m \tag{7}$$

where R is the rotation matrix with angle θ and where μ is a dilatation factor. μ and θ are given by the differential equations

$$\frac{\partial x'_1}{\partial x_1} = \frac{\partial x'_2}{\partial x_2} = \mu \cos\theta \qquad \frac{\partial x'_1}{\partial x_2} = -\frac{\partial x'_2}{\partial x_1} = \mu \sin\theta \tag{8}$$

Formula (5) becomes

$$\mathcal{O}'(x'_1, x'_2) = \mu^{-d} e^{-iJ\theta}\, \mathcal{O}(x_1, x_2) \tag{9}$$

Since μ and θ are the local dilatation and rotation parameters respectively, d is the dimension and J is the spin of the quantity considered.
For critical systems in two dimensions, x_1 and x_2 are the two coordinates. It is well known that a stastistical system at a point of transition of second order becomes

scale invariant. The corresponding rescaling of x_1 and x_2 is a particular case of (4). Polyakov has proposed that critical systems are invariant under the full conformal group (4).
One thus sees thas that both string theories and critical systems are based on conformally invariant field theories, with however different descriptions. σ and τ are string parameters while x_1, x_2 are the coordinates of the critical system.
In a conformally invariant field theory the improved energy momentum tensor is symmetric traceless and conserved. For two dimensional field theories in real σ, τ space this leads to

$$\left(\frac{\partial}{\partial\tau} \mp \frac{\partial}{\partial\sigma}\right)\left(T_0^0 \pm T_0^1\right) = 0 \tag{10}$$

Due to the periodicity in σ one can write

$$\begin{aligned} 2\pi\left(T_0^0 + T_0^1\right) &= \sum_m L_m z^{-m} \ ; \quad z = e^{i(\tau+\sigma)} \\ 2\pi\left(T_0^0 - T_0^1\right) &= \sum_m \bar{L}_m \bar{z}^{-m} \ ; \quad \bar{z} = e^{i(\tau-\sigma)} \end{aligned} \tag{11}$$

The operators L_m and $\bar{L}_m$ are the infinitesimal generators of conformal transformations. For a conformally covariant field which satisfies equation (5) one has

$$\begin{aligned} [L_m, \theta] &= z^m\left[z\frac{\partial}{\partial z} + (m+1)\delta\right]\theta \\ [\bar{L}_m, \theta] &= \bar{z}^m\left[\bar{z}\frac{\partial}{\partial \bar{z}} + (m+1)\bar{\delta}\right]\theta \end{aligned} \tag{12}$$

The operators L_m and $\bar{L}_m$ each satisfy the Virasoro algebra

$$\begin{aligned} [L_n, L_m] &= (n-m)L_{n+m} + \frac{c}{12}(n^3-n)\delta_{n,-m} \\ [\bar{L}_n, \bar{L}_m] &= (n-m)\bar{L}_{n+m} + \frac{c}{12}(n^3-n)\delta_{n,-m} \end{aligned} \tag{13}$$

where the central charge C depends on the model considered. Its actual value is a key point. We shall have more to say about this below. In general the field theories we are discussing are characterized by C and by the set of conformally covariant fields θ_α together with the set of conformal weights δ_α. Since the two Virasoro algebras (13) have the same properties we only consider explicitely the algebra of the L_m's most of the time.
As a first simple example, let us recall the essential

features of the standard bosonic (Veneziano) model. In this case one only considers massless two dimensional free fields X_μ. We shall denote by Λ_n the associated Virasoro generator. As it is well known they satisfy eq.(13) with

$$C = D \tag{14}$$

the vertex for the emission of the lightest string state is simply given by

$$V_k(z) = :e^{ik^\mu X_\mu(z)}: \tag{15}$$

where k^μ is the energy momentum of the emitted particle. It is well known that V_k satisfies condition (12) with

$$\delta = \frac{k^2}{2} \tag{16}$$

The present discussion of string is in the covariant formalim where one has to make sure that the time like components of X_μ decouple from the physical S matrix. Such a ghost killing mechanism requires first of all that the vertex V_k have dimension 1. From eq.(16) this leads to $k^2 = 2$ .the emitted particle is a tachyon with mass $m^2 = -2$.

For critical models the δ_α and $\bar{\delta}_\alpha$ are critical exponents. Indeed it is easy to see that the global dilatations, rotations ,and translations of x_1, x_2 are genrated by L_0 $\bar{L}_0$ L_{-1} and $\bar{L}_{-1}$. The vacuum state of the system must therefore be annihilated by these operators. As a result the two point function of any covariant operator can be computed up to a constant factor by means of equation (12). If $\delta = \bar{\delta}$ for instance, one finds

$$\langle 0 | \mathcal{O}(z,z^*), \mathcal{O}(z',z'^*) | 0 \rangle \sim \left(|z-z'|^2 \right)^{-2\delta} \tag{17}$$

Hence δ gives the power behaviour of the two point function at the critical point. It is quite obvious that δ must be positive for physical operators since the correlation functions must decrease when the separation increases.

As it is well known (2) (3) ,the derivative of a covariant operator is not covariant in general. An important exception is the case of an operator of vanishing weight. Its derivative with respect to has $\delta = 1$, $\bar{\delta} = 0$
Conversely, assume there exists an operator $I(z)$ with $\delta = 1$ Then it is obvious that

$$\left[L_n , \oint dz\, I(z) \right] = 0 \tag{18}$$

In stastistical mechanics such operators are called marginal. If they exist one has critical lines instead of

critical points since they can be added to the action with arbitrary coefficients without destroying conformal invariance. For the Veneziano model we have just recalled that $V_k(z)$ has weight 1. Indeed equation (18) for V_k is the basic ingredient for the decoupling of ghosts.
The notion of conformally covariant operator was introduced in string models [2] in order to dicuss posible generalizations of the Veneziano model. We now recall the essential points of this approach. Generalized string models involve other fields besides the free fields X_μ and the corresponding two dimensional field theory may have a non trivial interaction. Such is the case, for instance, if we have additional space components which are compactified. Quite generally we can consider that X_μ remains a free field which does not mix with the additional two dimensional fields. These will be characterized by the set of covariant operators $\mathcal{O}_\alpha$ together with the set of weights δ_α. The ghost killing condition now requires the emission vertex to have conformal weight one under the action of the total Virasoro generator

$$J_m = \Lambda_m + L_m \tag{19}$$

where L_m is the Virasoro generator of the additional two dimensional dynamics. This is realized by choosing

$$\mathcal{V}_\alpha = V_k \mathcal{O}_\alpha \qquad \frac{k^2}{2} + \delta_\alpha = 1 \tag{20}$$

We therefore see that the spectrum of lightest particles

$$m_\alpha^2 = 2(\delta_\alpha - 1) \tag{21}$$

will involve no tachyon provided that

$$\delta_\alpha \geqslant 1 \tag{22}$$

and this condition selects the conformally invariant field theories for which all covariant fields have weights larger than one. From the view point of critical systems this condition is unusual since it means that the corresponding two point function has a Fourier transform which has at most a logarithmic singularity at zero momentum. We shall come back to this later.
The last important general point about string theories is that they make sens only if the central charge of the total Virasoro algebra is equal to the critical value 26. This can be seen in many ways. The simplest one is to notice [4] that the central charge of the Faddev Popov ghosts is precisely -26 so that, with the above value, the central charge vanishes when all the fields are included and there is actually no breaking of conformal invariance. For the Veneziano model where only the X field enters this means that $\mathcal{D}$ =26. In the generalized models the total central charge is C+$\mathcal{D}$ where C is the central charge of the additional dynamics. Hence one must satisfy

$$\mathcal{D} = 26 - C \tag{23}$$

If C is an integer larger than one, this will effectively allow to lower the space time dimension.

At this point it is useful to recall some general properties of representations of the Virasoro algebra(13). From the group theory viewpoint it can be regarded as being in a Weyl Cartan basis, L_0 being the only operator of the commuting subalgebra, and L_m with n 0 being step operators. Hence a highest weight vector will be such that

$$L_0 |0,\mathcal{E}\rangle = \mathcal{E}|0,\mathcal{E}\rangle \quad ; \quad L_m|0,\mathcal{E}\rangle = 0 \;, \; m>0 \tag{24}$$

An irreducible representation is characterized by the values of $\mathcal{E}$ and C. The corresponding vector space which is called a Verma module, is spanned by all vectors of the form

$$|\{m\},\mathcal{E}\rangle = \prod_{k>0} (L_{-k})^{m_k} |0,\mathcal{E}\rangle \tag{25}$$

where k, m_k are arbitrary positive integers. They are eigenstates of L_0:

$$L_0 |\{m\},\mathcal{E}\rangle = (\mathcal{E}+N)|\{m\},\mathcal{E}\rangle \; ; \; N = \sum_k k\, m_k \tag{26}$$

All eigenvectors with the same eigenvalues are said to belong to the same level N. Kac [5] has considered the matrix of all inner products in a given module which is entirely determined from the Virasoro algebra together with the hermiticity condition

$$L_{-m} = L_m^{+} \tag{27}$$

Hence it is purely alebraic and only depends upon the values of $\mathcal{E}$ and C. It obviously factorizes into products of finite matrices at each level. Kac [5] has obtained a closed formula for each finite determinant. Define the quantity

$$\mathcal{E}(p,q) \equiv \frac{1}{48}\Big[(13-C)(p^2+q^2) - 24pq - 2(1-C) + (p^2-q^2)\sqrt{(C-1)(C-25)} \Big] \tag{28}$$

where p and q are arbitrary integers. If we consider a highest weight representation with $\mathcal{E} = \mathcal{E}(p,q)$ for some given p, q both larger than zero, the Kac determinant vanishes at the level N=pq. This vanishing shows that the metric of the Verma module need not be positive definite. The unitarity of the representation is thus in question. It is easy to show that the negative values of the highest weights are all excluded. For positive values we note that, for $C > 1$, $\varepsilon(p,q)$ is always negative for $p>1$, $q>1$, i.e. when it corresponds to a zero of a Kac determinant. Hence, in this region the Kac determinants never change sign for positive ε and one can show by explicit construction [6] that there exist a unitary representation for all $\varepsilon > 0, C > 1$. For $C < 1$, on the contrary, there are Kac zeroes for positive ε and the positivity of the metric is not assured. It has been shown that unitary representations only exist for [7]

$$C = 1 - 6/(r(r+1)) \quad , \; r \geq 3$$
$$\varepsilon = \varepsilon(p,q) \quad , \; 1 \leq p \leq q < r \tag{29}$$

where r, p, q are integers. Hence the allowed values of ε precisely coincide with zeroes of Kac determinants.

Going back to conformally invariant field thories we recall that, given a covariant operator with weight δ, it is easy to see that the state

$$\lim_{z \to 0} \mathcal{O}(z)|0\rangle \tag{30}$$

is a highest weight vector with $\varepsilon = \delta$. The spectrum of highest weights coincides with the set of conformal weights which, in general, involves more than one values. The representation of the Virasoro algebra is thus reducible since the Hilbert space is the sum of the corresponding Verma modules. The covariant operators are intertwening operators between the different irreducible representations. For arbitrary ε and C one can construct an infinite family of covariant operators with weights given by formula (28) for all p and q positive or negative as a natural byproduct of the exact quantum solution of quantum Liouville theory [8]. These operators are not all physical since formula (28) is not always positive. This shows nevertheless that the set of critical dimensions must coincide with Kac formula in general.

In view of formula (28), it is clear that one has to distinguish three regions for the possible values of C.

I-The region $C<1$

As we already pointed out, the Kac zeroes occur for positive highest weights. A systematic discussion has been given [3] which uses this fact, and shows the existence of special values of C

$$C = 1 + \frac{6(r+s)^2}{rs} \tag{31}$$

where r and s are integers of opposite signs. The virtue of this formula is that then

$$(C-1)(C-25) = 36(r^2-s^2)/r^2s^2 \tag{32}$$

is the ratio of squares of integers. In view of formula (29), unitarity is satisfied only for r+s=-1. This subseries remarkably reproduces a whole set of standard critical models[7]. In particular for r=3 and 5 one recovers the Ising (or 2 state Potts) model and the 3 state Potts model. A simple calculation shows that if we introduce

$$\sqrt{Q} = 2\cos\left[\frac{\pi}{12}(C-1)\left(1+\sqrt{\frac{(C-25)}{(C-1)}}\right)\right] \tag{33}$$

we obtain the correct number of spin components i.e. Q=2,3 for the Ising and the three state Potts model respectively. The Q state Potts model can be defined for continuous values of Q if we transform it into the random cluster model. For C<1 there exist various equivalent critical models. In particular the Q state critical Potts model is equivalent to a Coulomb gas model. For C<1 we have Q<4 and one is in a Coulomb phase. The point Q=4 corresponds to a point of transition of Kosterlitz-Thouless. Above Q=4 one enters into the plasma phase and the transition becomes first order. We shall come back to this below.

II-The region C>25

This region has some similarities with the region C<1 since in both cases the square root of formula (28) is real. A different approach is needed, however, since now the Kac determinants do not vanish for positive . The region C>1 is naturally covered by the quantum Liouville field theory since its central charge is given by[9]

$$C = 1 + 3/\hbar \tag{34}$$

where $\hbar$ is the Planck constant. The region C>25 corresponds to $\hbar$ <1/8 i.e. to the weak coupling regime of Liouville theory which is connected to the semi classical limit $\hbar \sim 0$. In the exact quantum solution[9], special values of C were again found

$$C = 1 + 6(N+1)^2/N \tag{35}$$

They can be put under the form of equation (31) continued to r=N and s=1. The spectrum of weights is again given by formula (28) with

$$\begin{aligned} \varepsilon &= \varepsilon(1, 2m-N) \quad , \quad 0 \leq m \leq \nu \, , \quad N = 2\nu+1 \\ \varepsilon &= \varepsilon(1, 2m-1-N) \quad , \quad 1 \leq m \leq \nu \, , \quad N = 2\nu \end{aligned} \tag{36}$$

It is easily checked that all this values are larger than 1, and condition (22) is satisfied. The associated string model has no tachyon. However formula (23) shows that $\mathfrak{D}$ <1! so that one has not gained much from this viewpoint. The existence of conformally invariant field theories for these values of C does however suggest that there are new critical models. These are models of a new type since all the known critical models have C<1, and a spectrum of conformal weights between 0 and 1. . The unusal feature of the new models is, as we already pointed out, that the two point functions are at most logarithmically divergent. The experimental feature of the transition are thus rather different from the standard ones.

III-The region $1<C<25$

In this case formula (28) gives complex values except when $p=\pm q$. The choice $p=q$ is unacceptable except for $p=q=1$, since it leads to negative ε. For three special values of C

$$C = 7 \ , \ 13 \ , \ 19 \tag{37}$$

local fields have been constructed[10] such that the spectrum of weights is given by formula (28) for

$$p=-q=1,2,3,\dots \tag{38}$$

Condition (22) is again satisfied and the associated string theory has no tachyon. Condition (23) leads to

$$\mathcal{D} = 19 \ , \ 13 \ , \ 7 \tag{39}$$

and there exist new string models for these values of $\mathcal{D}$. From the view point of statistical model one therefore predicts[8] isolated points of second order phase transition for the above values of C. This may be a bit surprising since for $C>1$ one is outside of the Coulomb phase. One can directly see, however that these points must enjoy special properties. Formula (33) when continued for $1<C<25$ leads to Q complex in general. For the values (37) one obtains

$$\sqrt{Q} = 2\cos\left(\frac{\pi}{2} + i\pi\frac{\sqrt{3}}{2}\right) \tag{40}$$

$$\sqrt{Q} = 2\cos(\pi + i\pi)$$

$$\sqrt{Q} = 2\cos\left(3\frac{\pi}{2} + i\pi\frac{\sqrt{3}}{2}\right)$$

and one can verify that the three special values are the only ones for which Q is real even though the argument of the cosine is complex.

As a conclusion it is clear that the study of conformally invariant field theories from the double view point of string theories and cratical systems has unravelled an interesting structure. The string theories discussed here at not based on free field theories in two dimensions and, hence, the dual amplitudes are difficult to determine. The common feature of all the new conformally invariant field theories discussed here is the appearence of operators of dimension 1. For the associated string theories, it corresponds to the existence of a massless string state. This fact should play a key role in the complete understanding of these models. On the other hand, we must say that, at the present time, the relevance of these new models to particle physics is not yet clear. The supersymmmetric version of the present discussion has been worked out in all details.[11]

LIOUVILLE MODEL ON THE LATTICE

L.D. Faddeev and L.A. Takhtajan
Steklov Mathematical Institute
Fontanka 27
Leningrad D11 - 191011 - USSR

Abstract :
Liouville equation is put on the lattice in a completely integrable way. The classical version is investigated in details and a lattice deformation of the Virasoro algebra is obtained. The quantum version still lacks a satisfactory definition of the Hamiltonian.

The Liouville equation

$$\frac{\partial^2\varphi}{\partial t^2} - \frac{\partial^2\varphi}{\partial x^2} + e^{2\varphi} = 0 \qquad (1)$$

on the finite interval $0 \leqslant X \leqslant 2\pi$ has attracted a considerable attention because of its possible role in the quantization of the string model |1|. In particular, the hamiltonian treatment of the corresponding classical dynamical system in cases of periodic and "open" boundary conditions was performed in |2| - |3|. The quasiclassical quantization obtained thereby is not completely satisfactory, especially in the periodic case. Indeed, the contribution of the zero modes to masses of the string excitations is continuous because of the non-compactness of the corresponding portion of phase space. The use of singular solutions |4| could give discrete spectrum of the zero modes which is however negative and so the tachyon problem becomes even worse. Another possibility is to abandon the quasiclassical approach completely with hope to get some kind of a nonequivalent quantization.

With these considerations in mind we have formulated some time ago a program to put Liouville model on the lattice in agreement with the general spirit of the treatment of the completely integrable models in 1+1 dimensional space-time |5|. As a first step we obtained the exactly soluble classical version of the model. Because of technical difficulties we were not able to develope the quantum version to the end. We advertized our partial results among our colleagues and realized that recently some became interested in such a program |6|. In this

REFERENCES

(1) J.-L.GERVAIS,B.SAKITA Nucl.Phys.B34(1971)832
(2) J.-L.GERVAIS,B.SAKITA Nucl.Phys.B34(1971)477
(3) A.A.Belavin,A.M.Polyakov ,A.B.Zamolodchikov Nucl.Phys. B241(1980)333
(4) D.Friedan Les Houches Lectures Notes 1982
(5) V.KAC. Proceeding of the International Congress of Mathematicians Helsinsky 1978;Lecture Notes in Physics vol.94 p.441 Springer Verlag
(6) J.-L.GERVAIS,A.NEVEU COM.MATH.PHYS.100(1985)15
(7) D.FRIEDAN,Z.QIU,S.SHENKER in Vertex Operator in Mathematics and Physics ed.J.LEPOWSKY et al.Springer; Phys.Rev.lett.52(1984)1575
(8) J.-L.GERVAIS,A.NEVEU Nucl.Phys.B257 FS14(1985)59
(9) J.-L.GERVAIS,A.NEVEU NucL.Phys.B224(1983)329;B238(1984) 125
(10) J.-L.GERVAIS,A.NEVEU Phys.lett.151B(1985) 271
(11) J.-F.ARVIS Nucl.Phys.B212(1983) 151; B218(1983) 303. O.BABELON Nucl.Phys.B258(1985)680

circumstances we decided to publish the results on the classical version before the completion of the quantum one. We think that the lattice deformation of the Virasoro algebra which naturally appears in our treatment is interesting by itself.

The paper is organized as follows. In sec. 1 we remind the results of |2| - |3| on the classical continuous model in the form which allows the natural lattice generalization. The latter will be described in sec. 2. Finally in sec. 3 we shall indicate a possible quantum generalization.

One of the authors (L.D.F.) is grateful to professors B. Diu and H.J. de Vega for their kind hospitality at the LPTHE. We thank Professors O. Babelon, J.L. Gervais, M. Jimbo, R. Marnelius and A. Neveu for interesting discussions.

1. Classical continuous model.

In what follows only periodic boundary conditions

$$\varphi(x+2\pi) = \varphi(x) \tag{2}$$

will be used. The main idea of |2| - |3| is to use the change of variables

$$\begin{pmatrix} \varphi \\ \pi \end{pmatrix} \longrightarrow \begin{pmatrix} u \\ v \end{pmatrix} \tag{3}$$

from the initial data $\varphi(x)$, $\pi(x) = \frac{\partial}{\partial t}\varphi(x)$ (time variable is supressed) to new fields with trival equations of motion

$$\frac{\partial}{\partial t} u = \frac{\partial}{\partial x} u \qquad \frac{\partial v}{\partial t} = -\frac{\partial v}{\partial x} \tag{4}$$

and some reasonably simple boundary condition. One variant of this map, described in detail in |7|, can be based on the auxillary linear problem (note the absence of the spectral parameter)

$$L\phi = \left(\frac{d}{dx} - Q(x)\right)\phi = 0 \tag{5}$$

where $Q(x)$ is a 2x2 matrix parametrized by the initial data

$$Q = \begin{pmatrix} \pi & e^{\varphi} \\ e^{\varphi} & -\pi \end{pmatrix} \tag{6}$$

Let $T(x,y)$ be a fundamental matrix solution normalized by $T(x,x) = I$. Then $T = T(2\pi, 0)$ is called a monodromy matrix. The particular form of $Q(x)$ guaranties that $\det T = 1$ and $\operatorname{tr} T \geqslant 2$, so that T is hyperbolic.

Using the notation

$$T(x,0) = T(x) = \begin{pmatrix} A(x) & B(x) \\ C(x) & D(x) \end{pmatrix} \tag{7}$$

We introduce functions $\mu(x)$ and $\nu(x)$

$$\mu(x) = \frac{B(x)}{A(x)} \qquad \nu(x) = \frac{D(x)}{C(x)} \tag{8}$$

with the following properties

$$\mu'(x) < 0 \quad , \quad \nu'(x) > 0$$

$$\mu(0) = \infty \quad , \quad \nu(0) = 0 \quad , \quad \mu > \nu \tag{9}$$

and

$$\mu(x+2\pi) = \mathcal{T}(\mu(x)) \quad , \quad \nu(x+2\pi) = \mathcal{T}(\nu(x)) \tag{10}$$

where for any f

$$\mathcal{T}(f) = \frac{Af + C}{Bf + D} \tag{11}$$

A, B, C, D being the matrix elements of the monodromy matrix. They are nontrivial observables so that the form (10) of the boundary conditions

seems not very transparent.
The Poisson structure

$$\{\varphi(x), \pi(y)\} = \gamma\,\delta(x-y) \tag{12}$$

where γ plays the role of the coupling constant, leads to the fundamental Poisson bracket relations

$$\{T(x) \overset{\otimes}{,} T(x)\} = [r, T(x) \otimes T(x)] \tag{13}$$

(we use already standard notation from |5|) where the 4×4 matrix r is given by

$$r = \begin{pmatrix} 0 & 0 & 0 & 0 \\ 0 & -\gamma & 2\gamma & 0 \\ 0 & 0 & -\gamma & 0 \\ 0 & 0 & 0 & 0 \end{pmatrix} \tag{14}$$

It follows from an evident relation

$$T(x) = T(x,y)\,T(y) \tag{15}$$

and commutativity of the matrix elements of $T(y)$ and $T(x,y)$ (ultralocality) that the following relations are true

$$\begin{aligned}\{u(x), u(y)\} = {} & \gamma\,\varepsilon(x-y)\,[u(x) - u(y)]^2 + \\ & + \gamma\,[\mu(x)^2 - \mu(y)^2]\end{aligned} \tag{15}$$

$$\begin{aligned}\{v(x), v(y)\} = {} & -\gamma\,\epsilon(x-y)\,[v(x) - v(y)]^2 + \\ & + \gamma\,[v(x)^2 - v(y)^2]\end{aligned} \tag{16}$$

$$\{u(x), v(y)\} = 2\gamma\,[\mu(x)v(y) - v(y)^2] \tag{17}$$

where $\epsilon(x)$ is a sign function and we confine ourselves to the fixed fundamental domain $0 < x, y < 2\pi$.

The following chain of Ansatze

$$\mu \longrightarrow \xi = \mu' \tag{18}$$

$$\xi \longrightarrow p = \frac{d}{dx} \ln \frac{1}{\sqrt{\mu'}} = -\frac{1}{2} \frac{\mu''}{\mu'} \tag{19}$$

$$p \longrightarrow s = p^2 + p' = \frac{1}{2}\left[\frac{3}{2}\left(\frac{\mu''}{\mu'}\right)^2 - \frac{\mu'''}{\mu'} \right] \tag{20}$$

and analogously for v is now introduced. The final object (the Schwartz derivative of μ) is known to be invariant under the transformation (11) and so the simple boundary conditions hold

$$s[u(x+2\pi)] = s[u(x)] \quad ; \quad s[v(x+2\pi)] = s[v(x)] \tag{21}$$

It is interesting that the variables introduced in (18) - (20) have beautiful Poisson brackets of their own. With notations $\xi(x) = \xi[u(x)]$, $\hat{\xi}(x) = \xi[v(x)]$ and so on we have the following list of formulae

$$\begin{aligned} \{\xi(x), \xi(y)\} &= -2\gamma\, \xi(x)\, \xi(y)\, \epsilon(x-y) \\ \{\hat{\xi}(x), \hat{\xi}(y)\} &= 2\gamma\, \hat{\xi}(x)\, \hat{\xi}(y)\, \epsilon(x-y) \\ \{\xi(x), \hat{\xi}(y)\} &= 2\gamma\, \xi(x)\, \hat{\xi}(y) \end{aligned} \tag{22}$$

so that for $\ln \xi$ the Poisson brackets are field independent. Continuing the differentiation we get

$$\begin{aligned} \{p(x), p(y)\} &= \gamma\, \delta'(x-y) \\ \{\hat{p}(x), \hat{p}(y)\} &= -\gamma\, \delta'(x-y) \end{aligned} \tag{23}$$

$$\{ p(x), \hat{p}(y) \} = 0$$

so that the p and $\hat{p}$ fields decouple. Finally for S we acquire the brackets

$$\{ s(x) , s(y) \} = \gamma [s(x) + s(y)] \delta'(x-y) + \gamma \delta'''(x-y) \quad (24)$$

and analogous relation for $\hat{S}(x)$, which are characteristic of the Virasoro algebra. Thus the phase space of the Liouville model is essentially the product of two such algebras.

It can be shown that the hamiltonian of our model

$$H = \frac{1}{2\gamma} \int_0^{2\pi} dx \, [\pi(x)^2 + \varphi'(x)^2 + e^{2\varphi}] \quad (25)$$

has a simple expression in terms of $S[u]$ and $S[v]$

$$H = \frac{1}{\gamma} \int_0^{2\pi} dx \, (S[u] + S[v]) \quad (26)$$

the equation of motion being linear

$$\dot{s} = \{ H, s \} = s'$$

This allows to call S (p^2, u, v) the angle action variables. The inverse map is given by the famous Liouville formula

$$e^{2\varphi} = - \frac{u' v'}{(u - v)^2} \quad (27)$$

with periodic $\varphi(x)$.

We finish this section with a comment on boundary conditions (10). One can simplify them by diagonalizing the monodromy matrix which is achieved by the transformation

$$u \longrightarrow \tilde{u} = \frac{u - z_1}{u - z_2} \;, \; v \longrightarrow \tilde{v} = \frac{v - z_1}{v - z_2} \tag{28}$$

where z_1, z_2 are the real fixed points of T

$$T(z_i) = z_i \; ; \; i = 1, 2 \; ; \; z_1 \geq z_2 \tag{29}$$

The new variables $\tilde{u}(x)$ and $\tilde{v}(x)$ have the following properties

$$\tilde{u}(0) = 1 \;, \quad \tilde{v}(0) = \frac{z_1}{z_2}$$

$$\tilde{u}' < 0 \;, \quad \tilde{v}' > 0 \;, \quad \tilde{u} > \tilde{v} \tag{30}$$

and satisfy simple boundary conditions

$$\tilde{u}(x + 2\pi) = e^{-2p}\, \tilde{u}(x) \;, \quad \tilde{v}(x + 2\pi) = e^{-2p}\, \tilde{v}(x) \tag{31}$$

where e^{-p} is the smallest eigenvalue of T, that means that $p \geq 0$. It is clear that p is the zero Fourier coefficient of the periodic function $p[\tilde{u}(x)]$

$$p = \int_0^{2\pi} dx \; p[\tilde{u}(x)] \tag{32}$$

We have calculated the Poisson brackets of these new variables with the following answer

$$\begin{aligned} \{ \tilde{u}(x), \tilde{u}(y) \} = {} & \gamma\, \epsilon(x-y)\, [u(x) - u(y)]^2 + \\ & + \gamma \frac{\tilde{u}(x) - \tilde{u}(y)}{\tilde{u}(2\pi) - \tilde{u}(0)} \Big\{ [\tilde{u}(x) - \tilde{u}(0)][\tilde{u}(y) - \tilde{u}(2\pi)] + \\ & + [\tilde{u}(x) - \tilde{u}(2\pi)][\tilde{u}(y) - \tilde{u}(0)] \Big\} \end{aligned} \tag{33}$$

$$\{\tilde{v}(x), \tilde{v}(y)\} = -\gamma\, \epsilon(x-y)\, [v(x) - v(y)]^2 +$$

$$+\gamma\, \frac{\tilde{\mu}(2\pi) + \tilde{\mu}(0)}{\tilde{\mu}(2\pi) - \tilde{\mu}(0)} \left[\tilde{v}(x)^2 - \tilde{v}(y)^2\right] \tag{34}$$

and

$$\{\tilde{\mu}(x), \tilde{v}(y)\} = 2\gamma\, \tilde{v}(y)\, [\tilde{\mu}(x) - \tilde{\mu}(0)]\, \frac{\tilde{\mu}(x) + \tilde{\mu}(2\pi)}{\tilde{\mu}(2\pi) - \tilde{\mu}(0)} \tag{35}$$

Note the invariance of the first relation with respect to transformation (11) with constant coefficients.

Fortunately the rather complicated transformation (28) is not really necessary and the original variables $\mu(x)$, $v(x)$ can be used to construct the generators of Virasoro algebra as well as Hamiltonian of Liouville model. It is for them that we are able to find generalization on the lattice both in classical and quantum case.

2. Classical model on the lattice.

Working in accordance with the general spirit of |5| we generalize Liouville model on the lattice beginning with the auxillary linear problem : instead of (5) we have now the equation

$$\phi_{n+1} = L_n\, \phi_n \tag{36}$$

and L_n must be constructed in terms of variables π_n , φ_n satisfying the discret form of Poisson brackets (12)

$$\{\pi_n, \varphi_m\} = \gamma\, \delta_{nm} \tag{37}$$

The matrix L_n is essentially uniquely defined by the requirements :

1. In the continuous limit $\pi_n = \int_{\Delta_n} dx\, \pi(x)$, $\varphi_n = \frac{1}{\Delta}\int_{\Delta_n} dx\, \varphi(x)$ where Δ_n is lattice with length Δ , L_n behaves as follows

$$L_n = \mathbb{1} + \Delta\, Q(x) + O(\Delta^2) \tag{38}$$

2. L_n exactly satisfies the fundamental Poisson relation of the form (13)

$$\{ L_n \overset{\otimes}{,} L_n \} = [r , L_n \otimes L_n] \tag{39}$$

with the r-matrix (14). The explicit formula is given by

$$L_n = \begin{pmatrix} \sqrt{1+\Delta^2 e^{2\varphi}}\; e^{\pi} & \Delta e^{\varphi} \\ \Delta e^{\varphi} & \sqrt{1+\Delta^2 e^{2\varphi}}\; e^{-\pi} \end{pmatrix} \tag{40}$$

Introducing the transport matrix

$$T_n = \overleftarrow{\prod_{\ell \le n}} L_\ell = \begin{pmatrix} A_n & B_n \\ C_n & D_n \end{pmatrix} \tag{41}$$

we let

$$\mu_n = \frac{B_n}{A_n} \quad , \quad \nu_n = \frac{D_n}{C_n} \tag{42}$$

The essential role of local relation (39) consists in the fact that it leads the same relation for T_n

$$\{ T_n \overset{\otimes}{,} T_n \} = [r , T_n \otimes T_n] \tag{43}$$

The latter in its turn gives the following Poisson bracket relations for u_n, v_n

$$\{u_n, u_m\} = \gamma \varepsilon_{nm} (u_n - u_m)^2 + \gamma (u_n^2 - u_m^2) \tag{44}$$

$$\{v_n, v_m\} = -\gamma \varepsilon_{nm} (v_n - v_m)^2 + \gamma (v_n^2 - v_m^2) \tag{45}$$

$$\{u_n, v_m\} = 2\gamma (u_n v_m - v_m^2) \tag{46}$$

Here the numbers n, m vary in the "fundamental domain" $1 \leqslant m,n \leqslant N$ where N is a length of the lattice and ϵ_{nm} is defined as follows

$$\epsilon_{nm} \begin{cases} 1 & n > m \\ 0 & n = m \\ -1 & n < m \end{cases} \tag{47}$$

The fortunate property of our lattice formulation is that the relations (44) - (46) look as the most naive generalization of the relations (15) - (17).

This luck continues in the construction of analogous of Ansatze (18) - (20).

$$u_n \longrightarrow \xi_n \longrightarrow p_n \longrightarrow s_n \tag{48}$$

We let

$$\xi_n = u_n - u_{n-1} \tag{49}$$

$$p_n = \frac{\xi_n - \xi_{n+1}}{\xi_n + \xi_{n+1}} = -\frac{u_{n+1} + u_{n-1} - 2u_n}{u_{n+1} - u_{n-1}} \tag{50}$$

$$s_n = (1-p_n)(1+p_{n-1}) = \frac{(u_{n+1}-u_n)(u_{n-1}-u_{n-2})}{(u_{n+1}-u_{n-1})(u_n-u_{n-2})} \tag{51}$$

and analoguously for σ_n.

Observe that s_n is invariant under the transformations (11).

The most striking property of the new variables consists in the relative simplicity of their Poisson brackets; straightforward calculations give the following formulae

$$\{\xi_n, \xi_m\} = -2\gamma\epsilon_{nm}\xi_n\xi_m \tag{52}$$

$$\{p_n, p_m\} = \frac{\gamma}{2}(\delta_{n+1,m} - \delta_{n,m+1})(1-p_n^2)(1-p_m^2) \tag{53}$$

and

$$\{s_n, s_m\} = \frac{\gamma}{2} s_n s_m \left[(4 - s_n - s_m) * (\delta_{n+1,m} - \delta_{n,m+1}) + s_{n-1}\delta_{n,m+2} - s_{n+1}\delta_{n+2,m} \right] \tag{54}$$

The last formula gives a lattice generalization of the Virasoro algebra, interesting in its own. In the continuous limit we have

$$s_n = 1 - \frac{\Delta^2}{4} s(u) \tag{55}$$

so that the hamiltonian

$$H = \frac{1}{\gamma} \sum_n \ln S_n(u) + \ln S_n(v) \tag{56}$$

is a natural generalization of (26).

The equations of motion

$$\dot{s}_n = \{ H, s_n \} = s_n (s_{n+1} - s_{n-1}) \tag{57}$$

generated by the hamiltonian are known to be completely integrable, as was shown by S. Manakov and M. Kac - P. van Moerbeke . In fact they apparently appeared first in the ecological papers of Volterra. So the variables S_n (and $\hat{S}_n$ corresponding to v_n) constitute the first step in constructing the angle-action variables for the Liouville model on the lattice. At this point we stop the discussion of the classical lattice model.

3. Partial quantum results.

Continuing to work in the spirit of |5| we get the quantum version of the lattice model via the construction of the L_n-operator in terms of the operators φ_n and π_n with the usual commutation relations

$$\{ \varphi_n, \pi_m \} = i \delta_{nm} \gamma \tag{58}$$

This L_n must turn to (40) in the classical limit and satisfy the fundamental commutation relation

$$R (L_n \otimes L_n) = (L_n \otimes L_n) R \tag{59}$$

with a particular C-number 4×4 matrix R .

The formula

$$L_n = \begin{pmatrix} e^{\pi/2}\sqrt{1+\Delta^2 e^{2\varphi}}\, e^{\pi/2} & \Delta e^{\varphi} \\ \Delta e^{\varphi} & e^{-\pi/2}\sqrt{1+\Delta^2 e^{2\varphi}}\, e^{\pi/2} \end{pmatrix} \quad (60)$$

gives such an object, the R-matrix being

$$R = \begin{pmatrix} 1 & 0 & 0 & 0 \\ 0 & 0 & e^{i\gamma} & 0 \\ 0 & e^{i\gamma} & 1-e^{2i\gamma} & 0 \\ 0 & 0 & 0 & 1 \end{pmatrix} \quad (61)$$

Now we literally repeat what was done before, namely introduce the transport matrix

$$T_n = \prod_{\ell \leq n} L_\ell = \begin{pmatrix} A_n & B_n \\ C_n & D_n \end{pmatrix} \quad (62)$$

and the operators

$$u_n = A_n B_n^{-1} \quad , \quad v_n = C_n D_n^{-1} \quad (63)$$

chosing a particular order of the factors. It is gratifying to check that these operators satisfy rather simple relations

$$u_n u_m = \theta^2 u_m u_n + (1-\theta^2) u_m^2 \, , \; n<m \quad (64)$$

$$v_m v_n = \theta^2 v_n v_m + (1-\theta^2) v_n^2 \, , \; n<m \quad (65)$$

$$u_n v_n = \theta^2 v_n u_n + (1-\theta^2) v_n^2 \qquad (66)$$

where $\theta = e^{i\gamma}$. Note that the coupling constant enters only through θ in analogy with the Sine-Gordon case.

Unfortunately we were not able to find natural generalizations of the ξ_n, p_n and S_n variables in the quantum case, so that the quantum version of the lattice Liouville model is still in a nonsatisfactory state. This was the reason why we did not publish our relatively old results on the classical version. However some new developments and in particular new results of M. Jimbo |8| on the tensor products of representations of Sklyanin algebras |9| which one of the authors (L.D.F.) learned during his visit to the LPTHE in april 1985 seem encouraging. With this optimistic note we finish this paper.

References.

|1| A.M. Polyakov, Phys. Lett. 103B (1981) 207.

|2| J.L. Gervais, A. Neveu, Nucl. Phys. B199 (1982) 59 ; B209 (1982) 125.

|3| A. Kihlberg, R. Marnelius, ITP - Göteborg preprint N82-2 (1982).

|4| C.P. Dzhozdrhadze, A.K. Pogrebkov, M.K. Polivanov, Theor. Math. Phys. 40 (1979) 221.

|5| L.D. Faddeev in Les Houches Lectures,vol.XXXIX North Holland(1984).

|6| J.L. Gervais, preprint ENS, 1985.

|7| A.K. Pogrebkov, M.K. Polivanov in Soviet Science Reviews, Sec. C (Math. Phys.) ed. S.P. Novikov, vol 3 (1985). Gordon and Breach.

|8| M. Jimbo, preprint RIMS (1985).

|9| E.K. Sklyanin, Funct. Anal. Appl., vol 17 (1983), P 34-48.

EXACT SOLVABILITY OF SEMICLASSICAL QUANTUM GRAVITY IN TWO DIMENSIONS AND LIOUVILLE THEORY

Norma SANCHEZ
ER 176, "Groupe d'Astrophysique Relativiste"
Département d'Astrophysique Fondamentale
Observatoire de Paris-Meudon
92195 Meudon Principal Cedex
FRANCE

The general solution of semiclassical two-dimensional Einstein equations is exactly found. It is given by a constant curvature metric parametrized by solutions ("wave functions") of a zero energy Schrodinger equation. Global, thermal and topological properties of the universe are analyzed as a function of its quantum matter content including the graviton contribution.

1 - Two dimensional field theories are useful in the understanding of more relativistif four dimensional theories and are in some cases, exactly solvable. Semiclassical approaches are helpful to get explicit (not merely formal) results and a qualitative understanding to quantization. Because of the difficulties to quantize Gravity, a semiclassical two dimensional treatment of the problem is interesting. Recently, two dimensional gravity has raised interest, mainly in connection with Polyakov's work on strings |1|. The *classical* Einstein equations do not describe the dynamics of the gravitational field in two dimensions because $\int \sqrt{g}\, R\, d^2_x$ is a topological invariant and because is $T^{\mu}_{\mu} = 0$ for a classical matter source. It has been proposed that Liouville theory which in geometric form reads $R + \Lambda = 0$, (Λ = const.) could governs the dynamics of two dimensional gravity |2|. In this paper we consider *semiclassical* Einstein equations in two dimensions and *derive* Liouville theory as *one* of the dynamical equations of the gravitational field, the other equation involved appears to be a Schrodinger's one. The trace anomaly $\langle \hat{T}^{\mu}_{\mu} \rangle \neq 0$ for a quantum matter source allows for a non trivial dynamics of the semiclassical Einstein equations in two dimensions. Semiclassical in this context means that matter fields $\hat{\phi}$ including the graviton are quantized to one-loop level and coupled to (c-number) gravity through the equations

$$R_{\mu\nu} - 2^{-1} R\, g_{\mu\nu} + \Lambda\, g_{\mu\nu} = 8\pi G \langle \hat{T}_{\mu\nu}(\hat{\Phi}, g_{\mu\nu}) \rangle \qquad (1)$$

$<\hat{T}_{\mu\nu}>$ is the expectation value of the stress tensor operator $\hat{T}_{\mu\nu}$ of quantum matter field, renormalized in such a way that is covariantly conserved $\nabla^{\nu}<\hat{T}_{\mu\nu}> = 0$.
Eqs (1) for $g_{\mu\nu}$ are highly complicated and need to be treated within some type of self consistent framework. $<\hat{T}_{\mu\nu}>$ depends on the geometry and on the choice of the quantum state $|>$, that is on the choice of the boundary conditions of matter fields. Therefore $<\hat{T}_{\mu\nu}>$ is not a local geometrical object. In two dimensions, the semiclassical eqs (1) reduce to

$$\Lambda g_{\mu\nu} = 8\pi G < \hat{T}_{\mu\nu} > \tag{2}$$

which are non-trivial because $<\hat{T}_{\mu}^{\mu}> \neq 0$. The metric can be always written in the conformally that form

$$dS^2 = C(u,v)\, du\, dv \tag{3}$$

where $u = x - t$, $v = x + t$. The geometry is uniquely characterized by the curvature scalar

$$R = 4 C^{-1} \partial_u \partial_v \ln C = 4 C^{-3} [C \partial_u \partial_v C - \partial_u C \partial_v C] \tag{4}$$

$<\hat{T}_{\mu\nu}>$ is uniquely determined by the trace anomaly value

$$\langle T_{\mu}^{\mu} \rangle = - \gamma \, (24\pi)^{-1} R \,, \tag{5}$$

and explicitely given by |3|

$$\langle T_{\mu\nu} \rangle = \Theta_{\mu\nu} \quad \gamma (48\pi)^{-1} R\, \delta_{\mu\nu} + P_{\mu\nu}$$

$$\langle T_{uu} \rangle = - \gamma (12\pi)^{-1} \sqrt{C}\; \partial_u^2 (\sqrt{C})^{-1} + \tilde{U}(u) \tag{6.a}$$

$$\langle T_{vv} \rangle = - \gamma (12\pi)^{-1} \sqrt{C}\; \partial_v^2 (\sqrt{C})^{-1} + \tilde{V}(v) \tag{6.b}$$

$$\langle T_{uv} \rangle = - \gamma (48\pi)^{-1} R\, g_{uv} \,. \tag{6.c}$$

$P_{\mu\nu}$ is any conserved traceless tensor taking into account the dependence of $<\hat{T}_{\mu\nu}>$ on the quantum state of matter fields. Its represents the non local part of $<\hat{T}_{\mu\nu}>$: $P_{uu} = \tilde{U}(u)$, $P_{vv} = \tilde{V}(v)$, $P_{uv} = P_{vu} = 0$. $\tilde{U}$ and $\tilde{V}$ are arbitrary functions of the indicated variables. The coefficient γ takes into account the spin(s) dependence and the number of degrees of freedom of the fields. The total value of γ is discussed in section 4. The semiclassical eqs (2) give

$$\langle T_{uu} \rangle = 0 \quad , \quad \langle T_{vv} \rangle = 0 \tag{7.a}$$

$$R + \tilde{\Lambda} = 0 \quad , \qquad \tilde{\Lambda} = 6\Lambda^{-1}\gamma \tag{7.b}$$

Eq. (7.b) is the Liouville equation in geometrical form. In terms of the conformal factor C it reads

$$4\,\partial_u \partial_v \ln C + \tilde{\Lambda} C = 0 \tag{8}$$

or

$$4\,\partial_u \partial_v \phi + \tilde{\Lambda}\beta^{-1} e^{\beta\phi} = 0 \;, \quad C = e^{\beta\phi} \;, \quad \beta = \text{const.}$$

As it is well known, the general solution is

$$\phi = \beta^{-1} \ln \frac{f'(u)\, g'(v)}{[1 + (\tilde{\Lambda}/8)\, f(u)\, g(v)]^2} \tag{9}$$

Here f and g are not totally arbitrary functions but determined in terms of $\tilde{U}$ and $\tilde{V}$ by eqs (7.a) :

$$\sqrt{f'}\; d_u^2 \left(\sqrt{f'}\right)^{-1} - 12\pi\gamma^{-1}\, \tilde{U}(u) = 0 \tag{10.a}$$

$$\sqrt{g'}\; d_v^2 \left(\sqrt{g'}\right)^{-1} - 12\pi\gamma^{-1}\, \tilde{V}(v) = 0 \tag{10.b}$$

That is to say, the solution to the back-reaction problem in two dimensions is determined by a constant curvature metric (eq. 3)

$$C = \frac{f'(u)\, g'(v)}{[1 - (R/8)\, f(u)\, g(v)]^2} \;, \tag{11}$$

parametrized by solutions of a zero-energy Schrodinger equation

$$d_u^2\, \tilde{X}_u(u) - 12\pi\gamma^{-1}\, \tilde{U}(u)\, \tilde{X}_u(u) = 0 \tag{12.a}$$

$$d_v^2\, \tilde{X}_v(v) = 12\pi\gamma^{-1}\, \tilde{V}(v)\, \tilde{X}_v(v) = 0 \tag{12.b}$$

By giving the "potentials" $\tilde{U}(u)$ and $\tilde{V}(v)$, i.e. by specifying the quantum state of the matter fields, eqs (12) determine the "wave functions"

$$\tilde{X}_u = (\sqrt{f'})^{-1} \quad , \qquad \tilde{X}_v = (\sqrt{g'})^{-1} \tag{13}$$

To know the geometry configuration as a function of the quantum state of matter fields, we consider the transformations

$$u_K = f(u) \quad , \qquad v_K = g(v) \tag{14}$$

The O(2,2) group of bilinear transformations is the invariance group for both Liouville equation (9) and the Schrodinger eq. (12). The first term of eq. (12) is the Schwarzian derivative (D [f]) of f : $D[f] = \sqrt{f'}\, d_u^2 \left(\frac{1}{\sqrt{f'}}\right) = \frac{f'''}{f'} - \frac{3}{2}\left(\frac{f''}{f'}\right)^2$. Under the Möbius or bilinear transformations, f becomes a new function, but D[f] is invariant, determinying the same vacuum state of the fields. Eq. (14) can be considered as the mapping relating some manifold covered by the coordinates u, v to its global analytic extension (realized in the coordinates u_K, v_K). These are monotonic increasing functions satisfying the conditions |4|.

$$u_{K\pm} = f(\pm\infty) \quad , \quad v_{K\pm} = g(\pm\infty) \tag{15}$$

u_{K+} (u_{K-}) can take finite or infinite values allowing for one, two or none event horizons in the space time. Same considerations hold for the mappings g. In particular, f = g. Properties of the Schrodinger eqs (12) can be derived from the asymptotic properties of these mappings. At an event horizon, $f'(-\infty) = 0$ and the "wave function" is $\tilde{X} = \infty$ there. On the contrary at the infinity, if for instance $f'(+\infty) = +\infty$, then $\tilde{X}(+\infty) = 0$ and $\tilde{U}(+\infty) = +\infty$. In particular, the values $\tilde{U} = 0$, $\tilde{V} = 0$ in eqs (10), determine f(g) as

$$\tilde{X} = \text{const}, \qquad f = (\alpha u + \beta) / (\sigma u + \delta) \tag{16}$$

with $(\alpha\delta - \beta\sigma) = 1$ and α, β, σ, δ, constant parameters in accordance with the invariance properties discussed above. The corresponding vacuum state ($|>_K$) can be considered as a reference or "minimal" vacuum at zero temperature, respect to which, states corresponding to non-zero potentials $\tilde{U}$ and $\tilde{V}$, appear as excited or thermal ones. A constant potential $\tilde{U}(u) = \tilde{U}_0$ such that $\tilde{U}_0/\gamma > 0$ (fig. 1) gives

$$\tilde{X} = A\, e^{-\tilde{\mathcal{H}} u} \quad , \qquad f = (2\tilde{\mathcal{H}} A^2)^{-1}\, e^{2\tilde{\mathcal{H}} u} \tag{17}$$

where A is a normalizing constant (we will choose $A = (\sqrt{2\tilde{\mathcal{H}}})^{-1}$) and $\tilde{\mathcal{H}}$, is the zero-energy transmission coefficient

$$\tilde{\mathcal{H}} = \sqrt{12\pi\, \gamma^{-1}\, \tilde{U}_0} \tag{18}$$

The solution $\tilde{X}$ has been choosen in order to have f as an increasing function. The mapping

$$u_K = e^{2\tilde{\mathcal{H}}u} \qquad 0 \leqslant u_K \;,\; v_K \leqslant +\infty \quad ,$$
$$v_K = e^{2\tilde{\mathcal{H}}v} \qquad -\infty \leqslant u \quad , \quad v \leqslant +\infty \tag{19}$$

defines an event horizon at $u_K\, v_K = 0$ $(uv = -\infty)$ and carries an intrinsic temperature

$$T = \pi^{-1}\tilde{\mathcal{H}} = \sqrt{12\,(\pi\gamma)^{-1}\;\tilde{U}_0} \tag{20}$$

as it can be seen by putting $t = i\tau (u = x - i\tau)$ and so $0 \leqslant \tau \leqslant \pi/\tilde{\mathcal{H}}$. On the contrary, if $(\tilde{U}_0/\gamma) < 0$, there is no transmission coefficient ($\tilde{\mathcal{H}}$ becomes imaginary) and no event horizon is formed. The geometry does not carry an intrinsic temperature in this case. More generally, each positive discontinuity in the "effective" potential $\tilde{U}_0/\gamma$ gives rise to an event horizon in the space time, the transmission coefficient $\tilde{\mathcal{H}} = \left|\frac{\tilde{X}'}{\tilde{X}}\right|_{\text{horizon}}$ playing the role of the "surface gravity" $\mathcal{H} = 2\tilde{\mathcal{H}}$ of the horizon.

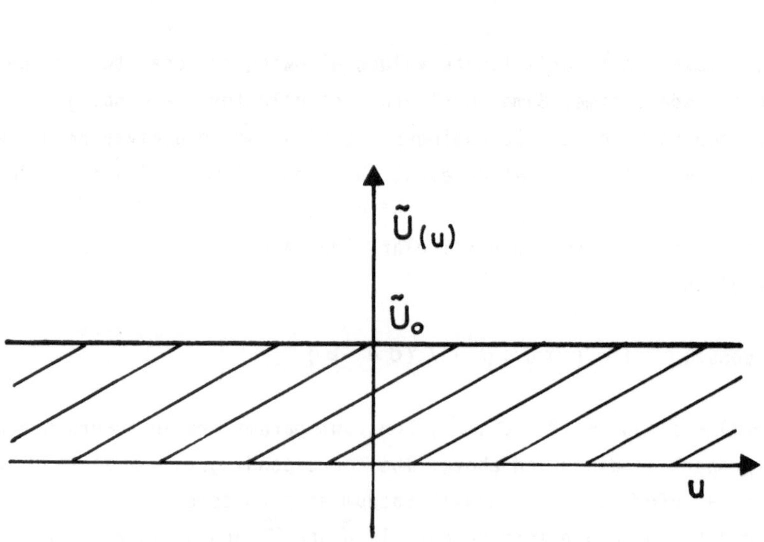

Fig. 1.a

Constant effective potential $(\tilde{U}_0/\gamma) > 0$ corresponding to the zero-energy Schrodinger eq. (12). The wave function $Ae^{-\tilde{\mathcal{H}}u}$, $\tilde{\mathcal{H}} = \sqrt{12\pi\gamma^{-1}\tilde{U}_0}$, determines a mapping $u_K = (2\tilde{\mathcal{H}}A^2)^{-1}\, e^{2\tilde{\mathcal{H}}u}$. γ is the trace anomaly factor.

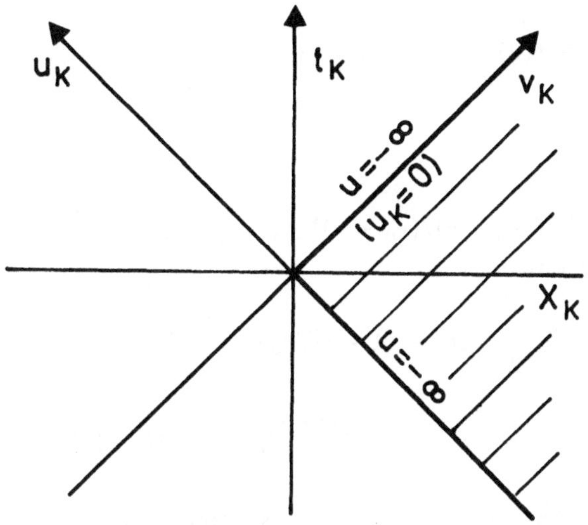

Fig. 1.b

Space time diagramm corresponding to the potential of fig.(1.a). The zero-energy transmission coefficient ($\mathfrak{K}$) is twice the surface gravity of the horizon ($\mathcal{H}$) ; $T = \pi^{-1}\mathcal{H}$ the associated temperature. $u_K = x_K - t_K$, $v_K + t_K$ are "Kruskal" (global) type coordinates, $u = v - t$, $v = x + t$ are of "Schwarzschild's" type.

2 - *Global properties of the semiclassical geometry.* It is convenient to rescale coordinates ($q = R/8$)

$$U = \sqrt{|q|}\,u_K \quad , \quad U = f(u) \quad ,$$
$$V = \sqrt{|q|}\,v_K \quad , \quad V = g(v) \tag{21}$$

such that $$dS^2_{(\mp)} = \frac{1}{|q|}\,\frac{1}{(1\mp UV)^2}\,dU\,dV \tag{22}$$

$$= \frac{1}{|q|}\,\frac{f'(u)\,g'(v)}{[1\mp f(u)g(v)]^2}\,du\,dv \tag{23}$$

The sign -(+) correspond here to $q > 0$ ($q < 0$) respectively. The case $q > 0$ describes a semiclassical de Sitter geometry. By defining

$$U = e^{\left(\frac{r^*-t}{r_H}\right)} = f(u) \quad ,$$
$$V = e^{\left(\frac{r^*+t}{r_H}\right)} = f(v) \qquad (24)$$

where $r^* = r_H\, 2^{-1} \ln[(r_H - r)/(r_H + r)]$, $r_H = 1/2\sqrt{|q|}$, the metric (22) can be written in the static form

$$dS^2_{(-)} = -(1 - r^2/r_H^2)\, dt^2 + (1 - r^2/r_H^2)^{-1}\, dr^2 \qquad (25)$$

which has an event horizon at $r = r_H = \sqrt{\gamma/3\Lambda}$. (See fig. (2)).

$\xi_{a\,;\,b}\, \xi^b = \mathcal{H} \xi_a$ ($\xi = \partial/\partial t$ is the Killing vector such that $|\xi| = 1$ at $r = 0$) defines the surface gravity as

$$\mathcal{H} = \sqrt{\Lambda_{eff} / 3} = \sqrt{3\Lambda/\gamma} \quad .$$

$\mathcal{H}$ is twice the "transmission coefficient" eq.(18) for $\tilde{U}_0 = \Lambda/(16\pi)$. The temperature is $T = (2\pi)^{-1}\mathcal{H} = (2\pi)^{-1}\sqrt{3\Lambda/\gamma}$ involving besides Λ the trace anomaly coefficient γ. The case $q < 0$ describes a semiclassical anti-de-Sitter geometry, obtained from the above situation by the analytic continuation $r_H \to -i\, r_H$. The mapping (24) becomes

$$U = e^{i\left(\frac{r^*-t}{r_H}\right)} \quad ,$$
$$V = e^{i\left(\frac{r^*+t}{r_H}\right)} \qquad (26)$$

for real time t and coordinate $r^* / r_H = -\,\mathrm{arctg}\,(r_H/r)$. The metric is real

$$dS^2_{(+)} = -\,(1 + r^2/r_H^2)\, dt^2 + \frac{dr^2}{(1+r^2/r_H^2)} \text{ , without event horizon.} \qquad (27)$$

The geometry does not carry an intrinsic (real) temperature (T becomes imaginary). The mapping eq. (26) in this case is not strictly increasing, which is associated to the fact that (AdS) is oscillatory in time and not globally hyperbolic.

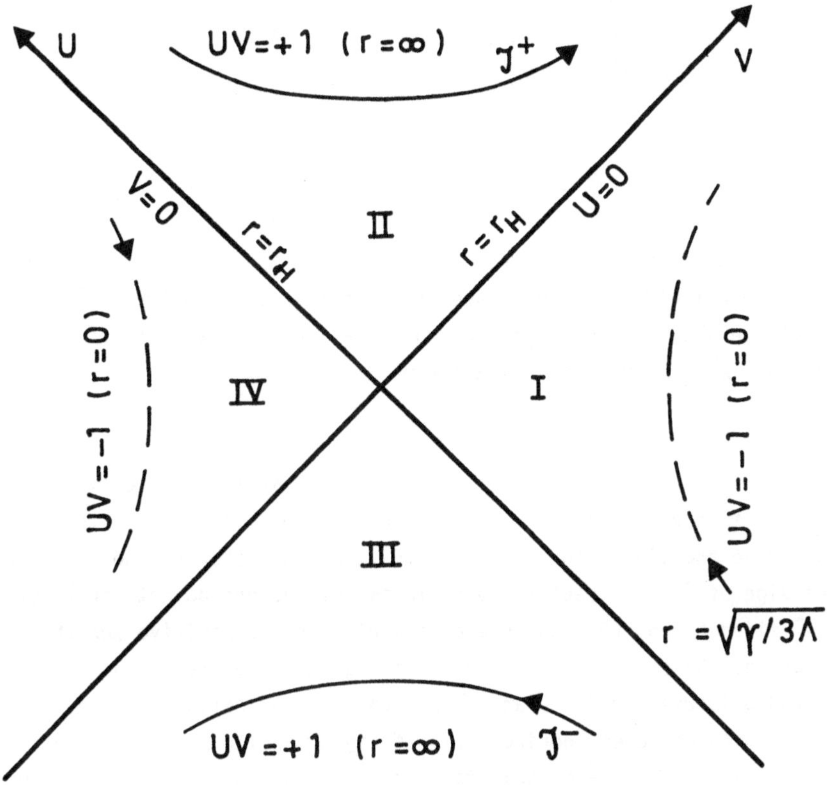

Fig. 2

Global structure of the space-time for the case $(\Lambda/\gamma) > 0$. The semiclassical geometry is of the de-Sitter type with one event horizon and intrinsic temperature $T = \sqrt{\Lambda_{eff}/3}$, $\Lambda_{eff} = 9\Lambda/\gamma$. For $(\Lambda/\gamma) < 0$, the geometry is anti-de-Sitter.

3 - *Instantons*. The analytic continuation $t = i\tau$ (τ real) maps the metric (22) onto a definite positive metric

$$ds^2_{(\mp)} = (\partial_Z \partial_{\bar{Z}} \Phi_{(\mp)}) \, dZ \, d\bar{Z} \tag{28}$$

Here $Z = X + iT$, $\bar{Z} = X - iT$. $\Phi_{(\mp)} = |q|^{-1} \ln [1 \mp Z \bar{Z}]$ is the solution of the Euclidean Liouville equation. For $q > 0$ Eq. (28) is the projective complex line ($\mathbb{C}P^1$); $1/2 \sqrt{|q|}$ is the curvature radius of the space.
We can consider $\mathbb{C}P^1$ as a gravitational instanton |6| of two dimensional Gravity : complete, non-singular and definite positive solution of the semiclassical Einstein equations in two dimensions . The Euler number is given by

$$\chi = (4\pi)^{-1} \int_{\mathcal{M}} \sqrt{g}\, R\, d^2x + (2\pi)^{-1} \int_{\partial\mathcal{M}} \sqrt{\sigma}\; K\, dy \tag{29}$$

and the euclidean action is

$$\hat{I} = 4^{-1} (\chi - 1 - \Lambda\Omega) \quad , \qquad \Omega = (2\pi)^{-1} \int \sqrt{g}\, d^2x \tag{30}$$

g and σ are the determinants of the metrics over the manifold $\mathcal{M}$ and over its boundary $\partial\mathcal{M}$, respectively. K is the trace of the extrinsic curvature. $\chi = 1$ and $\hat{I} = 0$ for flat Minkowski space ; $\chi = 0$ and $\hat{I} = -1/4$ for flat Rindler space. For the CP^1 instanton, $\chi = -1$ and $\hat{I} = -1/4(2 + \gamma/6)$.

4 - *Cosmological configurations and "critical dimensions"*. The cosmological constant $\bar{\Lambda} = 6\Lambda/\gamma$ in the Liouville eq. (9) is modified with respect to the classical one by the trace anomaly factor γ of eq. (5). The character of the solution depends on the sign of Λ/γ . Vector fields in two dimensions do not contribute to γ . For γ fixed by eq. (5), the scalar contribution is positive and that of gravitons is negative. Therefore : I) If sign $\Lambda \neq$ sign γ, i.e. $\Lambda > 0$ and $\gamma < 0$ (graviton dominated universe) or $\Lambda < 0$ and $\gamma > 0$ (matter dominated universe), the geometry has $R > 0$ with one event horizon. II) If sign $\Lambda =$ sign γ, the geometry has $R < 0$ without horizon. This means that for a given sign of Λ, the presence or absence of event horizons depends on the number of matter fields. The Universe could change from an Anti-de-Sitter to a de-Sitter phase (or vice-versa). The graviton contribution is crucial here to arise these possibilities. This contrasts with the standard classical situation (in four dimensions) in which R and the presence or not of event horizon only depends on Λ. If N (the number of matter fields) $\rightarrow \infty$ then the Hawking temperature $T \rightarrow 0$ and the semiclassical geometry is flat even if $\Lambda \neq 0$. If $\gamma = 0$ the dynamics is not determined by the semiclassical Einstein equations. In ref.7 the Liouville equation has been derived in the semiclassical context but the graviton contribution so crucial to this problem has been overlooked. The total value of γ as calculated in refs. 8 and 9 (denoted $\gamma_{(GKT)}$ and $\gamma_{(CD)}$ following notation of ref. 10) is

$$\gamma_{(GKT)} = (N_0 - 1 + N_{1/2} - \frac{15}{2} N_{3/2}) \; , \quad \gamma_{(CD)} = (N_0 - 1 - N_{1/2} + N_{3/2}) \tag{31}$$

Here, the graviton interacts with N_s massless fields of spins s, $s \leq 3/2$. The graviton contribution to γ was also obtained equal to -1 in ref. (10). In the context of quantized strings |1|, the trace anomaly coefficient for a theory with N matter fields coupled to two dimensional gravity was obtained equal to |1|

$$\gamma_{(P)} = N - 26 \text{ for bosons} \quad ,$$
$$\gamma_{(P)} = N - 10 \text{ for fermions (with supersymmetric coupling)} \qquad (32)$$

We denote it $\gamma_{(P)}$ because of ref. (1). (See also refs. 11-14 for a review). These values were calculated at the one loop level in the conformal gauge $g_{\mu\nu} = e^{\phi}\eta_{\mu\nu}$. The "critical dimension" 26 (10) in eq. (32) is only the ghost part (Faddeev-Popov determinant) of the graviton contribution. It does not take into account the quantization of the conformal factor (the Liouville field ϕ) that remains fixed. This should explain the difference between the values 1 in eq. (31) and 26 (10) in eq. (32). The value of γ that should be considered in the Liouville equation (9) of two dimensional gravity is that given by eq. (31) and not that of eq. (32). Understanding in connection with the quantization of the Liouville theory in this context desserves future investigation. It would be interesting to connect the results found here with those obtained from a semiclassical limit of the Hawking "wave function approach" |15| and of the Jackiw model |16|.
More details about this work are given elsewhere |17|.

References

1 - A.M. Polyakov, Phys. Lett. 103B, 207 and 211 (1981).
2 - R. Jackiw, Nucl.Phys. 252B, 343 (1985) and Refs therein.
3 - See e.g. N.D Birrell and P.C.W. Davies, "Quantum fields in curved spacetime" (Cambridge U.P., U.K., 1982).
4 - N. Sánchez, Phys.Rev. 24D, 2100 (1981) ; Phys.Lett. 81A, 424 (1981).
5 - G.W. Gibbons and S.W. Hawking, Phys.Rev. 15D, 2739 (1977).
6 - G.W. Gibbons and S.W. Hawking, Comm.Math.Phys. 66, 291 (1979).
7 - R. Balbinot and R. Floreanini, Phys.Lett. 151B, 401 (1985).
8 - R. Gastmans, R. Kallosh, C. Truffin, Nucl.Phys. B133, 417 (1978).
9 - S.M. Christensen, M.J. Duff, Phys.Lett. 79B, 213 (1978).
10 - S. Weinberg, in "General Relativity", S.W. Hawking and W. Israel, eds CUP (1979).
11 - D. Friedan in "Les Houches", session XXXIX - 1982, J.B. Zuber and R. Stora eds. pp 839-867, Elsevier Science Pub., 1984.
A. Neveu, ibid, pp 759-837.
12 - E.S. Fradkin, A.A. Tseytlin, Annals of Phys. 143, 413 (1982).
13 - O. Alvarez, Nucl.Phys. B216, 125 (1983).
14 - B. Durhuus, P. Olesen and J.L. Petersen, Nucl.Phys. B201, 176 (1982).
15 - S.W. Hawking, in "Les Houches" Session XL "Relativity, Groups and Topology II" B. de Witt and R. Stora eds., pp 333-379, North Holland (1984).
16 - M. Henneaux, Phys.Rev.Lett. 54, 959 (1985).
17 - N. Sánchez, Nucl.Phys. , B266, 487 (1986).

SOME FEATURES OF COMPLETE INTEGRABILITY IN SUPERSYMMETRIC GAUGE THEORIES

C. Devchand
Department of Mathematics,
University of Southampton.
and
Fakultat für Physik,
Universität Freiburg.

I. Introduction

The maximally supersymmetric gauge theories in four dimensions [1] have many interesting features. They are finite quantum field theories which may be obtained by dimensionally reducing the minimal (N=1) theory in ten space-time dimensions. They are also thought to be promising candidates for the realisation of tantalising duality conjectures [2,3], which generalise the duality between the Thirring model and the sine-Gordon model in two space-time dimensions [4] to four dimensional spontaneously broken gauge theories with monopoles. It has been suggested [3] that the latter feature is related to complete integrability. Moreover, the conformal invariance and the strongly constrained dynamics implied by the ultraviolet finiteness of the quantum theory suggest integrability of the field equations as a possible classical precursor underlying these features.

There has recently been much progress on the generalisation of the concept of complete integrability to field theories [5]. The origin of this concept is of course in hamiltonian dynamics, where for a hamiltonian system with a finite number N of degrees of freedom, the existence of N commuting integrals of motion means, by virtue of Liouville's theorem, that the system is fully integrable, i.e. that it is possible to completely separate the variables by a canonical transformation to action-angle variables. For infinite dimensional hamiltonian systems, the existence of an infinity of commuting integrals is thus a necessary (but not sufficient) condition of integrability. However a transformation to action-angle variables is implicit in the Inverse Scattering transform for such systems, since this incorporates what is effectively a nonlinear mapping to a free field theory. Similar nonlinear mappings are also the basis of methods which have been found to be useful for the solution of gauge theory type systems, namely the twistor methods [6] and the Riemann-Hilbert method[7]. These mappings are invertible and involve transforming differential equations to algebraic ones. They are therefore nonlinear analogues of the Fourier transform. Integrable systems are characterized by the absence of stochasticity, since there is no exchange of energy between degrees of freedom. This failure of equipartition of energy is what gives rise to the soliton phenomenon. Foremost amongst the properties characterizing the (possible) integrability of nonlinear systems is the possibility of writing the equations of motion in the Lax form:

$$\partial_0 L = [L,A] \tag{1}$$

(where L, A are linear differential operators), which we may rewrite as

$$[\partial_0+A,L] = 0 , \tag{2}$$

the compatibility condition for the set of equations:

$$(\partial_0+A)\psi = 0 = L\psi . \tag{3}$$

If the dimension of space-time is two, and if $L = \partial_1 + B$, for some B, eq. (2) is then just the condition for the vanishing of the curvature of the connection form C_μ with components $C_0 = A$, $C_1 = B$:

$$F_{\mu\nu} \equiv \partial_\mu C_\nu - \partial_\nu C_\mu + [C_\mu,C_\nu] = 0 . \tag{4}$$

If A and B are two components of a Lorentz vector, then the differential equation implied by (4) will be relativistic. Once the equation of motion has been cast into the form of (4), something important is guaranteed. For the corresponding linear system not only guarantees (formally) the existence of an infinity of conserved quantities [8], but also makes the equation of motion amenable to the algebraic methods of solution mentioned above. That it leads to an infinite number of conservation laws may be demonstrated, albeit only for a restricted class of models, using an argument [9-11] which is particularly instructive for gauge theory models. This considers a scattering problem

$$(\partial_\mu+C_\mu)\psi = 0 ; \quad \psi(-\infty,t) = 1 , \quad \dot{\psi}(+\infty,t) = 0 . \tag{5}$$

A solution to (5) exists (in any dimension) if (4) is satisfied. Now (4) need not necessarily be equivalent to the equations of motion. It could, for instance, be some identity in the problem. To proceed we need to invent a new combination of C_μ's depending on a parameter λ in such a way that a zero-curvature condition for $C'_\mu(\lambda) \equiv F(C_\mu,\lambda)$ now implies the equations of motion in addition to the previous identity. Now, identifying $C'_\mu(\lambda)$ with the potential of the scattering problem, it is clear that

$$\frac{dQ}{dt} = 0 ; \quad Q = P \exp \int_{-\infty}^{\infty} C'_1(\lambda)dx . \tag{6}$$

Expanding the path-ordered exponential in a power series in λ yields an infinity of non-local conserved charges. The crucial point about this argument is that the integrability condition is precisely a statement of the path-independence (i.e. integrability) of the phase factor of parallel transport [12]:

$$\psi_{X_1,X_2} = P \exp \int_{X_1}^{X_2} C'_\mu(\lambda)dx_\mu . \tag{7}$$

Indded, it is clear that the conservation law (6) stems from the boundary conditions of the scattering problem (5) for which $\psi_{-\infty,X}$ is a formal solution. The path-independence of (7) may be checked by considering its variation due to a variation of the path $x_\mu(t)$, parametrized by t [10]:

$$\frac{\delta}{\delta x_\mu(t)} \psi_{X_1,X_2} = C'_\mu(X_1)\psi_{X_1,X_2} - \psi_{X_1,X_2}C'_\mu(X_2)$$
$$+ \psi_{X_1 Y} \frac{dY_\nu}{dt} F_{\mu\nu}(Y) \psi_{Y,X_2} .$$

The first two terms are contributions of the end-points of the path; and we see that any path dependence (i.e. non-integrability) of the phase factor would be entirely encoded in the curvature $F_{\mu\nu}$ of the connection C'_μ (which in the present case is flat). However, it is, in general, not clear whether the integrability of the phase factor (7) and the consequent existence of the conservation laws (6), has any significance for the diagonalizability of the hamiltonian (i.e. complete integrability of the hamiltonian system). Although the phase factor $\psi_{-\infty\ +\infty}$ obeys a Yang-Baxter algebra for the d = 2 sigma models [13,33], the link of this fact with complete integrabitity is unclear. (Recall that for an infinite dimensional hamiltonian system to be completely integrable, one should require the existence of action-angle variables). In addition, there exist a related set of symmetry transformations of the field equations which generates an infinite dimensional non-abelian algebra [14,15]. The recent hope has been that just as the infinite dimensional abelian algebra of charges in involution characterizes the integrability of soliton theories, this non-abelian loop algebra characterizes the integrability of conformally invariant field theories with non-abelian symmetries. However, there has hitherto been scant evidence for the validity of this hope.

II. N = 4 supersymmetric gauge theories in superspace

Many integrability-related features of the principal chiral field equations, such as a Lax system, infinitely many 'hidden' symmetry transformations and related continuity equations, remarkably, also exist for the field equations of maximally supersymmetric gauge theories when these are written in terms of the N = 4 superfield potentials $A_A = (A_{\alpha\dot\beta}, A^s_\alpha, A_{\dot\beta t})$, where $\alpha,\dot\alpha = 1,2$; $s,t = 1,\dots,N$. (We follow the notation of [16].) In terms of these variables, which cannot be used to write down an off-shell theory, the equations of motion take the form of the following algebraic equations amongst the superfield Yang-Mills field strengths [17.18]:

$$F^{st}_{\alpha\beta} + F^{ts}_{\alpha\beta} = 0$$
$$F_{\dot\alpha s,\dot\beta t} + F_{\dot\alpha t,\dot\beta s} = 0 \qquad (8)$$
$$F^s_{\alpha,\dot\beta t} = 0 .$$

The θ-expansion of these equations yields the equations of motion for the component fields: the Yang-Mills field, 4 Majorana (or Weyl) spinors and 6 scalar fields [23]. There is no gauge-invariant action whose variation yields eq. (8). However, the algebraic nature of these equations is crucial for the features presented in refs. [16, 18-22]. It therefore seems quite likely that many of these features are merely artifacts of the dynamical variables used, and do not survive in any reformulation of the theory in terms of alternative variables (such as component fields or $N = 1$ or 2 superfields). Indeed it is by no means clear whether the infinite number of hidden symmetries of (8) and the corresponding spinor continuity equations have any significance for the integrability of the second order equations for the component fields. The starting point for the construction of these features is, as usual [8], a linear system for (8). Such a system formed the basis of a twistor-like transform for these equations [18]. The construction of [18] was based on the following equations governing the parallel propagation of a vector ϕ in the vector bundle along the direction in superspace given by $\zeta^A = (\mu^\alpha \lambda^{\dot\alpha}, \mu^\alpha, \lambda_{\dot\alpha})$:

$$\begin{aligned} &\mu^\alpha \nabla^s_\alpha \phi = 0 \\ &\lambda^{\dot\alpha} \bar\nabla_{\dot\alpha t} \phi = 0 \\ &\mu^\alpha \lambda^{\dot\alpha} \nabla_{\alpha\dot\alpha} \phi = 0 \; ; \end{aligned} \tag{9}$$

where $\mu^\alpha \lambda^{\dot\alpha}$ is a lightlike vector in MinKowski space and ζ^A is defined to be a lightlike vector in superspace. We may express (9) as the covariant constancy of ϕ along the $(1, \frac{1}{2}N)$ dimensional lightlike lines in superspace:

$$\zeta^A \nabla^s_A \phi = 0 \; ; \tag{10}$$

where $\nabla^s_A = (\nabla_{\alpha\dot\beta}, \nabla^s_\alpha, \bar\nabla_{\dot\beta s})$, as usual, if ϕ is propagated round a closed loop, it will not in general return to its original value; i.e. the propagation law (10) is not integrable. The condition for integrability (i.e. the path independence of ϕ) , is the condition that $\zeta^A A^s_A = (\mu^\alpha \lambda^{\dot\alpha} A_{\alpha\alpha}, \mu^{\dot\alpha} A^s_\alpha, \lambda^\alpha A_{\alpha s}) \equiv (W, U^s, V_s)$ is a flat connection. This is equivalent to the statement that

$$\begin{aligned} \phi &= P \exp \oint_\Gamma (W\mu^\alpha\lambda^{\dot\alpha} dx_{\alpha\dot\alpha} + U^s\mu^\alpha d\theta^s_\alpha + V_s\lambda^{\dot\alpha} d\theta_{\dot\alpha s}) \\ &= 0 \, , \end{aligned} \tag{11}$$

where Γ is a closed curve restricted to a lightlike line in superspace.

Now, using the covariant spinor derivatives

$$D^s \equiv \mu^\alpha D^s_\alpha = \mu^\alpha \frac{\partial}{\partial\theta^\alpha_s} + i \, \mu^\alpha \, \bar\theta^{\dot\beta s} \, \partial_{\alpha\dot\beta} \, ,$$

$$\bar D_s \equiv \lambda^{\dot\alpha} \bar D_{\dot\alpha s} = - \lambda^{\dot\alpha} \frac{\partial}{\partial\bar\theta^{\dot\alpha s}} - i \, \lambda^{\dot\alpha} \, \theta^\alpha_s \, \partial_{\alpha\dot\alpha} \, ,$$

which satisfy the algebra

$$\{D^s,\bar{D}_t\} = \mu^\alpha \lambda^{\dot\beta} \{D_\alpha,\bar{D}_{\dot\beta}\} = -2i\,\mu^\alpha \lambda^{\dot\beta}\, \partial_{\alpha\dot\beta} \equiv -2i\partial \tag{12}$$

$$\{D^s,D^t\} = 0 = \{\bar{D}_s,\bar{D}_t\} \tag{13}$$

we may write (10) in the form of a Lax system:

$$(D^s+U^s)\phi = 0 \tag{14a}$$

$$(\bar{D}_t+V_t)\phi = 0 \tag{14b}$$

$$(\partial + W)\phi = 0\ . \tag{14c}$$

The compatibility conditions for the spinor derivatives D^s and $\bar{D}_t$ following from (13):

$$D^sD^t\phi = -D^tD^s\phi\ ;\quad \bar{D}_s\bar{D}_t\phi = -\bar{D}_t\bar{D}_s\phi\ ,$$

yields the zero-curvature conditions

$$D^tU^s + D^sU^t + \{U^s,U^t\} = 0\ , \tag{15}$$

$$\bar{D}_tV_s + \bar{D}_sV_t + \{V_s,V_t\} = 0\ ; \tag{16}$$

whereas the compatibility condition following from (12):

$$D^s\bar{D}_t\phi + \bar{D}_tD^s\phi = -2i\delta^s_t\,\partial\phi$$

yields the equation

$$(D^sV_t+D_tU^s + \{U^s,V_t\})\phi = -2i\delta^s_t\,\partial\phi\ .$$

Now using (14c), we obtain a further zero-curvature condition

$$D^sV_t + \bar{D}_tU^s + \{U^s,V_t\} - 2i\delta^s_t\,W = 0\ . \tag{17}$$

Equations (15-17) are equivalent to the equations of motion (8). The space of all lightlike lines is the 'twistor space' of [18].

In order to proceed to the construction of continuity equations, it is convenient to write eqs. (9) for a particular curve in twistor space, parametrized by a complex parameter λ. We find it convenient to make the identifications $\mu^\alpha = (a,b)$, $\lambda^{\dot\alpha} = (a^2,b^2)$, $\lambda = b/a$. Eqs. (9) then take the form of a Lax-type system depending on one parameter [20]:

$$\begin{aligned}
&(\nabla^s_1 + \lambda\nabla^s_2)\chi(\lambda) = 0\\
&(\bar\nabla_{\dot 2t} + \lambda^{-2}\,\bar\nabla_{\dot 1t})\chi(\lambda) = 0\\
&(\nabla_{1\dot 2} + \lambda\nabla_{2\dot 2} + \lambda^{-1}\,\nabla_{2\dot 1} + \lambda^{-2}\,\nabla_{1\dot 1})\chi(\lambda) = 0\ .
\end{aligned} \tag{18}$$

We take the superfield functional $\chi(\lambda)$ to have analyticity properties in the complex λ-plane such that either

$$\chi(\lambda) = \psi(\lambda) = \sum_{n=0}^{\infty} \lambda^n \psi^{(n)} \tag{19}$$

or

$$\chi(\lambda) = \phi(\lambda) = \sum_{n=0}^{\infty} \lambda^{-n}\phi^{(n)} . \tag{20}$$

Writing out eqs. (8) explicitly:

$$F_{11}^{st} = 0 = F_{22}^{st} \; ; \; F_{\dot{1}s,\dot{1}t} = 0 = F_{\dot{2}s\dot{2}t} \tag{21a}$$

$$F_{1,\dot{1}t}^{s} = 0 = F_{2,\dot{2}t}^{s} \tag{21b}$$

$$F_{12}^{(st)} = 0 = F_{\dot{1}(s,\dot{2}t)} \tag{22a}$$

$$F_{1,\dot{2}t}^{s} = 0 = F_{2,\dot{1}t}^{s} \; , \tag{22b}$$

we note that the field equations may be partially integrated by writing the spinor potentials in the pure-gauge form

$$\begin{aligned} A_1^s &= g^{-1} D_1^s g \; , \quad A_{\dot{1}t} = g^{-1} \bar{D}_{\dot{1}t} g \\ A_2^s &= h^{-1} D_2^s h \; , \quad A_{\dot{2}t} = h^{-1} \bar{D}_{\dot{2}t} h \; , \end{aligned} \tag{23}$$

where the superfields g and h are given by $g^{-1} = \psi^{(0)}$, $h^{-1} = \phi^{(0)}$ in the expansions (19-20); and two components of the vector potential in the form:

$$A_{1\dot{1}} = g^{-1} \partial_{1\dot{1}} g \; , \quad A_{22} = h^{-1} \partial_{2\dot{2}} h \; . \tag{24}$$

The pure-gauges (23,24) solve the zero-curvature conditions (21), leaving eqs. (22) as the dynamical equations, with the remaining two components of the vector potential determined in terms of the spinor potentials by virtue of the traces of eqs. (22b).

The superfields g and h have transformation properties

$$g \to e^{-S(\lambda)} g e^{\alpha} \; , \quad h \to e^{R(\lambda)} h e^{\alpha} \tag{25}$$

where $\alpha = \alpha(x,\theta,\bar{\theta})$ is the parameter of local gauge transformations taking values in the lie algebra G of the gauge group, and

$$\begin{aligned} S(\lambda) &= \sum_{n=0}^{\infty} \lambda^n S^{(n)} (x,\theta,\bar{\theta}) \; , \\ R(\lambda) &= \sum_{n=0}^{\infty} \lambda^{-n} R^{(n)} (x,\theta,\bar{\theta}) \end{aligned} \tag{26}$$

are loop-algebra valued infinitesimal parameters satisfying the "killing" equations

$$[\zeta^A(\lambda)\nabla^S_A,\ S(\lambda)] = 0 = [\zeta^A(\lambda)\nabla^S_A,\ R(\lambda)] \tag{27}$$

whose solutions are functional symmetry generators. The coefficients in the expansions (26) span infinite dimensional vector spaces of symmetries. Eqs. (27) imply the field equations, and their general solution is given by S and R having the form of a similarity transformation of a generator of a global gauge transformation:

$$S^a = \psi\, T^a\, \psi^{-1}\ ,\quad R^a = \phi\, T^a\, \phi^{-1} \tag{28}$$

where we have expanded in a basis of G. Under S and R transformations, which we refer to as G^λ transformations, the spinor potentials (23) transform in the following fashion:

$$\begin{aligned}\delta(\lambda)\ A^S_A &= -\, g^{-1}\, D^s_A\, S(\lambda)g = -\ [\nabla^s_A, g^{-1}S(\lambda)g]\ , \quad \text{for } A = 1 \text{ or } \dot{1}\ ;\\ \delta(\lambda)\ A^s_A &= h^{-1}\, D^s_A\, R(\lambda)\ h \ = \ [\nabla^s_A, h^{-1}R(\lambda)h]\ , \quad \text{for } A = 2 \text{ or } \dot{2}\ .\end{aligned} \tag{29}$$

Implicit in the choice (23) is a restriction of the gauge freedom; and each coefficient in the power-series expansion of the G^λ transformation effects a nonlinearly realized local gauge transformation on the spinor potential respecting this gauge choice. The infinitesimal transformations (29) generate symmetries of the equations of motion (8); and they may be integrated to finite transformations since the generators of the infinitesimal transformations (29) close under the loop algebra $G \otimes \mathbb{C}[\lambda,\lambda^{-1}]$ with commutation relations

$$\left[L^m_a, L^n_b\right] = C_{abc}\ L^{m+n}_c \qquad m,n \in \mathbb{Z}\ , \tag{30}$$

where

$$L^n_a = \frac{d^n}{d\lambda^n} L_a(\lambda)\bigg|_{\lambda=0}\ ,\quad L_a = \int dz\ \delta_a\ A_A \frac{\delta}{\delta A_A}\ .$$

The proof may be found in [19]. If we further fix the gauge by writing

$$A_{1\dot{1}} = 0 = A^s_1 = A_{\dot{1}t}\ ,$$

in which gauge

$$A^s_2 = B\ D^s_2\ B^{-1}\ ,\quad A_{\dot{2}t} = B\ \bar{D}_{\dot{2}t}\ B^{-1}\ , \tag{31}$$

where B is the manifestly (local) gauge invariant superfield: $B = gh^{-1}$, which transforms covariantly under G^λ transformations:

$$B \to e^{-S}\ B\ e^{-R}\ . \tag{32}$$

In this gauge, the equations of motion (22) take the G^λ covariant form:

$$D_1^{(s}A_2^{t)} = 0$$

$$\bar{D}_{\dot{1}(s}A_{\dot{2}t)} = 0$$

$$D_1^s A_{\dot{2}t} + \delta_t^s \, 2i \, g \nabla_{1\dot{2}} g^{-1} = 0 \qquad (33)$$

$$\bar{D}_{\dot{1}s} A_2^t + \delta_s^t \, 2i \, g \nabla_{2\dot{1}} g^{-1} = 0$$

and the linear system (18) takes the form

$$\tau^A(\lambda) \; \mathcal{D}_A^s \, X(\lambda) = 0 \qquad (34)$$

where $\mathcal{D}_A^t = D_A^t + A_A^t$, and $X(\lambda)$ denotes either

(case A) $\quad g\Psi(\lambda) \equiv \Psi(\lambda) \; ; \; \Psi(\lambda=0) = \mathbb{1}$ $\quad$ or

(case B) $\quad g\phi(\lambda) \equiv \tilde{\Phi}(\lambda) \; ; \; \tilde{\Phi}(\lambda=\infty) = B$.

In the alternative gauge where

$$\tilde{A}_2^t = 0$$

$$\tilde{A}_{\dot{2}t} = 0$$

$$\tilde{A}_1^t = B^{-1} D_1^t B \qquad (35)$$

$$\tilde{A}_{\dot{1}t} = B^{-1} \bar{D}_{\dot{1}t} B \; ,$$

we have an alternative set of equations for the B-field:

$$D_2^{(s}\,\tilde{A}_1^{t)} = 0 = \bar{D}_{\dot{2}(s}\,\tilde{A}_{\dot{1}t)}$$

$$D_{\dot{2}s}\,\tilde{A}_1^t + \delta_s^t\, 2ih\, \nabla_{1\dot{2}}\, h^{-1} = 0 \qquad (36)$$

$$D_2^s\,\tilde{A}_{\dot{1}t} + \delta_t^s\, 2ih\, \nabla_{2\dot{1}}\, h^{-1} = 0 \ ,$$

and a linear system:

$$\zeta^A(\lambda)\, \tilde{\mathcal{D}}_A^s\, \tilde{X}(\lambda) = 0 \ , \qquad (37)$$

where $\tilde{\mathcal{D}}_A^t = D_A^t + \tilde{A}_A^t$, and $\tilde{X}(\lambda)$ denotes either (case B) $h\psi(\lambda) \equiv \tilde{\Psi}(\lambda)$; $\tilde{\Psi}(\lambda=0) = B^{-1}$,

or (case A) $\quad h\phi(\lambda) = \Phi(\lambda)$; $\Phi(\lambda=\infty) = 1$.

The superfield B also carries a representation of the loop algebra (30):

$$M_a^n = \int dz\, \delta_a^{(n)} B\, \frac{\delta}{\delta B} \ , \qquad (38)$$

where the generating function for the transformations $\delta^{(n)}B$ has the form

$$\delta B = \sum_{n=-\infty}^{\infty} \delta^{(n)} B = -\,(SB+BR) \ ,$$

with

$$S = \Psi(\lambda)\, T\Psi(\lambda)^{-1} \ , \quad R = \Phi(\lambda)\, T\Phi(\lambda)^{-1}$$

(for case A), or alternatively (case B)

$$S = \tilde{\Psi}(\lambda)\, T\tilde{\Psi}(\lambda)^{-1} \ , \quad R = \tilde{\Phi}(\lambda)\, T\tilde{\Phi}(\lambda)^{-1} \ .$$

The closure of the loop algebra follows by virtue of the equations of motion (33, 36), which are left invariant by the transformations (32), the infinitesimal forms of which are generated by (38). These infinitesimal transformations generate symmetries of (33, 36) since they leave the linear system invariant. Alternatively, it may be shown directly that eqs. (33, 36) are left invariant by virtue of eq. (27). Consider, for instance, the variation of (33a) under the transformation $\delta B = -\,SB$:

$$D_1^{(s}\,\delta(B\,D_2^{t)}\,B^{-1}) = D_1^{(s}\,D_s^{t)} S + D_1^{(s}[B\,D_2^{t)}\,B^{-1}, S]$$

$$= -\,D_2^{(t}\,D_1^{s)} S - \{B\,D_2^{(t}\,B^{-1},\, D_1^{s)} S\}$$

$= 0$, as a result of the consistency of (27).

Similarly, the variation of eq. (33c) yields [24]:

$$\delta\,[D_1^s(B\,\bar{D}_{\dot{2}t}B^{-1}) + 2i\,\delta_t^s\, g\,\nabla_{1\dot{2}}\, g^{-1}]$$

$$= D_1^s(\delta B\,\bar{D}_{\dot{2}t}B^{-1} + B\,\bar{D}_{\dot{2}t}\delta B^{-1}) + \delta_t^s\, 2i(\delta g\nabla_{1\dot{2}}g^{-1} + g\nabla_{1\dot{2}}\delta g^{-1} + g\delta A_{1\dot{2}}g^{-1})$$

$$= -\,\{\bar{\mathcal{D}}_{\dot{2}t}, D_1^s\, S\} + \delta_t^s\, 2i\, g\, \delta A_{1\dot{2}}\, g^{-1} \ ,$$

using (33c) and effecting the transformation $\delta B = - SB$ by the infinitesimal transformation: $\delta g = - Sg$, $\delta h = 0$. Now, since $F^s_{1,\dot{2}t} = 0$, we also have

$$- 2i\, \delta^s_t (g\, \delta A_{1\dot{2}}\, g^{-1}) = g\, \{\bar{\nabla}_{\dot{2}t},\, \delta A^s_1\} g^{-1}$$

$$= - \{\bar{\mathcal{D}}_{\dot{2}t},\, D^s_1\, S\}\ .$$

Eq. (33c) is therefore invariant under this transformation.

Corresponding to the above two sets of functional symmetry transformations, there exist two (completely equivalent) infinite sets of nonlocal spinor continuity equations, which may be represented compactly by the following expression for the nth continuity equation:

$$\left\{ D^{\alpha}_s [\epsilon^{\beta}_{\alpha}\, \alpha^s_t\, \mathcal{D}^{(n)t}_{\beta}\, \chi_{\beta}(\lambda)] - D^{\dot{\alpha}t} [\epsilon^{\dot{\beta}}_{\dot{\alpha}}\, \alpha^s_t\, \mathcal{D}^{(n)}_{\dot{\beta}s}\, \chi_{(\beta)}(\lambda)] \right\} \Bigg|_{\lambda=0} = 0\ ,$$

where

$$\alpha^s_t = 0\ ,\quad s = t$$
$$= 1\ ,\quad s \neq t\ ;$$

and (case A)

$$\mathcal{D}^{(n)t}_A = \frac{1}{\lambda^n} \mathcal{D}^t_A\ ,\quad \chi_{(A)}(\lambda) = \Psi(\lambda) = \sum_{n=0}^{\infty} \lambda^n\, \psi^{(n)}\ ,\quad A = 2, \dot{2}\ ,$$

$$\mathcal{D}^{(n)t}_A = \lambda^n\, \tilde{\mathcal{D}}^t_A\ ,\quad \chi_{(A)}(\lambda) = \Phi(\lambda) = \sum_{n=0}^{\infty} \lambda^{-n} \phi^{(n)}\ ,\quad A = 1, \dot{1}\ ;$$

or alternatively (case B)

$$\mathcal{D}^{(n)t}_A = \lambda^n\, \mathcal{D}^t_A\ ,\quad \chi_{(A)}(\lambda) = \tilde{\Phi}(\lambda) = \sum_{n=0}^{\infty} \lambda^{-n} \tilde{\phi}^{(n)}\ ,\quad A = 2, \dot{2}\ ,$$

$$\mathcal{D}^{(n)t}_A = \frac{1}{\lambda^n} \mathcal{D}^t_A\ ,\quad \chi_A\ (\lambda) = \tilde{\Psi}(\lambda) = \sum_{n=0}^{\infty} \lambda^n\, \tilde{\psi}^{(n)}\ ,\quad A = 1, \dot{1}\ .$$

Whether the features described here have any more than merely a formal significance, in particular, whether they are symptoms of an underlying classical structure constraining the quantum dynamics at high momenta, is still an open question. However, Witten has argued that since these features depend on a restriction of the data in twistor space, which destroys the invertibility of the twistor transform, they are merely artifacts of the parametrization used and cannot be expected to be of any physical relevance. It appears that this argument would also apply to some analogous features of the self-duality equations [25]. Indeed, no one has found a use for the infinite set of conserved currents and hidden symmetry transformations of the self-duality equations, although in this case it is clear that the equations <u>are</u> completely integrable.

III. Bogomolny equations for the bosonic sector of N = 4 super Yang-Mills

The features discussed in the previous section suggest the possible existence of non-trivial classical solutions to these equations. The bosanic sector of the N = 4 theory is given by the lagrangian density

$$L = \mathrm{tr}(-\frac{1}{4} F^2_{\mu\nu} - \frac{1}{2} (D_\mu A_i)^2 - \frac{1}{2} (D_\mu B_i)^2 - \frac{1}{4} [A_i,A_j]^2$$

$$- \frac{1}{4} [B_i,B_j]^2 - \frac{1}{2} [A_i,B_j]^2) \, , \tag{40}$$

$$\mu = 0,1,2,3 \; ; \quad i,j = 1,2,3 \; ,$$

in the notation of Gliozzi et al [1]. Conventionally [1], this lagrangian is viewed as a trivial reduction of ten-dimensional pure Yang-Mills, with the scalars A_i, B_j transforming as the self and anti-dual parts of a $\underline{6}$ of SU(4). To find critical points of the action (40) it is convenient to write down Bogomolny equations [26], which correspond to self-duality equations in four of the ten dimensions [27]. Alternatively, one may seek to find critical points corresponding to the generalized duality conditions [28]:

$$\lambda F_{\mu\nu} = T_{\mu\nu\rho\sigma} F_{\rho\sigma} \tag{41}$$

where T is a completely antisymmetric invariant of a subgroup of SO(d), the lorentz group in d euclidean dimensions. Applying the technology of [28] to the case of d = 10 Yang-Mills, with SO(10) breaking to a maximal subgroup, we find that an invariant T-tensor only exists for two maximal subgroups of SO(10) : SU(5) ⊗ U(1) and SO(4) ⊗ SU(4). The latter case corresponds to the conventional dimensional reduction yielding (40), with Bogomolny relations corresponding to self-duality in four dimensions together with conditions corresponding to a vanishing Higgs potential. This solution has been investigated by Osborn and also by Rossi [30]. There are three SU(5) × U(1) invariant sets of algebraic relations amongst the components of the field strength, corresponding to the three eigenvalues λ of the invariant T-tensor. Using complex coordinates, the most interesting set of equations may be written:

$$\begin{aligned} &F_{x\bar{x}} + F_{y\bar{y}} + F_{z\bar{z}} + F_{w\bar{w}} + F_{t\bar{t}} = 0 \\ &F_{ab} = 0 = F_{\bar{a}\bar{b}} \, , \quad a,b = x,y,z,w,t \, , \end{aligned} \tag{42}$$

a set of 21 equations. Although these relations saturate a Bogomolny-type bound on the action, their dimensional reduction does not yield lorentz-invariant relations in four dimensions. An alternative is to consider a T-tensor for a non-maximal subgroup of SO(10) having the desired reduction to SO(4) invariant relations. An example [31] is the following set of 17 equations left invariant under G_2 ⊗ SU(2) transformations:

$$01 + 32 + 56 + 89 = 0$$
$$02 + 13 + 64 + 97 = 0$$
$$03 + 21 + 45 + 78 = 0$$
$$04 + 26 + 53 = 05 + 34 + 61 = 0$$
$$06 + 42 + 15 = 14 + 25 + 36 = 0$$
$$07 + 83 + 29 = 08 + 37 + 91 = 0$$
$$09 + 72 + 18 = 17 + 39 + 28 = 0$$
$$48 + 57 = 0 = 59 + 68 = 67 + 49 \tag{43}$$

$$47 = 58 = 69 = 0 , \tag{44}$$

where a pair of numbers xy stands for the curvature two-form $xy \equiv F_{xy} = [D_x, D_y]$; $D_x = \frac{\partial}{\partial x} + A_x$. We use a notation in which, in the dimensionally reduced theory (40), the Higgs fields A_i and B_i are associated with the 4,5,6 and 7,8,9 components, respectively, of a ten-dimensional vector; the first four components being the vector potential. This set of equations has the disadvantage that the last three equations (44) imply a rather trivial solution, at least for low dimensional gauge groups. This problem was circumvented in ref. [31], where the last three equations (44) were replaced by the two relations

$$47 = 58 = 69 . \tag{45}$$

Remarkably, this modification yields a non-trivial lorentz invariant set of equations having the Bogomolny property of saturating a lower bound of the modified theory:

$$L' = L + \frac{1}{2} [A_i, B_i]^2 , \tag{46}$$

where L is the lagrangian density (40) and no sum over i is implied in the extra term. This modification of course breaks the $N = 4$ supersymmetry of the full theory and does not have a conventional pure Yang-Mills interpretation in higher dimensions. In ten dimensions, this theory has the interesting feature of not being fully SO(10)-covariant, and provides a further example of a higher dimensional theory not respecting the full space-time symmetry. Ward [29] has discovered some solvable systems of this genre. We note that a higher dimensional theory need not have any more than four dimensional Lorentz invariance; and our modified lagrangian (46) is invariant under the $SO(4) \otimes U(1) \otimes U(1) \otimes U(1) \otimes z_3$ subgroup of the original SO(10) . This feature is also reminiscent of recent discussions of Weinberg [32] on G-structures. The Bogomolny equations (43,45) have the further interesting feature that they incorporate in a nontrivial manner, the G_2 invariant $d = 7$ equations [28]

$$C_{abc} F_{bc} = 0 , \tag{47}$$

where the c's are the structure constants of the algebra of the imaginary octonions:

$$[e_a, e_b] = C_{abc}\, e_c , \quad a,b = 1,\dots,7 .$$

These equations correspond to the T-tensor

$$T_{abcd} = \phi_{abcd} = \frac{1}{3!} \epsilon_{abcdefg} C_{efg}$$

where the ϕ's are the structure constants of the associator of the imaginary octonions:

$$[e_a, e_b, e_c] = (e_a e_b)e_c - e_a(e_b e_c) = \phi_{abcd} e_d ,$$

which is completely antisymmetric.

The equations (47) may be obtained by dimensionally reducing the $d = 8$ $\tilde{SO}(7)$ invariant equations

$$F_{8a} = C_{abc} F_{bc} , \tag{48}$$

which have an interesting feature with regard to $d = 10$ supersymmetric Yang-Mills. The condition for the superinvariance of the lorentz invariant spinor ground state, $\langle\lambda\rangle = 0$, is given by

$$\delta\lambda = 0 = F_{\mu\nu} \gamma^{\mu\nu} \epsilon , \quad \mu,\nu = 1,\ldots..10 ,$$

where ϵ is the anticommuting supersymmetry parameter, which in light cone coordinates we may choose to write as

$$\epsilon = \zeta \otimes \eta$$

where ζ is a spinor on the light cone and η is a constant $SO(8)$ spinor in the transverse directions. It may be seen from this condition that a necessary condition for the satisfaction of this equation is that the transverse components of the gauge field satisfy

$$F_{ij} \gamma^{ij} \eta = 0 , \quad i,j = 1,\ldots,8 . \tag{49}$$

These equations are equivalent to eqs. (48), if η is taken to be purely left-handed or right-handed [28].

IV. Further remarks

Although I have explicitly considered (in Section II) the on-shell constraints of $N = 4$ super Yang-Mills theory, I should point out that the discussion is equally valid for all gauge field equations admitting linear systems which imply the existence of a set of 1-parameter family of flat connections

$$A^s(\lambda) = d^s(\lambda)\psi(\lambda)\cdot\psi(\lambda)^{-1} \tag{50}$$

linearly related to the gauge (vector or spinor) potentials of the theory:

$$A^s(\lambda,x) = \sum_a \alpha^s_a(\lambda)A_a(x) ,$$

(where the α's are constants depending on the spectral parameter λ), such that sufficiently many components of the curvature $F_{ab}(A_a)$ vanish in order that every A_a has the form of one of two (or more) independent pure-gauges:

$$A_{a_i} = g^{-1}\,\partial_{a_i}\,g\,, \quad A_{b_i} = h^{-1}\,\partial_{b_i}\,h$$

without loss of any generality (i.e. all the curvatures $F_{a_ia_j}$ and $F_{b_ib_j}$ are zero as a consequence of the consistency of (50). Both the self-duality equations and the constraint equations (8) are sets of equations of this genre; and so are some of the completely solvable higher dimensional equations recently discussed by Ward [29]. In particular his $Sp(2) \times Sp(1)$ invariant equations in 8 dimensions are amenable to the construction of infinite sets of symmetries and continuity equations. Using four complex coordinates y,z,w,t, the set of equations solving eight dimensional Yang-Mills may be written

$$A_a = g^{-1}\,\partial_a\,g\,, \quad A_{\bar a} = h^{-1}\,\partial_{\bar a}\,h\,, \quad a = y,z,w,t\,;$$

$$\begin{aligned} &F_{y\bar y} + F_{z\bar z} = 0 \\ &F_{w\bar w} + F_{t\bar t} = 0 \\ &F_{t\bar z} - F_{y\bar w} = F_{y\bar t} + F_{w\bar z} = 0\,; \end{aligned} \tag{51}$$

and they are compatibility conditions for the linear system:

$$\begin{aligned} &(D_y - \lambda^{-1}\,D_{\bar z})\Phi = 0 \\ &(D_{\bar y} + \lambda\,D_z)\Phi = 0 \\ &(D_w + \lambda^{-1}\,D_{\bar t})\Phi = 0 \\ &(D_{\bar w} - \lambda\,D_t)\Phi = 0\,. \end{aligned} \tag{52}$$

This linear system is all that is required to show the existence of an infinite set of hidden symmetries; the generators of which close under the loop algebra of the gauge group. To display the related continuity equations we have to fix the gauge. For instance, in the gauge $A_a = 0$, equations (51) reduce to

$$A_{\bar a} = B\,\partial_{\bar a}\,B^{-1}\,, \quad B = gh^{-1}\,;$$

$$\begin{aligned} &\partial_t\,A_{\bar z} - \partial_y\,A_{\bar w} = 0 \\ &\partial_y\,A_{\bar t} + \partial_w\,A_{\bar z} = 0 \\ &\partial_y\,A_{\bar y} + \partial_z\,A_{\bar z} = 0 \\ &\partial_w\,A_{\bar w} + \partial_t\,A_{\bar t} = 0\,. \end{aligned} \tag{53}$$

The hidden symmetry currents may now be derived by noting that the last two equations in (53) imply the continuity equation

$$\sum_a \partial_a J^{(1)}_{\bar a} = 0 \; ; \quad J^{(1)}_{\bar a} = A_{\bar a} \, ,$$

which is satisfied by virtue of the λ-independent parts of the linear system (52):

$$A_{\bar y} = -\partial_z X^{(1)} \, , \quad A_{\bar z} = \partial_y X^{(1)} \, , \quad A_{\bar t} = -\partial_w X^{(1)} \, , \quad A_{\bar w} = \partial_t X^{(1)} \, ,$$

where we have assumed that Φ has the expansion

$$\Phi = \sum_{n=0}^{\infty} \lambda^{-n} X^{(n)} \, .$$

The derivation of the higher continuity equations follows closely the analogous derivation for the chiral model in two dimensions. The significance of these continuity equations, as for those of the 4d self-duality equations, is unclear.

To conclude, we note that the dimensionally reduced version of equations (51) may be reformulated to take the form of a Dirac equation if the gauge group has an SU(2) subgroup. Define a matrix in an SU(2) basis

$$\Psi = \begin{pmatrix} \phi_{\bar w} & \phi_{\bar t} \\ -\phi_t & \phi_w \end{pmatrix} ,$$

and a Dirac operator

$$D = \begin{pmatrix} D_y & -D_{\bar z} \\ D_z & D_{\bar y} \end{pmatrix}$$

then eqs. (51) correspond to the conditions

$$\det \Psi = 0 = \Psi^\dagger \Psi \tag{54}$$

$$D\Psi = 0 \tag{55}$$

together with the consistency condition for (55):

$$D^\dagger D \Psi = 0 \, .$$

I should like to thank David Fairlie for discussions. Section III is based on work done in collaboration with him. I also thank Norma Sanchez and Hector de Vega for inviting me to give the talk on which this article is based, R. Stora and P. Ramond for hospitality at les Houches where this article was written up; and E. Witten for some comments.

References

[1] F. Gliozzi, J. Scherk and D. Olive, Nucl. Phys. B122 (1977) 253;
L. Brink, J. Schwarz and J. Scherk, Nucl. Phys. B121 (1977) 77.
[2] P. Goddard, J. Nuyts and D. Olive, Nucl. Phys. B125 (1977) 1;
C. Montoneu and D. Olive, Phys. Lett. 72B (1977) 117.
[3] D. Olive, in Monopoles in Quantum Field Theory (N. Craigie et al, Eds.) World Scientific, 1982.
[4] S. Coleman, Phys. Rev. D11 (1975) 2088.
[5] L.D. Faddeev, in Proceedings les Houches 1982, Recent Advances in Field Theory and Statistical Mechanics, eds. J.-B. Zuber and R. Stora.
[6] M. Atiyah and R.S. Ward, Comm. Math.Phys. 55 (1977) 117.
[7] V. Zakharov and S. Manakov, Sov. Sci. Revs. (Phys. Rev.) A1, 133 (1979).
[8] P. Lax, Comm. Pure Appl. Math. 21 (1968) 467.
[9] M. Luscher and K. Pohlmeyer, Nucl. Phys. B137 (1978) 46.
[10] A. Polyakov, Phys. Lett. 82B (1979) 247.
[11] B. Hasslacher and A. Neveu, Nucl. Phys. B151 (1979) 1.
[12] C.N. Yang, Phys. Rev. Lett. 33 (1974) 445.
[13] M. Davies, P. Houston, J. Leinaas, A. Macfarlane, Phys. Lett. 119B (1982) 187.
[14] L. Dolan, Phys. Rev. Lett. 47 (1981) 1371; Phys. Rep. 109 (1984) 1.
[15] C. Devchand and D.B. Fairlie, Nucl. Phys. B194 (1982) 232;
Ge M.-L. and Y.-S. Wu, Phys. Lett. 108B (1982) 411;
K. Ueno and Y. Nakamura, Phys. Lett. 117B (1982) 208.
[16] C. Devchand, Nucl. Phys. B238 (1984) 333.
[17] M. Sohnius, Nucl. Phys. B136 (1978) 461.
[18] E. Witten, Phys. Lett. 77B (1978) 394.
[19] C. Devchand, Phys. Rev. D (1985).
[20] I.V. Volovich, Lett. Math. Phys. 7 (1983) 517; Theor. Math. Phys. 57 (1983) 1269, I. Aref'eva and I.V. Volovich, Phys. Lett. 149B (1984) 131.
[21] L.-L. Chau, Ge M.-L. and Z. Popowicz, Phys. Rev. Lett. 52 (1984) 1940.
[22] I.V. Volovich, Phys. Lett. 129B (1983) 429.
[23] J. Harnad, J. Hurtubise, M. Legare, and S. Shnider, Nucl. Phys. B256 (1985) 609.
[24] This proof of the invariance of eq. (34c) was written down erroneously in ref. [16]. I thank I.V. Volovich for drawing my attention to this error.
[25] See e.g. L.-L. Chau, Ge M.-L., A. Sinha and Y.-S. Wu, Phys. Lett. 121B (1983) 391 (and references therein).
[26] E.B. Bogomolny, Sov. J. Nucl. Phys. 24 (1976) 449.
[27] P. Goddard and D. Olive, Rep. Prog. Phys. 41 (1978) 91;
E. Witten and D. Olive, Phys. Lett. 78B (1978) 97;
M.A. Lohe, Phys. Lett. 70B (1977) 325;
D. Olive, Nucl. Phys. B153 (1979) 1.
[28] E. Corrigan, C. Devchand, D.B. Fairlie and J. Nuyts, Nucl. Phys. B214 (1983).
[29] R.S. Ward, Nucl. Phys. B236 (1984) 381.
[30] H. Osborn, Phys. Lett. 83B (1979) 321;
P. Rossi, Phys. Lett. 99B (1981) 229.
[31] C. Devchand and D.B. Fairlie, Phys. Lett. 141B (1984) 73.
[32] S. Weinberg, Phys. Lett. 138B (1984) 47.
|33| H.J. de Vega, H. Eichenherr and J.M. Maillet, Nucl.Phys. B240, 377 (1984).
H. Eichenherr, in "Non Linear Equations in Classical and Quantum Field Theory", Ed. by N. Sánchez, Lect. Notes in Phys. 226, 171-195, Springer-Verlag (1985).

MONOPOLES AND RECIPROCITY

E. Corrigan

Department of Mathematical Sciences
Durham, UK.

1. Introduction.

One of the more remarkable features of the theory of self-dual non abelian monopoles is the way in which they may be constructed using the Atiyah-Drinfeld-Hitchin-Manin-Nahm (ADHMN) procedure[1,2]. Moreover, whilst it is possible to think of the construction rather abstractly in terms of algebraic geometry[3] it is also possible to cast it into a concrete and memorable form where it appears to embody an interesting principle[4,5], possibly having wider implications.

Since the ADHM construction is now widely reviewed[1] it might be useful to sketch the notion of 'reciprocity' for monopoles[4] and then to discuss a couple of ideas which are more speculative and less developped. These concern static, but non self dual, monopoles whose existence is known[6], but whose properties are hardly explored.

2. Self-dual monopoles and reciprocity.

Over the last few years the mathematical investigation of monopoles has largely focussed on the description of static solutions to the SU(2) Yang-Mills-Higgs equations in the Bogomolny-Prasad-Sommerfield (BPS) limit[7], (vanishing Higgs' potential), for which the gauge field, A_μ, and scalar field, Φ, satisfy

$$F_{jk} = \partial_j A_k - \partial_k A_j + i\left[A_j, A_k\right] = \epsilon_{jkl}\, D_l \Phi\,, \quad j,k,l = 1,2,3$$

$$A_0 = 0 = F_{0i} \tag{2.1}$$

More succinctly, if we set

$$A_4 = \Phi \,, \qquad F_{4i} = -D_i\Phi \tag{2.2}$$

we may write,

$$F_{\alpha\beta} = \tfrac{1}{2}\epsilon_{\alpha\beta\gamma\delta}F_{\gamma\delta}\,, \qquad \alpha,\beta,\gamma,\delta = 1,..,4, \tag{2.3}$$

and recognise the Bogomolny equations, (2.1), as self-duality equations in four dimensions reduced to three, since no field depends on x_4.

Additionally, the scalar field satisfies asymptotically

$$\Phi \sim \begin{pmatrix} 1/2 & 0 \\ 0 & -1/2 \end{pmatrix}\left(1-\frac{k}{r}\right), \qquad r=|\underline{x}| \to \infty, \tag{2.4}$$

locally, (but not globally), over the 'sphere at infinity'. Here, k is a positive integer , or zero. The tilde denotes gauge equivalence rather than equality since it must be remembered that generally Φ cannot be gauge rotated to a constant over the whole sphere; it has non-trivial topology associated with it.

The interesting solutions are everywhere regular with an energy

$$E = \mathrm{tr}\int d^3x\, \{\underline{B}^2 + (\underline{D}\Phi)^2\} = 4\pi k\,, \qquad B_i = \tfrac{1}{2}\epsilon_{ijk}F_{jk}\,, \tag{2.5}$$

and a magnetic flux

$$M = 2\,\mathrm{tr}\int_{S_\infty} d\underline{S}\cdot(\Phi\,\underline{B}) = 4\pi k. \tag{2.6}$$

For each choice of k there is a solution set to eqs. (2.3) satisfying conditions (2.4) and (2.5) and depending upon 4k-1 parameters[8]. Ansatze to produce these solutions were first discussed by Ward[9] for the case of two monopoles, extended to the general case by Corrigan and Goddard[10], and proved to be complete by Hitchin[11].

Given a solution to the monopole problem for a given k, we may consider the Dirac equation

$$\left(\frac{\partial}{\partial x_4} + i A_4 - i \underline{\sigma} . \underline{D} \right) \psi = 0 \tag{2.7}$$

for a left-handed spinor ψ. (For a self-dual monopole the 'right-handed' partner to (2.7) has no normalisable solutions at all.) There is a very nice theorem due to Callias[12] concerning the normalisable solutions to eq.(2.7). They are all of the form:

$$\psi(x_4, \underline{x}) = e^{i x_4 z} \, \Psi(\underline{x}, z), \tag{2.8}$$

and, there are precisely k linearly independent solutions provided $|z| \leq 1/2$, and none otherwise. The spinor function $\Psi(\underline{x},z)$ can be picked in such a way that

$$\int d^3\underline{x} \;\; \Psi_i^+ \, \Psi_j = \delta_{ij} \qquad\qquad i,j = 1,\dots,k, \tag{2.9}$$

by an orthonormalisation process.

Let us denote by $\underline{T}$ the 'expectation value' of the coordinate $\underline{x}$ with respect to the wave function Ψ. Thus,

$$\underline{T}_{ij} = \int d^3\underline{x} \;\; \Psi_i^+ \, \underline{x} \, \Psi_j , \tag{2.10}$$

and, in addition, set

$$(T_4)_{ij} = \int d^3\underline{x} \;\; \Psi_i^+ \left(i \frac{\partial}{\partial z} \right) \Psi_j . \tag{2.11}$$

Then, the remarkable property of the $\underline{T}$'s is that they satisfy

$$\frac{d\underline{T}}{dz} = i \, \underline{T} \wedge \underline{T} + i \left[T_4, \underline{T} \right] \tag{2.12}$$

which are just the self-duality equations reduced from four to one dimension by deleting dependence on three variables, (z is the fourth). Note that with respect to z dependent changes of basis for the Ψ's, (i.e. U(k) transformations), $\underline{T}$ transforms homogeneously while T_4 transforms as a gauge field,

$$T_4 \rightarrow T_4 + i\, U^{+} \frac{dU}{dz} \,. \tag{2.13}$$

Picking the gauge $T_4 = 0$, eqs.(2.12) reduce to Nahm's equations[2] on the interval $|z| \leqslant 1/2$. It is tempting to think of the z variables as 'momentum-like' variables and to think of eqs.(2.12) as a set complementary to eqs.(2.3).

Consider the Dirac equation again, this time in z-space. That is,

$$\left[\left(\frac{\partial}{\partial z_4} + i T_4 \right) - i \underline{\sigma} \cdot \left(\underline{\nabla}_z + i \underline{T} \right) \right] \tilde{\psi}\, (z_4, \underline{z}) = 0 \,, \tag{2.14}$$

where we shall think of z_4 as what we called z previously, and set $T_4 = 0$ by a choice of gauge, given the other components $\underline{T}(z)$ are independent of the other three variables. Eq.(2.14) has a set of solutions of the form

$$\tilde{\psi}\, (z, \underline{z}) = e^{-i \underline{x} \cdot \underline{z}} \; \tilde{\Psi}\, (z, \underline{x}) \,, \tag{2.15}$$

and, the interesting ones, by analogy with eqs.(2.8) and (2.9) are those satisfying

$$\int_{-1/2}^{1/2} dz \;\; \tilde{\Psi}_i^{+} (z, \underline{x})\, \tilde{\Psi}_j (z, \underline{x}) = \delta_{ij} \tag{2.16}$$

That is, the normalisable solutions over the interval $|z| \leqslant 1/2$.

In the previous situation, the number of solutions to eq.(2.7) was determined essentially by the asymptotic behaviour of the Higgs field, Φ. Here, the number of normalisable solutions might be expected to be governed by the asymptotic (i.e. as $z \to \pm 1/2$) behaviour of the three $\underline{T}$'s, since they are, like Φ, the components of a gauge field corresponding to deleted variables in the dimensional reduction. That is indeed the case. Nahm[2] argued (and Hitchin[3] and Donaldson[13] proved) that provided the three matrices $\underline{T}$ satisfy

$$\underline{T} \sim \frac{\underline{t}}{\frac{1}{2} \pm z} \tag{2.17}$$

at the end points, where $\underline{t}$ is an irreducible representation of the SU(2) Lie algebra, then there are precisely two normalisable solutions to eq.(2.14), normalisable in the sense of eq.(2.16). It ought to be possible to demonstrate eq.(2.17) using the properties of the solutions to the original Dirac equation, eq.(2.7), but we have not been able to do so. It would be nice to find a direct argument connecting these facts.

Finally, we can make use of the two normalisable solutions to eq.(2.14) to construct the original, three-dimensional, gauge and Higgs fields. We set,

$$\Phi_{ij} = (A_4)_{ij} = \int_{-1/2}^{1/2} dz \; \tilde{\Psi}_i^{+} \, z \, \tilde{\Psi}_j \,, \tag{2.18}$$

$$\underline{A}_{ij} = \int_{-1/2}^{1/2} dz \; \tilde{\Psi}_i^{+} \left(i \underline{\nabla}_x \right) \tilde{\Psi}_j \,. \tag{2.19}$$

In other words, the Higgs field is the 'expectation' value of z, compare eq.(2.10), and $\underline{A}$ is defined in much the same way as T_4 in eq.(2.11). These definitions of Φ and $\underline{A}$ are, of course the necessary ones according to the ADHMN construction[2]. What is surprising, and striking, is the way in which the T and A gauge fields are each defined by the zero mode solutions to the Dirac operator in the background of the other and what is more, according to the same prescription, suitably interpreted. For more details and related facts, see refs.(4,5). Nahm's version of the ADHM construction thus consists of solving eqs.(2.12) for the $\underline{T}$'s (in the gauge $T_4 = 0$) subject to the end conditions (2.17), solving eq.(2.14) subject to the condition (2.16) and computing the monopole fields A, Φ via eqs.(2.18) and (2.19). These are then guaranteed to satisfy the Bogomolny equations, and to be a solution of energy $4\pi k$. Moreover, all solutions to the Bogomolny equations can be obtained using this method, at least in principle if not in practice.

3. Non self dual monopoles.

With regard to the question of non self dual monopoles most of the effort to date has been concentrated on proving existence, in the static case[6], and in developping an idea of Manton in the time dependent or scattering situation[14]. Manton's idea was to explore the motion of slowly moving monopoles (of like magnetic charge) using the fact that for vanishingly small velocities the monopoles follow geodesics in the parameter space of the self dual solutions. That does not mean their trajectories in space-time are necessarily simple, however. The metric on parameter space is defined as follows. Suppose A_μ ($\mu = 1,..,4$) and $A_\mu + \delta A_\mu$ are two neighbouring solutions of the Bogomolny equations for the vector potential $\underline{A}$ and the scalar field Φ then the metric is defined by

$$\langle \delta A_\mu, \delta A_\mu \rangle = \int d^3\underline{x} \ \mathrm{tr}\left(\delta A_\mu \delta A_\mu \right), \tag{3.1}$$

(in the background gauge).

Atiyah and Hitchin[15] have computed the metric in the case of two monopoles and used the geodesics to discover interesting facts about two monopole scattering in this adiabatic approximation.

Besides the important question of the scattering solutions containing monopoles of like charge there are also questions about solutions containing both monopoles and antimonopoles. One would reasonably expect these to be time dependent, but there are also some that are static, a fact as we have already mentioned proved by Taubes a couple of years ago[6]. Necessarily, these solutions although static fail to satisfy the Bogomolny equations, but instead satisfy the full Yang-Mills-Higgs equations, (still with vanishing potential for the Higgs field),

$$D_\mu F_{\mu\nu} = 0, \qquad \nu = 1, .., 4. \tag{3.2}$$

Presumably, for the simplest of these solutions the Higgs field approaches a constant, say $\sigma_3/2$, globally over the sphere at infinity. The solution has no nett magnetic charge. However, there must still be a $1/r$ term (cf. eq.(2.4), otherwise the Higgs field contributes nothing to the energy and satisfies $\underline{D}\Phi = 0$ everywhere, a disallowed situation if Φ is to be singularity free but not everywhere constant. Indeed, examining eqs.(3.2) asymptotically, one finds

$$\Phi \sim \sigma_3/2 - A/r, \qquad r \to \infty, \tag{3.3}$$

where $\mathrm{tr}(A\sigma_3) > 0$ and constant, so that the contribution to the energy of the Higgs field is,

$$E_\Phi = \mathrm{tr} \int d^3\underline{x} \; (\underline{D}\,\Phi)^2 \;= 4\pi\, \mathrm{tr}\,(A\,\Phi_\infty) \tag{3.4}$$

There appears to be nothing simple to be said about the contribution to the energy from the gauge potential.

With a non self dual background the Dirac equation will have both left and right handed solutions, only the difference $n_R - n_R$ being related to the magnetic charge. Nevertheless, it might be interesting to try to construct the quantities $\underline{T}$, via eq.(2.10), for the solutions of each type. For static solutions, $\underline{T}$ and T_4 must still be gauge fields dependent upon the single variable z. The question is what equation do they satisfy? Do they have any interesting properties at all? Even for the time dependent case in the adiabatic approximation mentioned earlier, it would be interesting to know how the $\underline{T}$'s develop in time, given the moving monopoles described by fields whose parameters have a time development determined by the geodesic motion in the parameter space itself. One is tempted to speculate that the nice properties of monopoles may arise precisely because they are transforms, in the Nahm sense, of a two dimensional theory. One is also tempted to speculate that the two dimensional theory might be another Yang-Mills theory, though there is no evidence for that at the moment. The idea did suggest, however, that it might be useful to tackle one (and two) dimensional gauge theories (reduced from four (and five) dimensions as above) with particular regard to solutions on a finite interval with singularities at the endpoints(16), since these are the useful ones in the self dual case. The same equations (apart from a sign) have been discussed by Nikolaevski and Schur, Savviddy and Chang(17), and probably others, in the context of Yang-Mills 'mechanics' and chaotic behaviour. Before elaborating a little more in the next section, there's another idea perhaps worth mentioning.

Din and Zakrzewski(18) discovered some years ago a method of generating non self dual solutions to CP^N models in two Euclidean dimensions. These solutions can be regarded as carefully balanced combinations of instantons and antiinstantons which are static solutions with respect to a theory in three dimensions (two space - one time), for which the instantons are static 'lumps'. These solutions are rather like monopoles, but easier to construct owing to their two dimensional nature. Indeed, the instantons of charge k are basically generated by polynomials of degree k in the complex variable $z = x_1 + x_2$. Thus, in the CP^N model, the N + 1 complex functions p_i, i = 0,..,N, satisfying

$$\sum_0^N p_i^* p_i = 1, \tag{3.5}$$

are constructed by setting

$$p_i(z,z^*) = f_i(z) / |\underline{f}| \,, \qquad i = 0,..,N \tag{3.6}$$

where $|\underline{f}| = (\Sigma f_i^* f_i)^{1/2}$, and f_i is a polynomial function of z only, of degree k. All the instantons are obtained this way[19], the antiinstantons by taking functions of z^* instead.

Remarkably, all the non dual solutions to the CP^N model are obtained via a projection procedure applied to instantons in CP^M models for $M = \binom{N}{r}$, r = 2,...,N-1,each of the latter being generated naturally from the fundamental CP^N instanton. Set,

$$F_{i_0 i_1 .. i_r} = f_{[i_0} f^{(1)}_{i_1} f^{(2)}_{i_2} \cdots f^{(r)}_{i_r]} \tag{3.7}$$

where $f^{(q)}$ denotes $\frac{d^q f}{dz^q}$, and the square brackets denote antisymmetrization. Then a solution to the CP^N model is found by normalising the quantities,

$$F_i = \sum_{i_0 \cdots i_{r-1}} F^*_{i_0 \cdots i_{r-1}} F_{i_0 \cdots i_{r-1} i} \,. \tag{3.8}$$

What is more, all the solutions are obtainable this way, starting from the complete set of instantons $f_i(z)$, (including the corresponding set of antiinstantons, by the way).

Maybe, in the monopole case too, there is the possibility of obtaining non dual solutions by projecting self dual solutions belonging to larger gauge groups down to SU(2). It is encouraging to note that Nahm's method of construction works for larger gauge groups as well,[2] subtly putting together elementary 'SU(2)-like' contributions. Unfortunately, the construction has only been carried out explicitly for spherically symmetric self dual monopoles so far[20], and they do not appear to be helpful. That there is a sort of projection producing non self dual solutions can be illustrated by producing a singular solution as follows.

Start with the trivial solution to Nahm's equations, $\underline{T} = 0$, and solve (2.14). I.e.,

$$\left(\frac{d}{dz} - \underline{\sigma}.\underline{x}\right)\tilde{\Psi}(z,\underline{x}) = 0 . \qquad (3.9)$$

Thus,

$$\tilde{\Psi}(z,\underline{x}) = e^{rz}\,\tfrac{1}{2}(1+\underline{\sigma}.\hat{\underline{x}})\,v_1 + e^{-rz}\,\tfrac{1}{2}(1-\underline{\sigma}.\hat{\underline{x}})\,v_2 , \qquad (3.10)$$

where $v_{1,2}$ are independent of z. Suppose we consider the interval $a \leqslant z \leqslant b$, select just half of (3.10), say the part containing v_1, and compute the vector potential according to the prescription (2.18),(2.19) and the new range of z. We shall find, for the various components of the vector potential in spherical polar coordinates:

$$A_r = A_\theta = 0 , \qquad A_\phi = \frac{1-\cos\theta}{2r\sin\theta} \qquad (3.11)$$

and

$$A_4 = \Phi = \frac{1}{2}\frac{d}{dr}\ln\left(\frac{e^{2rb}-e^{2ra}}{r}\right). \qquad (3.12)$$

Since we projected out half of the Dirac solutions we end up with a U(1) potential which is in fact a Dirac potential for a singular monopole of charge 1/2. What about eq.(3.12)? Setting aside the singularity problems, (3.11) and (3.12) do not in general satisfy Maxwell's equations, except in two special limits:

(i) $a \to -\infty$, $b \to 0$, $\Phi \to -1/2r$

(ii) $a \to b$.

In case (i), mentioned by Nahm several years ago[21], we obtain a singular self dual solution. In case (ii) we obtain another solution to Maxwell's equations, this time non self dual. The idea then is to try to extend case (ii) to a non abelian situation and a non singular solution. Attempts so far have been unsuccessful, however. Other ideas, such as reducing directly from the tensor product of two monopoles, or a monopole tensored with an antimonopole seem not to be helpful either.

4. Second order Nahm equations.

The one dimensional Yang-Mills equations reduced from four (euclidean) dimensions are, in the gauge $T_4 = 0$,

$$\frac{d^2 \underline{T}}{dz^2} - \sum_i \left[T_i, \left[T_i, \underline{T} \right] \right] = 0 \qquad (4.1)$$

and

$$\left[\underline{T}_{\cdot}, \frac{d\underline{T}}{dz} \right] = 0. \qquad (4.2)$$

We[16] have studied these to a limited extent, concentrating on the possible pole residues which, if $\underline{T} \sim \frac{\underline{t}}{\lambda - z}$, must satisfy,

$$\sum_i \left[t_i, \left[t_i, \underline{t} \right] \right] = 2\underline{t} \qquad (4.3)$$

We supposed, originally, that these algebraic equations would be relatively easy to solve. However, eqs.(4.3) turn out to be quite interesting in themselves, permitting a variety of possibilities other than $\pm\underline{t}$ being a representation of the SU(2) algebra. Indeed, there is no need to restrict the discussion to three dimensions (i.e. the t's) as afar as the algebraic problem is concerned. The tougher problem of cataloguing all solutions to eqs.(4.1) and (4.2) with a given behaviour at two poles $\underline{I} \sim \frac{\underline{t}^{\pm}}{\frac{1}{2} \mp z}$, seems to be beyond us at present.

Perhaps the most interesting solution we discovered to eq.(4.3) was in the case of an SU(3) gauge group, (and appears to be unique to that case in three dimensions). There we found a solution interpolating continuously between the two 'dual' possibilities. In other words, there is a $\underline{t}(\theta)$ such that for $\theta = 0$, $\underline{t}(0)$ satisfies the SU(2) algebra, $i\underline{t} = \underline{t}\wedge\underline{t}$, while for $\theta = \pi$, $-\underline{t}(\pi)$ satisfies the SU(2) algebra. This result was unexpected.

It is perhaps worth remarking that every solution we found has a distinctive behaviour under a 'commutator mapping'.

$$\underline{t} \rightarrow \underline{t}' = -i\,\underline{t}\wedge\underline{t} \tag{4.4}$$

Clearly the solutions corresponding to SU(2) algebras are fixed points of this mapping. However, other solutions correspond to 'solvable' algebras in the sense that the iterated commutator mapping eventually yields zero. For the interpolating solution above the commutator mapping always yields another usually inequivalent solution, corresponding in fact to the mapping,

$$\theta \rightarrow -2\theta \mod 2\pi \tag{4.5}$$

in terms of θ.

This is probably not a place to produce a catalogue of known possibilities, particularly as we do not understand the systematics of the solutions. Further details can be found in refs.(16,22). We were led to a study of these equations in an attempt to discover an application of Nahm's transformation to Taubes' solution. We still need to be able to tackle the appropriate Dirac equations in order to be able to work through a tractable case. At the moment it looks hard. However, to end with a personal comment -- I also thought in 1978 that progress on self dual monopoles was unlikely to be made analytically!

Acknowledgements.

I am grateful to Hector de Vega, Norma Sanchez and the CNRS for the opportunity to air these rather rough ideas.

References.

1. M.F. Atiyah, V.G. Drinfeld, N.J. Hitchin and Yu.I. Manin, Phys. Letts.65A (1978) 195.

V. G. Drinfeld and Yu.I. Manin, Commun. Math. Phys. 63 (1978) 177.

M.F. Atiyah, "Geometry of Yang-Mills Fields", Lezione Fermioni, Pisa 1979.

2. W. Nahm, Phys. Letts. 90B (1980) 413.

"Multimonopoles in the ADHM construction." In: Proceedings of the Symposium on Particle Physics. Z. Horvath et al., eds., Visegrad 1981.

"Construction of all self-dual monopoles by the ADHM method." In: Monopoles and Quantum Field Theory. N.S. Craigie et al., eds., Singapore: World Scientific 1982.

"The algebraic geometry of multimonopoles", Group Theoretical Methods in Physics, Istanbul 1982.

3. N.J. Hitchin, Commun. Math. Phys. 89 (1983) 145.

4. E. Corrigan and P. Goddard, Ann. Phys. (NY) 154 (1984) 253.

5. W. Nahm, "Self-dual monopoles and calorons." Talk at the XII Colloquium on Group Theoretical Methods in Physics. Treste, 1983.

6. C. Taubes, Commun. Math. Phys. 86 (1982) 257.

7. E.B. Bogomolny Sov. J. Nuclear Phys. 24 (1976) 449.

M.K. Prasad and C.M. Sommerfield, Phys. Rev. Letts. 35 (1975) 760

8. E. Weinberg, Phys. Rev. 20 (1979) 936.

9. R.S. Ward, Commun. Math. Phys. 79 (1981) 317.

Phys. Letts. 102B (1981) 136.

10. E. Corrigan and P. Goddard, Commun. Math. Phys. 80 (1981) 575.

11. N.J. Hitchin, Commun. Math. Phys. 83 (1982) 579.

12. C. Callias, Commun. Math. Phys. 62 (1978) 213.

13. S. Donaldson, Commun. Math. Phys. 93 (1984) 453.

Commun. Math. Phys. 96 (1984) 387.

14. N. Manton, Phys. Letts. 110B (1982) 54.
"Monopole interactions at long range", Cambridge preprint 1985.

15. M.F. Atiyah and N.J. Hitchin, Phys. Letts 107A (1985) 21.

16. E. Corrigan, P.R. Wainwright and S.M.J. Wilson, Commun. Math. Phys. 98 (1985) 259.

17. E.S. Nikolaevski and L.N. Schur, JETP Letts. 36 (1982) 218.
G.K. Savviddy, Phys. Letts. 130B (1983) 303.
S.J. Chang, Phys. Rev. D20 (1984) 259.

18. A.M. Din and W.J. Zakrzewski, Nucl. Phys. 174B (1980) 397.
W.J. Zakrzewski, Classical solutions to CP^{n-1} models and their generalisations", Lecture Notes in Physics 151 (Springer-Verlag, 1982).

19. A. d'Adda, P. diVecchia and M. Luscher, Nucl. Phys. 146B (1978) 63.

20. M.C. Bowman, E. Corrigan, P. Goddard, A. Puaca and A. Soper, Phys. Rev. D12 (1984) 3100.

21. W. Nahm, Phys. Letts. 93B (1980) 42.

22. E. Corrigan, "Some comments on a cubic algebra", invited talk, Srni Czechoslovakia, 1985, to appear in Circolo Matematico di Palermo.

NON-LOCAL CONSERVATION LAWS FOR NON-LINEAR SIGMA MODELS WITH FERMIONS

Michael Forger
Theory Division, CERN
1211 Geneva 23, Switzerland

1. - INTRODUCTION

The history of the non-linear σ models begins in 1960, when Gell-Mann and Lévy proposed their original σ model as an effective field theory describing the strong interactions between nucleons and pions at low energies [1-3]. It was soon realized, however, that there seemed to be no way to extend this model to a somewhat higher energy range in such a way as to incorporate strong interactions with strange particles, and the idea was therefore not pursued very vigorously. In the last few years, however, the subject has been revived, though in a rather different physical context, by Buchmüller, Peccei and Yanagida: they propose supersymmetric σ models as effective field theories describing, e.g., the weak interactions between leptons and quarks at low energies [4]. Finally, we should mention the observation by Cremmer and Julia that supersymmetric non-linear σ models appear naturally in extended supergravity [5].

In two-dimensional space-time, non-linear σ models have come to play an even more outstanding rôle. On the one hand, this is mostly due to the striking analogies that exist between two-dimensional non-linear σ models and four-dimensional non-Abelian gauge theories. At the classical level, for example, both are of geometric nature ("geometry dictates interactions"), are conformally invariant and admit topologically non-trivial Euclidean solutions (instantons) [6,7]; moreover, both can be coupled naturally to fermionic matter and to gravity. (For other, more intricate connections, see [8].) At the quantum level, we mention asymptotic freedom [9,10] and, as examples of known properties of σ models that are believed, but still not proved, to be shared by gauge theories, dynamical mass generation and confinement [11]. On the other hand, a large class of non-linear σ models provides examples of integrable systems in two-dimensional space-time, and it is this very attractive feature that shall be dealt with in the present notes.

Since "integrability" is a rather flexible notion that tends to be given different meanings by different people, let us specify in what sense we shall use this term here. At the classical level, the model is said to be integrable if its non-linear equations of motion are equivalent to (or at least imply) the compatibility conditions for a certain linear system of first-order partial differential equations (Lax pair) containing a spectral parameter; this hidden symmetry then gives rise to infinite series of local as well as non-local conservation laws. At the quantum level, we shall call the model integrable if it admits a conserved quantum non-local charge -- the quantum counterpart of the (first) conserved classical non-local charge. Existence of such a charge puts strong constraints on the dynamics: namely, particle production is suppressed, and the S-matrix factorizes into two-body amplitudes [12], which can often be calculated exactly [13].

In the case of two-dimensional non-linear σ models, the basic integrability condition is that the σ model field takes values in a Riemannian symmetric space $M = G/H$. More specifically, it has been known for some time that the purely bosonic non-linear σ model is classically integrable if $M = G/H$ is Riemannian symmetric [14,15] (for reviews, see [16,17]) and is quantum integrable if in addition H is simple [18]. Analogous results for non-linear σ models with fermions are more recent [19-21] and, in the quantized case, still incomplete.

The subject of this review will be to explain what is presently known about integrability of non-linear σ models with fermions. More specifically, we shall present the results of Ref. [19], in a somewhat abbreviated form, in Section 2 (classical theory) and Section 4 (quantum theory), while Section 3 contains as yet unpublished material that establishes the equivalence between the two apparently different formulations of Grassmannian models, with minimally or supersymmetrically coupled fermions, that have been used in Refs. [19] and [21].

It is a pleasure for me to acknowledge the productive collaboration with E. Abdalla, in the course of which the major part of what follows has been developed. For supplementary reading, I recommend his review in the same series [22].

2. - CLASSICAL NON-LINEAR σ MODELS WITH FERMIONS

Let us begin by briefly reviewing the general method of coupling matter fields to non-linear σ models. From a global point of view, σ model fields are maps q from space-time to a given Riemannian manifold M, called the target space, while matter fields are sections Φ of a certain Hermitian complex (or Riemmanian real) vector bundle $S \otimes q^*V$ over space-time: this bundle arises by taking the tensor product of an appropriate spinor or tensor bundle S over space-time with the pull-back q^*V to space-time, via the σ model field q, of a given Hermitian complex (or Riemannian real) vector bundle V over M, called the target bundle. Generically, M and V do not admit any symmetries (a symmetry of a Riemannian manifold is an isometry), and one must resort to describing the fields q and Φ in terms of their components with respect to (arbitrarily chosen) local co-ordinates for M and local trivializations for V. Here, however, we shall assume that the target space is a Riemannian homogeneous ($\equiv$ coset) space $M = G/H$, and that the target bundle is an associated vector bundle $V = G \times_H V_0$, derived from a given unitary (or orthogonal) representation of the stability group H on a given finite-dimensional complex (or real) vector space V_0. Then instead of σ model fields q taking values in M, we can (at least locally) use σ model fields g taking values in G, defined modulo H, and instead of matter fields Φ that are sections of $S \otimes q^*V$, we can (at least locally) use matter fields ϕ whose spinor or tensor components are functions taking values in V_0, defined modulo H. In the following, our matter fields will be Dirac spinor fields χ, and so the basic fields g and χ in our class of models transform according to

$$g \longrightarrow gh \quad , \quad \chi \longrightarrow h^{-1} \cdot \chi \tag{2.1}$$

under gauge transformations (the gauge group being H) and according to

$$g \longrightarrow g_0 g \quad , \quad \chi \longrightarrow \chi \tag{2.2}$$

under global symmetry transformations (the global symmetry group being G).

Before going on, let us fix some notation. First of all, the manifold M, being Riemannian homogeneous, is the quotient space $M = G/H$ of a connected Lie group G, with Lie algebra $\mathfrak{g}$, modulo a compact subgroup $H \subset G$, with Lie algebra $\mathfrak{h} \subset \mathfrak{g}$, such that the following holds: there exists an Ad(H)-invariant subspace $\mathfrak{m}$ of $\mathfrak{g}$ such that $\mathfrak{g}$ is the (vector space) direct sum of $\mathfrak{h}$ and $\mathfrak{m}$:

$$\mathfrak{g} = \mathfrak{h} \oplus \mathfrak{m} . \tag{2.3}$$

Thus we have the commutation relations

$$[\mathfrak{h}, \mathfrak{h}] \subset \mathfrak{h} \quad , \quad [\mathfrak{h}, \mathfrak{m}] \subset \mathfrak{m} . \tag{2.4}$$

The decomposition of elements $X \in \mathfrak{g}$ corresponding to (2.3) will be written

$$X = X_{\mathfrak{h}} + X_{\mathfrak{m}} . \tag{2.5}$$

We also assume that the given left-invariant Riemannian metric $(.,.)$ on M can be obtained, by restriction, from some bi-invariant pseudo-Riemannian metric $(.,.)$ on G for which the direct decomposition (2.3) is orthogonal. (This amounts essentially to requiring that M be naturally reductive; we refer to Refs. [7,17] for a detailed discussion.) Finally, as we are ultimately interested in the question of integrability, we shall assume throughout that M is not only Riemannian homogeneous but in fact Riemannian symmetric: otherwise, the model would not be integrable even in the pure model limit (when all matter fields are required to vanish). Essentially, this means that in addition to (2.4), we also have the commutation relation

$$[\mathfrak{m},\mathfrak{m}] \subset \mathfrak{h}. \tag{2.6}$$

Riemannian symmetric spaces are completely classified; see [23, pp. 516 and 518] for a list.

Returning to the non-linear σ model, we proceed to define composite fields made up from the basic fields g and χ. For example, the bosonic sector provides vector fields A_μ and k_μ, taking values in $\mathfrak{h}$ and in $\mathfrak{m}$, respectively, defined as in the pure model [15-17]:

$$A_\mu = (g^{-1}\partial_\mu g)_{\mathfrak{h}} \quad , \quad k_\mu = (g^{-1}\partial_\mu g)_{\mathfrak{m}} \tag{2.7}$$

(cf. (2.5)). On the other hand, the fermionic sector gives rise to various fields which are bilinear in the spinors and are built by inserting either generators of the representation or operators which commute with all such generators, as well as γ_μ or 1 or $\gamma_5 = \gamma_0\gamma_1$. The one of interest to us here is the vector field B_μ, taking values in $\mathfrak{h}$, defined by the requirement that for all generators $T \in \mathfrak{h}$,

$$(B_\mu, T) = -\tfrac{i}{2}\,\overline{\chi}\gamma_\mu T\cdot\chi \ . \tag{2.8}$$

More explicitly, in terms of an arbitrary basis of generators $T_j \in \mathfrak{h}$, with $g_{jk} = (T_j,T_k)$ and $(g^{jk}) = (g_{jk})^{-1}$,

$$B_\mu = B^j_\mu T_j \quad , \quad B^j_\mu = -\tfrac{i}{2}\,g^{jk}\,\overline{\chi}\gamma_\mu T_k\cdot\chi \ . \tag{2.9}$$

Note that A_μ is a gauge potential, while k_μ and B_μ are gauge covariant ($A_\mu \to h^{-1}A_\mu h + h^{-1}\partial_\mu h$, $k_\mu \to h^{-1}k_\mu h$ and $B_\mu \to h^{-1}B_\mu h$ under gauge transformations (2.1)); all of them are globally invariant (invariant under global symmetry transformations (2.2)). We therefore introduce the gauge field

$$F_{\mu\nu} = \partial_\mu A_\nu - \partial_\nu A_\mu + [A_\mu, A_\nu] \tag{2.10}$$

and the covariant derivatives

$$D_\mu g = \partial_\mu g - g A_\mu \quad , \quad D_\mu D_\nu g = \partial_\mu D_\nu g - D_\nu g \, A_\mu \, , \ \ldots \tag{2.11}$$

$$D_\mu X = \partial_\mu X + A_\mu \cdot X \quad , \tag{2.12}$$

$$D_\mu k_\nu = \partial_\mu k_\nu + [A_\mu, k_\nu] \quad , \tag{2.13}$$

$$D_\mu B_\nu = \partial_\mu B_\nu + [A_\mu, B_\nu] \quad . \tag{2.14}$$

Conjugating the gauge covariant and globally invariant fields k_μ, B_μ and $F_{\mu\nu}$ by means of the bosonic field g, we obtain the gauge invariant and globally covariant composite fields

$$j_\mu = - g k_\mu g^{-1} \tag{2.15}$$

$$j_\mu^M = g B_\mu g^{-1} \tag{2.16}$$

and

$$G_{\mu\nu} = g F_{\mu\nu} g^{-1} \quad . \tag{2.17}$$

Note that

$$k_\mu = g^{-1} D_\mu g \quad , \quad j_\mu = - D_\mu g \, g^{-1} \quad . \tag{2.18}$$

Moreover, as a consequence of the symmetric space structure of M, we have the identities

$$[k_\mu, k_\nu] = - F_{\mu\nu} \quad . \tag{2.19}$$

$$D_\mu k_\nu - D_\nu k_\mu = 0 \quad . \tag{2.20}$$

(This is proved by taking the $\mathfrak{h}$-component and the $\mathfrak{m}$-component of the identity

$$\partial_\mu (g^{-1}\partial_\nu g) - g^{-1}\partial_\mu g \, g^{-1}\partial_\nu g - \partial_\nu (g^{-1}\partial_\mu g) + g^{-1}\partial_\nu g \, g^{-1}\partial_\mu g = 0 \ ,$$

inserting (2.7), and using the commutation relations (2.4) and (2.6).) After conjugation by g, these identities take the form

$$[j_\mu, j_\nu] = -G_{\mu\nu} \,. \tag{2.21}$$

$$\partial_\mu j_\nu - \partial_\nu j_\mu + 2[j_\mu, j_\nu] = 0 \,. \tag{2.22}$$

Having fixed our notation, we can now write down the Lagrangian of our model: it reads

$$L = \frac{1}{2} g^{\mu\nu} (D_\mu q, D_\nu q) + \frac{i}{4} \bar{\chi} \overleftrightarrow{\not{D}} \chi + \frac{a}{2} g^{\mu\nu} (B_\mu, B_\nu) \;, \tag{2.23}$$

where a is a coupling constant, to be determined later. The equations of motion are

$$g^{\mu\nu} (D_\mu k_\nu - D_\mu B_\nu + [B_\mu, k_\nu]) = 0 \,. \tag{2.24}$$

$$\not{D} \chi = a \not{B} \cdot \chi \,. \tag{2.25}$$

The latter implies the following equations of motion for the composite field B_μ:

$$g^{\mu\nu} D_\mu B_\nu = 0 \,. \tag{2.26}$$

$$D_\mu B_\nu - D_\nu B_\mu - 2a [B_\mu, B_\nu] = 0 \,. \tag{2.27}$$

Indeed, in terms of an arbitrary basis of generators $T_j \in \mathfrak{h}$, with $g_{jk} = (T_j, T_k)$, $(g^{jk}) = (g_{jk})^{-1}$, $c_{jkl} = g_{jm} c^m_{kl}$ and $[T_k, T_l] = c^m_{kl} T_m$, we have

$$\begin{aligned}
(g^{\mu\nu} D_\mu B_\nu, T_j) &= -\frac{i}{2} g^{\mu\nu} \overline{D_\mu \chi} \gamma_\nu T_j \cdot \chi - \frac{i}{2} g^{\mu\nu} \bar{\chi} \gamma_\nu T_j \cdot D_\mu \chi \\
&= \frac{ia}{2} g^{\mu\nu} \bar{\chi} \gamma_\nu [B_\mu, T_j] \cdot \chi \\
&= \frac{a}{4} g^{\mu\nu} c_{lkj} g^{km} g^{ln} (\bar{\chi} \gamma_\mu T_m \cdot \chi)(\bar{\chi} \gamma_\nu T_n \cdot \chi) \\
&= 0 \,,
\end{aligned}$$

$$\begin{aligned}
(\varepsilon^{\mu\nu} D_\mu B_\nu, T_j) &= -\frac{i}{2} \overline{D_\mu \chi} \gamma^\mu \gamma_5 T_j \cdot \chi + \frac{i}{2} \bar{\chi} \gamma_5 \gamma^\mu T_j \cdot D_\mu \chi \\
&= \frac{ia}{2} \bar{\chi} \gamma^\mu \gamma_5 [B_\mu, T_j] \cdot \chi \\
&= -a \varepsilon^{\mu\nu} c_{lkj} B^k_\mu B^l_\nu \\
&= a \varepsilon^{\mu\nu} ([B_\mu, B_\nu], T_j) \,,
\end{aligned}$$

where we have used $\varepsilon^{\mu\nu}\gamma_\nu = \gamma^\mu\gamma_5 = -\gamma_5\gamma^\mu$ and the total antisymmetry of the c_{jkl}. After conjugation by g, (2.24) and (2.26) become

$$g^{\mu\nu}(\partial_\mu j_\nu - [j_\mu, j_\nu^M]) = 0 \quad , \tag{2.28}$$

$$g^{\mu\nu}(\partial_\mu j_\nu^M + [j_\mu, j_\nu^M]) = 0 \quad , \tag{2.29}$$

while (2.27) takes the form

$$\partial_\mu j_\nu^M - \partial_\nu j_\mu^M + [j_\mu, j_\nu^M] - [j_\nu, j_\mu^M] - 2a\,[j_\mu^M, j_\nu^M] = 0 \ . \tag{2.30}$$

Finally, we note that the gauge invariant and globally covariant composite field

$$J_\mu = j_\mu + j_\mu^M \tag{2.31}$$

is precisely the Noether current of the theory, corresponding to the global symmetry of the Lagrangian (2.23) under G (cf. (2.2)); this explains the notation j_μ^M, standing for "matter field contribution to the Noether current ". In particular, adding (2.28) and (2.29) shows that the field J_μ is a conserved current, as it must be:

$$g^{\mu\nu}\partial_\mu J_\nu = 0 \ . \tag{2.32}$$

Our main statement about the model defined by the Lagrangian (2.23) is now that if the coupling constant a for the quartic fermion self-interaction term in (2.23) takes the special value

$$a = -\frac{1}{2} \ , \tag{2.33}$$

then the model is integrable. More specifically, this means that given a solution of the equations of motion, there exists a one-parameter family of gauge invariant and globally covariant G-valued fields $U^{(\lambda)}$, determined uniquely up to normalization by the requirement that they satisfy the following linear system of first-order differential equations:

$$\begin{aligned}\partial_\mu U^{(\lambda)} = U^{(\lambda)} \{ &((1 \mp \cosh\lambda)\, j_\mu - \sinh\lambda\ \varepsilon_{\mu\kappa} j^\kappa) \\ &+ \tfrac{1}{2}((1-\cosh 2\lambda)\, j_\mu^M \mp \sinh 2\lambda\ \varepsilon_{\mu\kappa} j^{M\kappa})\}\end{aligned} \tag{2.34}$$

(either sign is possible, so we should really write $U_\pm^{(\lambda)}$). In fact, a rather tedious but straightforward computation shows that the Frobenius integrability conditions for the linear system (2.34) are precisely the equations (2.22), (2.28), (2.29) and (2.30) with (2.33). One of the interesting features of this construction is that the fields $U^{(\lambda)}$ serve as the generating functional for an infinite series of non-local conservation laws. Namely, we can define a one-parameter family of gauge invariant and globally covariant $\mathfrak{g}$-valued fields $J_\mu^{(\lambda)}$ by setting

$$J_\mu^{(\lambda)} = U^{(\lambda)} \{ (\pm \cosh\lambda \; j_\mu + \sinh\lambda \; \varepsilon_{\mu\kappa} j^\kappa) + (\pm \cosh 2\lambda \; j_\mu^M + \sinh 2\lambda \; \varepsilon_{\mu\kappa} j^{M\kappa}) \} U^{(\lambda)-1} \tag{2.35}$$

(either sign is possible, so we should really write $J_{\mu\pm}^{(\lambda)}$). Then as a consequence of (2.22), (2.28), (2.29), (2.30) with (2.33) and (2.34), the fields $J_\mu^{(\lambda)}$ are conserved currents:

$$g^{\mu\nu} \partial_\mu J_\nu^{(\lambda)} = 0 \; . \tag{2.36}$$

Expanding this around $\lambda = 0$ gives an infinite series of $\mathfrak{g}$-valued conservation laws which (except for the very first: $J_\mu^{(\lambda=0)} = J_\mu$) are non-local. In particular, the first non-local charge is given by

$$Q^{(1)} = \int dy_1 \, dy_2 \, \theta(y_1 - y_2) \, [J_0(t,y_1), J_0(t,y_2)] - \int dy \, (J_1 + j_1^M)(t,y) . \tag{2.37}$$

Its conservation (i.e., time-independence) can also be checked directly from (2.32) and the equation

$$\partial_\mu (J_\nu + j_\nu^M) - \partial_\nu (J_\mu + j_\mu^M) + 2 [J_\mu, J_\nu] = 0 \tag{2.38}$$

which follows from combining (2.22) and (2.30) with (2.33).

We conclude by collecting some notation to be used in Section 4. There, we shall assume without further notice that the quotient space M = G/H is an irreducible Riemannian symmetric space of the compact type; in particular, G is a compact semisimple Lie group. Then in the orthogonal Ad(H)-invariant direct decomposition (2.3) of $\mathfrak{g}$, the complementary space $\mathfrak{m}$ is irreducible but the stability algebra $\mathfrak{h}$ may be reducible, and will in general decompose into its centre $\mathfrak{h}_0$ and r simple ideals $\mathfrak{h}_1, \ldots, \mathfrak{h}_r$. (More concretely, we may suppose dim $\mathfrak{h}_0 \leq 1$ and $r \leq 2$ [23, p. 518].) Thus we arrive at a further orthogonal Ad(H)-invariant direct decomposition

$$\mathfrak{h} = \mathfrak{h}_0 \oplus \mathfrak{h}_1 \oplus \ldots \oplus \mathfrak{h}_r \tag{2.39}$$

with commutation relations

$$[\mathfrak{h}_i, \mathfrak{h}_j] = \{0\} \quad \text{for } i \neq j \text{ or } i = 0 = j \;. \tag{2.40}$$

Moreover, all $\mathfrak{h}$-valued fields are further decomposed according to (2.39), i.e.,

$$A_\mu = A_\mu^{(0)} + A_\mu^{(1)} + \ldots + A_\mu^{(r)} \;, \tag{2.41}$$

$$F_{\mu\nu} = F_{\mu\nu}^{(0)} + F_{\mu\nu}^{(1)} + \ldots + F_{\mu\nu}^{(r)} \;, \tag{2.42}$$

$$B_\mu = B_\mu^{(0)} + B_\mu^{(1)} + \ldots + B_\mu^{(r)} \;, \tag{2.43}$$

and in analogy to (2.16) and (2.17), we set

$$j_\mu^{M(i)} = g\, B_\mu^{(i)}\, g^{-1} \;. \tag{2.44}$$

$$G_{\mu\nu}^{(i)} = g\, F_{\mu\nu}^{(i)}\, g^{-1} \;. \tag{2.45}$$

3. - EXAMPLE: THE GRASSMANNIAN MODEL

For the case of the complex Grassmannians, the bosonic sector of the model is specified by taking, e.g., G = U(N) and H = U(p) × U(q) with N = p+q. The Lie algebra $\mathfrak{g}$ is the Lie algebra u(N) of all anti-Hermitian complex (N×N)-matrices, and it carries the standard scalar product given by

$$(Z_1, Z_2) = \frac{1}{2}\,\mathrm{tr}(Z_1^+ Z_2) = -\frac{1}{2}\,\mathrm{tr}(Z_1 Z_2) \quad \text{for } Z_1, Z_2 \in u(N) \;. \tag{3.1}$$

We shall in the following use the block matrix notation

$$Z = \begin{pmatrix} A & B \\ C & D \end{pmatrix} \begin{matrix} \updownarrow p \\ \updownarrow q \end{matrix} \;, \quad \text{(columns: } \overset{\leftrightarrow}{p},\ \overset{\leftrightarrow}{q}\text{)} \tag{3.2}$$

The orthogonal direct decomposition (2.3) is also standard:

$$\mathfrak{h} = \left\{ \begin{pmatrix} A & 0 \\ 0 & D \end{pmatrix} \middle|\, A^+ = -A ,\, D^+ = -D \right\} \;, \quad \mathfrak{m} = \left\{ \begin{pmatrix} 0 & -R^+ \\ R & 0 \end{pmatrix} \right\} . \tag{3.3}$$

It is convenient to decompose the bosonic field g in the form $g = (X,Y)$, where X and Y are matrix fields which have N rows but p and q columns, respectively. The fact that $g^+g = 1_N$, $gg^+ = 1_N$ then expresses itself in subsidiary conditions on X and Y:

$$X^+X = 1_p \ , \quad X^+Y = 0 \ , \qquad XX^+ + YY^+ = 1_N \ . \qquad (3.4)$$
$$Y^+X = 0 \ , \quad Y^+Y = 1_q \ ,$$

The gauge potential becomes

$$A_\mu = \begin{pmatrix} A_\mu^X & 0 \\ 0 & A_\mu^Y \end{pmatrix} \quad \text{with} \quad \begin{array}{l} A_\mu^X = X^+\partial_\mu X \ , \\ A_\mu^Y = Y^+\partial_\mu Y \ , \end{array} \qquad (3.5)$$

and we introduce covariant derivatives

$$D_\mu X = \partial_\mu X - X A_\mu^X \ , \quad D_\mu D_\nu X = \partial_\mu D_\nu X - D_\nu X A_\mu^X \ , \ \ldots$$
$$D_\mu Y = \partial_\mu Y - Y A_\mu^Y \ , \quad D_\mu D_\nu Y = \partial_\mu D_\nu Y - D_\nu Y A_\mu^Y \ , \ \ldots \qquad (3.6)$$

so that $D_\mu g = (D_\mu X, D_\mu Y)$, $D_\mu D_\nu g = (D_\mu D_\nu X, D_\mu D_\nu Y)$, ... Moreover, differentiating the constraints (3.4), we obtain

$$X^+ D_\mu X = 0 \ , \quad X^+ D_\mu Y + D_\mu X^+ Y = 0 \ .$$
$$Y^+ D_\mu X + D_\mu Y^+ X = 0 \ , \quad Y^+ D_\mu Y = 0 \ . \qquad (3.7)$$

Due to (3.4), this means that we can actually solve for $D_\mu Y$ in terms of $D_\mu X^+$ or for $D_\mu X$ in terms of $D_\mu Y^+$:

$$D_\mu Y = - X D_\mu X^+ Y \ , \quad D_\mu X = - Y D_\mu Y^+ X \ . \qquad (3.8)$$

Next, we have

$$k_\mu = \begin{pmatrix} 0 & X^+ D_\mu Y \\ Y^+ D_\mu X & 0 \end{pmatrix} \ . \qquad (3.9)$$

Conjugating by g, and using (3.8), we obtain, according to (2.15),

$$j_\mu = X D_\mu X^+ - D_\mu X X^+ = Y D_\mu Y^+ - D_\mu Y Y^+ \ . \qquad (3.10)$$

Similarly, it follows from (3.1) and (3.9) that

$$(k_\mu, k_\nu) = \mathrm{Re}\,\mathrm{tr}(D_\mu X^+ D_\nu X) = \mathrm{Re}\,\mathrm{tr}(D_\mu Y^+ D_\nu Y) \quad . \qquad (3.11)$$

In order to fix the fermionic sector of the model, we have to specify the fermion representation of the stability group $H = U(p) \times U(q)$ and can then compute the composite field

$$B_\mu = \begin{pmatrix} B_\mu^X & 0 \\ 0 & B_\mu^Y \end{pmatrix} \qquad (3.12)$$

from the relation

$$\mathrm{tr}(B_\mu^X A) + \mathrm{tr}(B_\mu^Y D) = \mathrm{tr}(B_\mu T) = -2(B_\mu, T) = i\bar{\chi}\gamma_\mu T\cdot\chi \qquad (3.13)$$

which must hold for any generator

$$T = \begin{pmatrix} A & 0 \\ 0 & D \end{pmatrix} \qquad (3.14)$$

in $\mathfrak{h}$. We shall here concentrate on two particular possibilities which, for reasons to be commented upon later, are known as the minimal model and the supersymmetric model, respectively.

A. - Minimal Model

The representation space is $V_0 = \mathbb{C}^p$, on which $U(p)$ acts by the fundamental representation (or rather its dual) and $U(q)$ acts trivially. More specifically, the fermionic field ψ (we shall write ψ instead of χ) can be viewed as a $(1\times N)$-matrix (row vector) of Dirac spinor fields. In particular, the gauge transformation law (2.1), in this representation, takes the form

$$g \to gh \quad , \quad \psi \to \psi h^X \quad , \qquad (3.15)$$

where the field h is block diagonal,

$$h = \begin{pmatrix} h^X & 0 \\ 0 & h^Y \end{pmatrix} \quad , \qquad (3.16)$$

and so the appropriate covariant derivative becomes

$$D_\mu \psi = \partial_\mu \psi - \psi A_\mu^X \quad . \qquad (3.17)$$

For the remainder of this subsection, we write $\psi = (\psi^a)$, where indices a (or b) run from 1 to p, while indices i (or j or k) run from 1 to N. Then from (3.13) and (3.14), we have, in this representation

$$\mathrm{tr}(B_\mu^X A) + \mathrm{tr}(B_\mu^Y D) = -i\bar{\psi}^a \gamma_\mu \psi^b A^{ba} .$$

This shows that

$$(B_\mu^X)^{ab} = -i\bar{\psi}^a \gamma_\mu \psi^b .$$
$$B_\mu^Y = 0 . \tag{3.18}$$

Conjugating by g, we obtain, according to (2.16),

$$(j_\mu^M)_{ij} = -i\, X_i^a\, \bar{\psi}^a \gamma_\mu \psi^b\, \bar{X}_j^b . \tag{3.19}$$

Similarly, it follows from (3.1) and (3.12) that

$$(B_\mu, B_\nu) = \frac{1}{2} (\bar{\psi}^a \gamma_\mu \psi^b)(\bar{\psi}^b \gamma_\nu \psi^a) . \tag{3.20}$$

Finally, the kinetic + minimal coupling term for the fermions in the Lagrangian (2.23) takes the form

$$\bar{\chi} \overleftrightarrow{\not{\partial}} \chi = \bar{\psi}^a \overleftrightarrow{\not{\partial}} \psi^a . \tag{3.21}$$

Hence in this representation, the Lagrangian (2.23) is (apart from a total factor of 2) precisely the Lagrangian for the minimal model written down in Ref. [21]. There, it is also explained in what sense this model is minimal.

B. - Supersymmetric Model

The representation space is $V_0 = \mathfrak{m}$, on which U(p) × U(q) acts by conjugation. More specifically, the fermionic field χ can be viewed as an (N×N)-matrix of Dirac spinor fields, subject to constraints as expressed by (3.3). In particular, the gauge transformation law (2.1), in this representation, takes the form

$$g \to gh \quad , \qquad \chi \to h^{-1} \chi h \quad , \tag{3.22}$$

so that the appropriate covariant derivative becomes

$$D_\mu \chi = \partial_\mu \chi + [A_\mu, \chi] . \tag{3.23}$$

It is however more convenient to work with a shifted fermionic field ψ which is again an (N×N)-matrix of Dirac spinor fields, defined as follows:

$$\psi = g\chi \ . \tag{3.24}$$

From (3.24), (2.1) -- or rather (3.22) -- and (2.2), it is obvious that ψ and g have the same transformation law both under gauge transformations and under global symmetry transformations, and so the appropriate covariant derivative becomes, as in (2.11),

$$D_\mu \psi = \partial_\mu \psi - \psi A_\mu \ . \tag{3.25}$$

Then obviously,

$$D_\mu \psi = D_\mu g \, \chi + g \, D_\mu \chi \ . \tag{3.26}$$

Once again, it is convenient to decompose the fermionic field ψ in the form $\psi = (\psi^X, \psi^Y)$, where ψ^X and ψ^Y are matrices of Dirac spinor fields which have N rows but p and q columns, respectively. The fact that

$$\chi = g^+ \psi = \begin{pmatrix} X^+\psi^X & X^+\psi^Y \\ Y^+\psi^X & Y^+\psi^Y \end{pmatrix} \tag{3.27}$$

lies in $\mathfrak{m}$ then expresses itself in subsidiary conditions on ψ^X and ψ^Y:

$$\begin{aligned} X^+\psi^X = 0 \quad &, \quad X^+\psi^Y + \psi^{X+}Y = 0 \ . \\ Y^+\psi^X + \psi^{Y+}X = 0 \quad &, \quad Y^+\psi^Y = 0 \ . \end{aligned} \tag{3.28}$$

Due to (3.4), this means that we can in fact solve for ψ^Y in terms of ψ^{X+} or for ψ^X in terms of ψ^{Y+}:

$$\psi^Y = -X\psi^{X+}Y \quad , \quad \psi^X = -Y\psi^{Y+}X \ . \tag{3.29}$$

Moreover, we introduce covariant derivatives

$$\begin{aligned} D_\mu \psi^X &= \partial_\mu \psi^X - \psi^X A_\mu^X \ , \\ D_\mu \psi^Y &= \partial_\mu \psi^Y - \psi^Y A_\mu^Y \ , \end{aligned} \tag{3.30}$$

so that $D_\mu \psi = (D_\mu \psi^X, D_\mu \psi^Y)$. For the remainder of this subsection, we write $\psi^X = ((\psi^X)_i^a)$ and $\psi^Y = ((\psi^Y)_i^c)$, where indices a (or b) run from 1 to p and indices c (or d) run from 1 to q, while indices i (or j or k) run from 1 to N. Then from (3.13) and (3.14), we have, in this representation,

$$\mathrm{tr}(B_\mu^X A) + \mathrm{tr}(B_\mu^Y D) = \frac{i}{2}(\gamma_0\gamma_\mu)_{\alpha\beta}\, \mathrm{tr}(X_\alpha^+[T,X_\beta])$$

$$= \frac{i}{2}(\gamma_0\gamma_\mu)_{\alpha\beta}\,\mathrm{tr}\left(\begin{pmatrix} 0 & \psi_\alpha^{X+}Y \\ \psi_\alpha^{Y+}X & 0\end{pmatrix}\begin{pmatrix} 0 & AX^+\psi_\beta^Y - X^+\psi_\beta^Y D \\ DY^+\psi_\beta^X - Y^+\psi_\beta^X A & 0\end{pmatrix}\right)$$

$$= -\frac{i}{2}(\gamma_0\gamma_\mu)_{\alpha\beta}\{\mathrm{tr}(Y^+\psi_\beta^X\psi_\alpha^{X+}YD + \psi_\alpha^{X+}YY^+\psi_\beta^X A) + \mathrm{tr}(X^+\psi_\beta^Y\psi_\alpha^{Y+}XA + \psi_\alpha^{Y+}XX^+\psi_\beta^Y D)\}$$

$$= -i(\gamma_0\gamma_\mu)_{\alpha\beta}\{\mathrm{tr}(\psi_\alpha^{X+}\psi_\beta^X A) + \mathrm{tr}(\psi_\alpha^{Y+}\psi_\beta^Y D)\},$$

where in the third respectively fourth equality, we have used that the components of ψ are anticommuting c-numbers respectively that $\gamma_0\gamma_\mu$ is a symmetric (2×2)-matrix. This shows that

$$(B_\mu^X)^{ab} = -i\,\overline{(\psi^X)_i^a}\,\gamma_\mu(\psi^X)_i^b\ .$$
$$(B_\mu^Y)^{cd} = -i\,\overline{(\psi^Y)_i^c}\,\gamma_\mu(\psi^Y)_i^d\ . \tag{3.31}$$

Conjugating by g, and using (3.29), we obtain, according to (2.16),

$$(j_\mu^M)_{ij} = -iX_i^a\,\overline{(\psi^X)_k^a}\,\gamma_\mu(\psi^X)_k^b\,\overline{X}_j^b + i\,\overline{(\psi^X)_j^a}\,\gamma_\mu(\psi^X)_i^a$$
$$= +i\,\overline{(\psi^Y)_j^c}\,\gamma_\mu(\psi^Y)_i^c - iY_i^c\,\overline{(\psi^Y)_k^c}\,\gamma_\mu(\psi^Y)_k^d\,\overline{Y}_j^d\ . \tag{3.32}$$

Similarly, it follows from (3.1) and (3.12) that

$$(B_\mu,B_\nu) = \frac{1}{2}((\overline{(\psi^X)_i^a}\gamma_\mu(\psi^X)_i^b)(\overline{(\psi^X)_j^b}\gamma_\nu(\psi^X)_j^a)) + \frac{1}{2}((\overline{(\psi^X)_i^a}\gamma_\mu(\psi^X)_j^a)(\overline{(\psi^X)_j^b}\gamma_\nu(\psi^X)_i^b))$$
$$- \frac{1}{2}((\overline{(\psi^Y)_i^c}\gamma_\mu(\psi^Y)_i^d)(\overline{(\psi^Y)_j^d}\gamma_\nu(\psi^Y)_j^c)) + \frac{1}{2}((\overline{(\psi^Y)_i^c}\gamma_\mu(\psi^Y)_j^c)(\overline{(\psi^Y)_j^d}\gamma_\nu(\psi^Y)_i^d)) \tag{3.33}$$

Finally, we compute

$$\overline{X}\overleftrightarrow{\not{B}}X = \frac{1}{2}g^{\mu\nu}(\gamma_0\gamma_\mu)_{\alpha\beta}\,\mathrm{tr}\Big(\begin{pmatrix} 0 & \psi_\alpha^{X+}Y \\ \psi_\alpha^{Y+}X & 0\end{pmatrix}\begin{pmatrix} 0 & D_\nu X^+\psi_\beta^Y + X^+D_\nu\psi_\beta^Y \\ D_\nu Y^+\psi_\beta^X + Y^+D_\nu\psi_\beta^X & 0\end{pmatrix}$$
$$-\begin{pmatrix} 0 & D_\nu\psi_\alpha^{X+}Y + \psi_\alpha^{X+}D_\nu Y \\ D_\nu\psi_\alpha^{Y+}X + \psi_\alpha^{Y+}D_\nu X & 0\end{pmatrix}\begin{pmatrix} 0 & X^+\psi_\beta^Y \\ Y^+\psi_\beta^X & 0\end{pmatrix}\Big)$$

$$= \frac{1}{2}g^{\mu\nu}(\gamma_0\gamma_\mu)_{\alpha\beta}\{\mathrm{tr}(\psi_\alpha^{X+}D_\nu\psi_\beta^X - D_\nu\psi_\alpha^{X+}\psi_\beta^X) + \mathrm{tr}(\psi_\alpha^{Y+}D_\nu\psi_\beta^Y - D_\nu\psi_\alpha^{Y+}\psi_\beta^Y)\},$$

and

$$
\begin{aligned}
(\gamma_0\gamma_\mu)_{\alpha\beta}\, tr(\psi_\alpha^{Y+} D_\nu \psi_\beta^Y - D_\nu\psi_\alpha^{Y+}\psi_\beta^Y)
&= (\gamma_0\gamma_\mu)_{\alpha\beta}\, tr(Y^+\psi_\alpha^X X^+ D_\nu(X\psi_\beta^{X+}Y) - D_\nu(Y^+\psi_\alpha^X X^+)X\psi_\beta^{X+}Y) \\
&= (\gamma_0\gamma_\mu)_{\alpha\beta}\, tr(Y^+\psi_\alpha^X D_\nu\psi_\beta^{X+}Y - Y^+D_\nu\psi_\alpha^X\psi_\beta^{X+}Y) \\
&= (\gamma_0\gamma_\mu)_{\alpha\beta}\, tr(\psi_\alpha^X D_\nu\psi_\beta^{X+} - D_\nu\psi_\alpha^X\psi_\beta^{X+}) \\
&= (\gamma_0\gamma_\mu)_{\alpha\beta}\, tr(\psi_\alpha^{X+} D_\nu\psi_\beta^X - D_\nu\psi_\alpha^{X+}\psi_\beta^X) \ ,
\end{aligned}
$$

where in the last equality, we have used that the components of ψ are anticommuting c-numbers and that $\gamma_0\gamma_\mu$ is a symmetric (2×2)-matrix; moreover, we have repeatedly used the fact that

$$
\begin{aligned}
D_\mu Y^+\psi^X &= 0 \quad , \quad D_\mu X^+\psi^Y = 0 \ , \\
\psi^{X+} D_\mu Y &= 0 \quad , \quad \psi^{Y+} D_\mu X = 0 \ ,
\end{aligned}
\tag{3.34}
$$

which follows from (3.7) and (3.28) by inserting $1_N = XX^+ + YY^+$ in the middle. This shows that the kinetic + minimal coupling term for the fermions in the Lagrangian (2.23) takes the form

$$
\bar{\chi}\overleftrightarrow{\not{D}}\chi = \overline{(\psi^X)_i^a}\,\overleftrightarrow{\not{D}}(\psi^X)_i^a = \overline{(\psi^Y)_i^c}\,\overleftrightarrow{\not{D}}(\psi^Y)_i^c \ . \tag{3.35}
$$

Hence in this representation, the Lagrangian (2.23) is (apart from a total factor of 2) precisely the Lagrangian for the supersymmetric model written down in Ref. [21]. That this is indeed the supersymmetric model is well known (see, e.g., Ref. [24]); for a proof in the more general context of Section 2, see Ref. [19].

4. - QUANTIZATION AND ANOMALIES

When the transition from classical to quantum non-linear σ models is performed, the algebraic structure of the model may change, and has to be re-examined. This happens because the fields are now operator-valued distributions, and products of operators at the same point are therefore a priori ill-defined. The standard solution of this problem is to introduce normal products $\mathcal{N}[\ldots]$, e.g., by first splitting points and then subtracting appropriate terms in such a way as to cancel all the divergences that occur when the points are made to recoalesce. One of the things that has to be checked, then, is whether this procedure of subtracting infinities does or does not preserve a given symmetry of the model: if not, this shows up as an anomaly in the conservation law (Ward identity) for the corresponding Noether current.

A related question that arises, in view of the integrability properties of the classical non-linear σ models discussed in Sections 2 and 3, is whether, and in what sense, the quantum non-linear σ models are also integrable. Namely, one may wonder whether the classical non-local symmetry, expressed through Eqs. (2.34)-(2.36), persists at the quantum level. One feature of such a quantum integrability would be the existence of a conserved quantum non-local charge, which should be the quantum counterpart of the conserved classical non-local charge defined in (2.37). To find such a counterpart, note first that at the quantum level, the classical expression (2.37) itself is ill-defined, due to the short-distance ($y_1-y_2 \to 0$) singularity in the (matrix) commutator $[J_0(t,y_1),J_0(t,y_2)]$. In fact, it is the precise nature of this singularity that provides the key to a correct definition of the quantum non-local charge, and we must therefore analyze the Wilson short-distance expansion for the (matrix) commutator of two currents at nearby (spacelike separated) points:

$$[J_\mu(x+\varepsilon), J_\nu(x-\varepsilon)] \sim \sum_k C_{\mu\nu}^{(k)}(\varepsilon)\, \mathcal{N}[\mathcal{O}_k(x)] \qquad (\varepsilon^2 < 0)\,. \tag{4.1}$$

Here, $\sim$ means equality up to terms that go to zero as $\varepsilon \to 0$, and k labels a complete set of independent composite local operators of (canonical) dimension $\leqslant 2$ which take values in $\mathfrak{g}$ and are both gauge invariant and globally covariant. The standard examples for composite operators of this type are

$$\begin{aligned} &\text{in dimension 1:} && \dot{j}_\mu\,,\quad \dot{j}_\mu^{M(i)}\,.\\ &\text{in dimension 2:} && \partial_\mu \dot{j}_\nu\,,\quad \partial_\mu \dot{j}_\nu^{M(i)}\,,\\ & && G_{\mu\nu}^{(i)}\,,\quad \mathcal{N}[[\dot{j}_\mu,\dot{j}_\nu^{M(i)}]]\,,\quad \mathcal{N}[[\dot{j}_\mu^{M(i)},\dot{j}_\nu^{M(i)}]]\,. \end{aligned} \tag{4.2}$$

In general, depending on the relation between the various representations of the stability group H involved (as expressed through the existence or non-existence of appropriate intertwining operators), there may or may not exist other composite operators of this type, but for the following we shall assume that this is not the case. In particular, this means that the only composite local operators of dimension 1 appearing on the right-hand side of (4.1) are the contributions j_μ and $j_\mu^{M(i)}$ to the current J_μ, which implies that the transition from the classical to the quantum non-local charge can be achieved by a simple renormalization of the second term in (2.37). More specifically, the Wilson expansion (4.1) takes the form

$$[J_\mu(x+\varepsilon), J_\nu(x-\varepsilon)] \sim C^{\rho}_{\mu\nu}(\varepsilon)\, J_\rho(x) + \sum_{i=0}^{r} \hat{C}^{(i)\rho}{}_{\mu\nu}(\varepsilon)\, j_\rho^{M(i)}(x)$$
$$+ D^{\sigma\rho}_{\mu\nu}(\varepsilon)\, (\partial_\sigma J_\rho)(x) + \sum_{i=0}^{r} \hat{D}^{(i)\sigma\rho}{}_{\mu\nu}(\varepsilon)\, (\partial_\sigma j_\rho^{M(i)})(x)$$
$$+ \sum_{i=0}^{r} E^{(i)}\, \varepsilon_{\mu\nu}\, \varepsilon^{\sigma\rho}\, G^{(i)}_{\sigma\rho}(x) \tag{4.3}$$
$$+ \sum_{i=0}^{r} F^{(i)}\, \varepsilon_{\mu\nu}\, \varepsilon^{\sigma\rho}\, \mathcal{N}[[j_\sigma, j_\rho^{M(i)}](x)]$$
$$+ \sum_{i=0}^{r} G^{(i)}\, \varepsilon_{\mu\nu}\, \varepsilon^{\sigma\rho}\, \mathcal{N}[[j_\sigma^{M(i)}, j_\rho^{M(i)}](x)] \qquad (\varepsilon^2 < 0),$$

and the charge $Q^{(1)}$ is defined as the limit

$$Q^{(1)} = \lim_{\delta \to 0} Q^{(1)}_\delta \tag{4.4}$$

of a cut-off charge $Q^{(1)}_\delta$, which reads

$$Q^{(1)}_\delta = \int_{|y_1 - y_2| \geq \delta} dy_1\, dy_2\, \theta(y_1 - y_2)\, [J_0(t,y_1), J_0(t,y_2)]$$
$$- Z(\delta) \int dy\, J_1(t,y) - \sum_{i=0}^{r} \hat{Z}^{(i)}(\delta) \int dy\, j_1^{M(i)}(t,y)\,. \tag{4.5}$$

Moreover, general principles of field theory imply that the coefficients $C^{\rho}_{\mu\nu}(\varepsilon)$, $D^{\sigma\rho}_{\mu\nu}(\varepsilon)$, $Z(\delta)$ respectively $\hat{C}^{(i)\rho}{}_{\mu\nu}(\varepsilon)$, $\hat{D}^{(i)\sigma\rho}{}_{\mu\nu}(\varepsilon)$, $\hat{Z}^{(i)}(\delta)$ appearing in (4.3) and (4.5) can all be expressed in terms of a function $D_1(-\varepsilon^2)$ respectively $\hat{D}_1^{(i)}(-\varepsilon^2)$ which, just like $Z(\delta)$ resp. $\hat{Z}^{(i)}(\delta)$, has a logarithmic singularity at the origin. Finally, it can be shown that if the remaining coefficients $E^{(i)}$, $F^{(i)}$, $G^{(i)}$ happen to vanish, then the charge $Q^{(1)}$, as defined in (4.4) and (4.5), is conserved. For more details, we refer the reader to Ref. [19].

In the case of the complex Grassmannian models and, in particular, the $\mathbb{C}P^{N-1}$ models, more information is available. Namely, an explicit group-theoretical analysis of the representations involved shows that the composite local operators listed in (4.2) are indeed the only ones that may appear on the right-hand side of (4.1). Moreover, and this is certainly a most remarkable feature, the coefficients $E^{(i)}$, $F^{(i)}$, $G^{(i)}$ can all be calculated perturbatively within the 1/N-expansion, and turn out to vanish to all orders! In other words, both the minimal model and the supersymmetric model admit a conserved quantum non-local charge, and as a result, particle production is suppressed, so that the S-matrix factorizes into two-body amplitudes which, at least in the $\mathbb{C}P^{N-1}$ case, can be calculated exactly. This is in sharp contrast with the situation in the pure model, where the quantum non-local charge develops an anomaly, and conclusions about the S-matrix cannot be drawn. For more information on this subject, we refer the reader to Ref. [21].

5. - OPEN PROBLEMS

To conclude, let me briefly resume the present state of the art and comment on further prospects. First of all, I hope to have convinced the reader that the issue of classical integrability for non-linear σ models can be regarded as completely settled -- both for the purely bosonic models and for the models with fermionic matter fields. Second, however, I believe that the origin of quantum integrability is not really understood. It is true that for the purely bosonic models, there exists a simple and general criterion for deciding whether they are quantum integrable [18], but for the models with fermionic matter fields, no such criterion has been found, and it seems very likely that the method employed so far (short-distance expansions for (matrix) commutators of currents) is in fact unable to solve the problem. A more promising approach would be to realize that since quantum integrability can be viewed as absence of a certain anomaly, it should find a natural, non-perturbative explanation in terms of an appropriate index theorem. At the moment, this is just a conjecture, motivated by the complete analogy between our anomaly and the standard axial anomaly. Namely, in the case of the complex Grassmannian models, explicit calculations within the 1/N-expansion show that both anomalies satisfy a non-renormalization theorem à la Adler-Bardeen: the non-zero contributions come from lowest order in 1/N, and suffer no radiative corrections. But to my knowledge, the only raison d'être for such a non-renormalization theorem is that the coefficient of the integrated anomaly is controlled by an index theorem. The new aspect here is that although the two anomalies are identical, and in particular, local, the anomalous currents themselves are very different: our current is non-local while the axial current is of course local. This might mean that the index problem to be formulated involves a (non-local) pseudo-differential operator, rather than a (local) differential operator such as the Dirac operator.

From a physical point of view, the systematic mechanism behind quantum integrability of non-linear σ models is also still somewhat mysterious. In the case of the complex Grassmannian models, it is clearly related to confinement. In fact, and this is a general feature, the model contains a gauge potential A_μ which will in general become dynamical and hence give rise to a confining long-range force between the partons (fundamental bosonic quanta): only gauge invariant composite fields are physical. In the presence of fermions, however, vacuum polarization screens this long-range force, which in certain special circumstances will disappear altogether (i.e., A_μ ceases to be a propagating field). It is just this total screening effect that takes place in the minimal and supersymmetric Grassmannian models, leading to deconfinement of the partons and factorizability of the corresponding S-matrix. A systematic and non-perturbative understanding of exactly when and how total screening does take place is however an open question. Hopefully, the "fermionization" procedure of Polyakov and Wiegmann [25], suitably generalized, will help to gain more insight into all these problems.

REFERENCES

1) M. Gell-Mann and M. Lévy, Nuovo Cim. 16 (1960) 705-726.

2) S. Weinberg, Phys. Rev. 166 (1968) 1568-1577.

3) B.W. Lee, "Chiral Dynamics", Gordon and Breach, New York (1972).

4) W. Buchmüller, S.T. Love, R.D. Peccei and T. Yanagida, Phys. Lett. 115B (1982) 233-236;
W. Buchmüller, R.D. Peccei and T. Yanagida, Phys. Lett. 124B (1983) 67-73;
W. Buchmüller, R.D. Peccei and T. Yanagida, Nucl. Phys. B227 (1983) 503-546; B231 (1984) 53-64; B244 (1984) 186-206.

5) E. Cremmer and B. Julia, Nucl. Phys. B159 (1979) 141-212.

6) M. Forger, "Instantons in Non-Linear σ Models, Gauge Theories and General Relativity", in: Differential Geometrical Methods in Mathematical Physics, Proceedings, Clausthal, Germany 1978, ed. by H.D. Doebner, Lecture Notes in Physics, Vol. 139 (Springer Verlag, Berlin-Heidelberg-New-York, 1981).

7) M. Forger, "Differential Geometric Methods in Non-Linear σ Models and Gauge Theories", Ph.D. Thesis, Freie Universität Berlin, unpublished (1980).

8) M. Dubois-Violette and Y. Georgelin, Phys. Lett. 82B (1979) 251-254;
A.M. Polyakov, Phys. Lett. 82B (1979) 247-250.

9) A.M. Polyakov, Phys. Lett. 59B (1975) 79-81;
E. Brézin, S. Hikami and J. Zinn-Justin, Nucl. Phys. B165 (1980) 528-544.

10) D.J. Gross and F. Wilczek, Phys. Rev. Lett. 30 (1973) 1343-1346;
H.D. Politzer, Phys. Rev. Lett. 30 (1973) 1346-1349.

11) A. D'Adda, P. Di Vecchia and M. Lüscher, Nucl. Phys. B146 (1978) 63-76; B152 (1979) 125-144.

12) M. Lüscher, Nucl. Phys. B135 (1978) 1-19.

13) A.B. Zamolodchikov and Al.B. Zamolodchikov, Nucl. Phys. B133 (1978) 525-535.

14) K. Pohlmeyer, Commun. Math. Phys. 46 (1976) 207-221;
M. Lüscher and K. Pohlmeyer, Nucl. Phys. B137 (1978) 46-54;
V.E. Zakharov and A.V. Mikhailov, Sov. Phys. JETP 47 (1978) 1017-1027.

15) H. Eichenherr and M. Forger, Nucl. Phys. B155 (1979) 381-393; B164 (1980) 528-535; Commun. Math. Phys. 82 (1981) 227-255.

16) H. Eichenherr, "Geometric Analysis of Integrable Non-Linear σ Models", in: Integrable Quantum Field Theories, Proceedings, Tvärminne, Finland 1981, ed. by J. Hietarinta and C. Montonen, Lecture Notes in Physics, Vol. 151 (Springer Verlag, Berlin-Heidelberg-New York, 1982).

17) M. Forger, "Non-linear σ Models on Symmetric Spaces", in: Non-Linear Partial Differential Operators and Quantization Procedures, Proceedings, Clausthal, Germany 1981, ed. by S.I. Andersson and H.D. Doebner, Lecture Notes in Mathematics, Vol. 1037 (Springer Verlag, Berlin-Heidelberg-New York, 1983).

18) E. Abdalla, M. Forger and M. Gomes, Nucl. Phys. B210 [FS6] (1982) 181-192.

19) E. Abdalla and M. Forger, "Integrable Non-Linear σ Models with Fermions, CERN preprint TH.4238/85 (1985), to appear in Commun. Math. Phys.

20) E. Abdalla, M.C.B. Abdalla and M. Gomes, Phys. Rev. D27 (1983) 825-836.

21) E. Abdalla, M. Forger and A. Lima-Santos, Nucl. Phys. B256 (1985) 145-180.

22) E. Abdalla, "Non-Linear σ Models. A Geometrical Approach to Quantum Field Theory", in: Non-linear Equations in Classical and Quantum Field Theory, Proceedings, Meudon-Paris, France 1983/84, ed. by N. Sanchez, Lecture Notes in Physics, Vol. 226 (Springer Verlag, Berlin-Heidelberg-New York, 1985).

23) S. Helgason, "Differential Geometry, Lie Groups and Symmetric Spaces", (Academic Press, New York, 1978).

24) S. Aoyama, Nuovo Cim. 57A (1980) 176-184.

25) A.M. Polyakov and P.B. Wiegmann, Phys. Lett. 131B (1983) 121-126;
P.B. Wiegmann, Phys. Lett. 141B (1984) 217-222; 152B (1985) 209-214.

INVERSE SCATTERING TRANSFORM IN ANGULAR MOMENTUM AND APPLICATIONS TO NON-LOCAL EFFECTIVE ACTIONS

J. Avan

L.P.T.H.E. Jussieu

Tour 16, 1er étage

4 place Jussieu

75230 PARIS CEDEX 05

I. INTRODUCTION

The problem of finding saddle-points of non-local actions, and more generally of trying to deal with non-local effective field theories, often arises in quantum theory. The one-loop effective action which contains a functional determinant depending on the classical field, describing quantum corrections to the energy of a classical configuration at first order in $\hbar$, is amongst the best-known and oldest [1] examples of non-local actions. We shall here concentrate on these "one-loop effective actions" which we choose to define more generally as a functional of classical fields, consisting of

a) a non-local term, which is the functional determinant of a differential operator containing the fields as "potentials" of a linear problem;

b) a local term, polynomial in the fields and their derivatives.

"One-loop effective actions" often appear in the formalism of functional integral, after one has integrated over bosonic or fermionic fields on which the classical action depends quadratically. This can follow from introducing auxiliary variables, as it is the case when one formulates 1/N expansions of field theories:

1) 1/N expansion of $(\Phi^2)^2$ in any number of dimensions [2] $\nu \leq 4$ is obtained by a Hubbard-Stratanovitch transformation:

$$\exp \int_{-\infty}^{+\infty} (\vec{\Phi}^2)^2 \, d^\nu x = \iint \mathcal{D}v \; \exp - \int_{-\infty}^{+\infty} (v^2 - 2v.\vec{\Phi}^2) \, d^\nu x \quad (1.1)$$

Once it is applied to the classical lagrangian, this transformation makes it into a quadratic function of -variables, and Gaussian integration leads to the field theory described by the following path integral:

$$\mathcal{Z}(J) = \iint \mathcal{D}v \; \exp\left\{-\frac{N}{2} S_{eff} - \int J^a . \langle x| \frac{1}{\partial^2 + m^2 + v(.)} |x\rangle J^b.\right\} \quad (1.2)$$

where $$S_{eff} = \mathrm{Log\ det} \left(\frac{-\partial^2 + m^2 + v}{-\partial^2 + m^2} \right) + \frac{m^{\nu-2}}{(4\pi)^{\nu/2}} \Gamma(1 - \nu/2) \int_{-\infty}^{+\infty} (v(x). \, d^\nu x) - \frac{m^{\nu-4}}{8 g_B} \int_{-\infty}^{+\infty} v^2(x) \, d^\nu x \quad (1.3)$$

The action that generates 1/N expansion of the theory containing the auxiliary field v is a "one-loop effective action".

2) 1/N expansion of 2-dimensional fermionic models is also obtained by introducing one or two auxiliary field(s), to replace quadratic expressions ($\bar{\psi}\psi$) or ($\bar{\psi}\gamma_5\psi$) [3]. In this way, one gets the 1/N expansion for the Gross-Neveu model [14]

$$\mathcal{L} = \bar{\Psi}_a . \not{\partial} . \Psi_a + g_B (\bar{\psi}_a \psi_a)^2 \quad (1.4)$$

or the Chiral-Gross-Neveu model

$$\mathcal{L} = \bar{\Psi}_a \not{\partial} . \Psi_a + g_B . \left\{ (\bar{\psi}_a . \psi_a)^2 - (\bar{\psi}_a . \gamma_5 . \psi_a)^2 \right\} \quad (1.5)$$

or any quartic-coupled scalar + pseudoscalar model.

$$\mathcal{L} = \bar{\Psi}_a . \not{\partial} . \Psi_a + g_S (\bar{\psi}_a . \psi_a)^2 + g_P (\bar{\psi}_a . \gamma_5 . \psi_a)^2 \quad (1.6)$$

One has to replace the quartic self-couplings by quadratic couplings to an auxiliary field, as in (1.1):

$$(\bar{\Psi}\Psi)^2 \longrightarrow \sigma^2 - \sigma\bar{\Psi}\Psi \tag{1.7a}$$

$$(\bar{\Psi}.\gamma_5.\Psi)^2 \longrightarrow w^2 - \bar{\Psi}\gamma_5 w \Psi \tag{1.7b}$$

The generating functional for 1/N expansion of such models reads, after gaussian-integration over the fermions:

$$\mathcal{Z}(J) = \int \mathcal{D}\sigma\,\mathcal{D}w \,.\, \exp\left\{- N.S_{eff} - \int J_a \,.\, \langle x | \frac{1}{-\partial\!\!\!/ + m + \sigma + i\gamma_5 w} | x\rangle J_a\right\} \tag{1.8}$$

J^a are the fermionic sources; S_{eff} reads:

$$. \; S_{eff} = \ln\det\left(\frac{-\partial\!\!\!/ + m + \sigma + i\gamma_5 w}{-\partial\!\!\!/ + m}\right) + \frac{2^{\nu/2-1}}{(4\pi)^{\nu/2}}\Gamma\left(1 - \nu/2\right)\int_{-\infty}^{+\infty}(\sigma^2 + 2m\sigma + w^2)\,d^\nu x$$

$$+ \, 4/g_R \int_{-\infty}^{+\infty} w^2 . \, d^\nu x \tag{1.9}$$

. $W \equiv 0$ for Gross-Neveu model

. $1/g_R \equiv 0$ for CGN-model

One gets again the 1-loop effective action structure. This also occurs when one formulates 1/N expansion of O(N) nonlinear sigma model and of CPN model [4,5].

One also gets this structure after integrating over the fermions in a theory containing matter fields (spinors) coupled to boson fields. We shall mention the effective action of Yukawa model in 4 dimensions [6]:

$$\mathcal{L} = \bar{\Psi}.\partial\!\!\!/.\Psi + m.\bar{\Psi}.\Psi + g.\bar{\Psi}(\phi_1 + i\gamma_5\phi_2)\Psi - \frac{1}{2}\Big\{$$

$$(\partial\Phi_1)^2 + (\partial\Phi_2)^2 + m_1^2\phi_1^2 + m_2^2\phi_2^2 + \lambda(\phi_1^2 + \phi_2^2)^2 \tag{1.10}$$

which generates a 1-loop effective action in ϕ_1 and ϕ_2 after integration over the Grassmann variables.

In general, computation of functional integrals which are <u>not</u> quadratic requires the use of saddle-point method. In particular, the study of non-constant, localized, finite-action euclidean saddle-points (instantons) is very important, to determine the behaviour of large orders of perturbative expansion [7] and the stability or instability (generated by tunnel effect) of the perturbative vacuum [8]. One therefore need to solve equations of the type:

$$\langle x| \frac{1}{\partial^2 + m^2 + v} |x\rangle = c_1 + c_2 \; v(x)$$

(c_1 , c_2 are numerical constants) (1.11a)

or (when Fermions are involved)

$$tr \left\{ \langle x| \frac{1}{\not\partial + m + v + i\gamma_5 w} |x\rangle - \langle x| \frac{1}{\not\partial + m} |x\rangle \right. = \text{Polynomial function of}$$

(v, w, $\partial_x v$, $\partial_x w$) (1.11b)

Such non-local and non-polynomial equations are of course very difficult to solve. More generally, one would like to be able to deal with these 1-loop effective actions, without worrying anymore about their non-locality. It is clear that one has to introduce auxiliary variables, in one-to-one correspondance with the fields, such that the action becomes local when expressed in term of those new variables. This non-trivial change of variables is provided by the inverse scattering transform. Here we present a brief exposition of the methods and results for 1-loop effective actions. A detailed account can be found in the original references [2, 3, 4, 6, 17, 30].

II. PRINCIPLES AND TECHNIQUES OF I.S.T.

We shall associate the linear differential operator Θ inside (ln det) (eqs. (1.3), (1.9)) with a diffusion problem $\Theta\Psi = 0$. It can be shown on general arguments [9] that the functional determinant of the differential operator Θ can be expressed in a <u>local form</u> as a function of suitable scattering data. However such a relation requires until now that the fields (considered as potentials in this diffusion problem) depend on one single variable. We shall be mainly interested in spherically symmetric solutions of the gap equations (1.11), since they can reasonably be assumed to be least-action configuration of the functional integral. (This is rigorously proved for local actions, see [10]). Other type of dependence have been used, according to the manifest symmetry of the problem: axially-dependent fields v(z) were used in [11] for effective-action studies in an external constant field.

Using only spherically-symmetric fields allows one to recast the functional determinant as a sum over partial waves:

$$\ln \det \left(\frac{-\partial^2 + m^2 + \sigma}{-\partial^2 + m^2} \right) = \sum_{l=0}^{+\infty} d(\nu, l) \ln \det \left[\frac{-d_r^2 + \frac{l^2 - 1/4}{r^2} + m^2 + \sigma}{-d_r^2 + \frac{l^2 - 1/4}{r^2} + m^2} \right] \tag{2.1}$$

where $d(\nu,l)$ is the degeneracy of l-angular momentum in ν-dimensions.

Similar expressions can be obtained for 2-dimensional fermionic actions [3], 4-dimensional Yukawa model [6]. Such functional determinants as:

$$\ln \det \frac{\begin{bmatrix} -m - v - iw & d_r + \frac{J + 1/2}{r} \\ d_r - \frac{J - 1/2}{r} & -m - v + iw \end{bmatrix}}{\begin{bmatrix} -m & d_r + \frac{J + 1/2}{r} \\ d_r - \frac{J - 1/2}{r} & -m \end{bmatrix}} \tag{2.2}$$

will appear in the corresponding partial wave expansion.

Anyhow, it is clear on (2.1) and (2.2) that one has to use as a spectral variable in the auxiliary linear problem the angular momentum l or J. The main steps are now:

1) introduction of the scattering data associated to the diffusion problem (2.1) or (2.2), with the angular momentum as a spectral parameter (l and J will be variables, m being a constant mass scale). This allows to express the functional determinant in a local form as a function of these S.D..

2) One has to replace as for as possible the local polynomials of the potentials and their derivatives, by local functions of the scattering data. This cannot always be done, in contrast to our first step. We shall see that, in two dimensions at least, this can only be done for integrable models: NLσ [12], GN [13], CGN [13]. The link between integrability and separability of $1/N$ effective action is here quite clear. It is not so clear in 4 dimensions, where the notion of integrability itself is not clear, anyway.

3) If one has succeeded to solve the stationary-point equations for the effective action, which is then (fully) written in terms of the S.D., it is possible to obtain the corresponding fields, by applying an inverse scattering transform to the critical S.D. (hence the name of the method). This I.S.T. amounts to solving a Gel'fand Levitan-Marchenko integral equation; this can be done in closed form, in some cases, in particular when the kernel is degenerate.

The whole method of defining scattering data, relating them to the functional determinant of the differential operator (step 1), obtaining relations between local functionals of fields and local functionals of the S.D. (called trace identities) (step 2), and getting back from S.D. to potential (step 3), has been formulated for Schrödinger equation with energy as spectral parameter [15], Dirac equation with energy as spectral parameter [16], Schrödinger equation with angular momentum [17], [2], Dirac equation with angular momentum in two [3] and four [6] dimensions, Schrödinger equation with potential-dependant spectral parameter [18], [19]. We shall now explain the procedure and technical details for Schrödinger equation, with angular momentum as spectral parameter. Other equations can be treated in exactly the same way.

Step 1 : Scattering data and trace identity

We are studying the equation:

$$\left\{ - d_r^2 + \frac{\ell^2 - 1/4}{r^2} + 1 + v(r) \right\} \varphi(r) = 0 \tag{2.3}$$

Considering ℓ as a complex variable, we shall define the following quantities:

<u>Jost solution</u> $\equiv$ normalized regular solution when $r \to 0$:

$$f(r, \ell) =_{(r\to 0)} r^{\ell + 1/2} \left(1 + \theta(r)\right), \quad \mathrm{Re}\ \ell > 0 \tag{2.4}$$

<u>Regular solution at $+\infty$</u>

$$\varphi(r, \ell) = e^{-r} \left(1 + O(1/r)\right), \quad \forall \ell \tag{2.5}$$

These definitions will only be meaningful if v(r) verify suitable boundary conditions. Namely $v(r) \ll r^{-2}$ when $r \to 0$ or $r \to +\infty$.

One now defines the Jost function as:

$$\varphi(r, \ell) = \frac{F(\ell)}{\ell} f(r, -\ell) - \frac{F(-\ell)}{\ell} f(r, \ell) \tag{2.6}$$

$F(\ell)$ is clearly defined only when $\mathrm{Re}\ \ell \geqslant 0$, and is analytic in this half-plane. One can show that F has an analytic continuation for $0 > \mathrm{Re}\ \ell > -1$ for regular enough potentials v. $F(\ell)$ more generally describes how a function , regular solution of (2.3) when $r \to +\infty$, becomes singular when $r \to 0$. In all the previously mentioned equations which can be treated in this way, this will be the meaning of $F(\ell)$. To give it a mathematical form, one defines the wronskian of two solutions of (2.3) as:

$$W(\Phi_1, \Phi_2) = d_r\Phi_1 . \Phi_2 - d_r \Phi_2 . \Phi_1 \tag{2.7}$$

and it easily follows that

$$2F(\ell) = W\left(f(\ell, r), \varphi(\ell, r)\right) \tag{2.8}$$

Now we define the scattering data of the operator in (2.3). It consists of

a) discrete eigenvalues: ℓ_K such that $F(\ell_K) = 0$. They correspond to the only fully regular solutions of (2.3)

b) normalization coefficients c_K of the φ-solutions corresponding to the eigenvalues. In the case of Schrödinger equation, it can be shown that ℓ_K and c_K are necessarily real when v is real. This will not be true in general for Dirac equation [3, 6] or Schrödinger equation with "potential-dependent" angular momentum:

$$\left(-d_r^2 + \frac{(\ell - r\theta)^2 - 1/4}{r^2} + 1 + v(r)\right) \varphi(r) = 0 \tag{2.9}$$

that one meets in 1/N-expansion of CP_N [19]

c) continuous S.D.

$D(\tau) = \left|\frac{F(i\tau)}{F_0(i\tau)}\right|$ where $F_0(\ell)$ is the Jost function for zero-potential:

$$F_0(i\tau) = 2^{i\tau}\,\Gamma(1 + i\tau) \tag{2.10}$$

It will be shown later that there is a one-to-one correspondence between the scattering data defined above, and the potential v. This will always be the case in all the linear problems mentioned above. We can now obtain, from the definition of the scattering data and the fact that $\frac{F(\ell)}{F_0(\ell)}$ goes to 1 when $|\ell|$ goes to $+\infty$ (which is true in all cases, except equation (2.9), which anyhow requires a special treatment, see [19]), an analytic representation of $F(\ell)$ for any ℓ, Re $\ell > 0$.

$$\frac{F(\ell)}{F_0(\ell)} = \prod_{K=1}^{N_B} \frac{\ell - \ell_K}{\ell + \ell_K} \exp\left\{\frac{2\ell}{\pi}\int_0^{+\infty} \frac{\ln D(\tau)}{\ell^2 + \tau^2}\, d\tau\right\} \tag{2.11}$$

Similar expressions will be obtained in the other cases. Now we can obtain the fundamental relation between the Jost function and the functional determinant, which is the key of the whole procedure:

$$\ln \det \left[\frac{-d_r^2 + 1 + v + \frac{\ell^2 - 1/4}{r^2}}{-d_r^2 + 1 + \frac{\ell^2 - 1/4}{r^2}} \right] = \ln \frac{F(\ell)}{F_0(\ell)} \tag{2.12}$$

This identity follows from derivating the l.h.s. with respect to , and using the value of the Green function of the operator as:

$$G(r, r', \ell) = \frac{f(r_<, \ell) . \varphi(r_>, \ell)}{2 F(\lambda)} \tag{2.13}$$

together with the wronskian property:

$$d_r W(\Phi_1(\ell_1, r), \Phi_2(\ell_2, r)) = \frac{1}{r^2} \frac{\Phi_1(\ell_1, r) \Phi_2(\ell_2, r)}{\ell_1^2 - \ell_2^2} \tag{2.14}$$

This identity (2.14) is very useful in that it allows one to compute scalar products standardly defined as:

$$< \Phi_1, \Phi_2 > = \int_0^{+\infty} \frac{dr}{r^2} \Phi_1(r) \Phi_2(r) \tag{2.15}$$

Fundamental trace identities such as (2.12) can be derived in all previously mentioned cases. We can now write the non-local functional determinant as a local function of the scattering data, from (2.11) and (2.12).

Step 2

It is now possible to expand both sides of the fundamental trace identity (2.12) in asymptotic series of $1/\ell$. This will lead to relations between local functionals of the fields and their derivatives, and local functions of the scattering data. Those relations can however be obtained in an easier way by introducing a change of variables in (2.3) as $\varphi = \varphi_0 \exp \int_{+\infty}^{r} \Psi dr$ where

. Ψ is the regular solution at $r \to \infty$ for $v \neq 0$

. φ_0 is the regular solution for v = 0. In fact φ_0 is equal to $\sqrt{2\pi r}\cdot K_\ell(r)$

It is clear that $F(\ell) = \exp \int_{+\infty}^{0} \Psi\, dr$. Inserting Ψ in (2.3) leads to the associated Ricatti equation :

$$d_r \Psi + \Psi^2 + 2\,\frac{\varphi_0'}{\varphi_0} + v = 0 \tag{2.16}$$

Using the asymptotic expansion of $\frac{\varphi_0'}{\varphi_0}$ in $1/\ell$, and solving (2.16) order by order in $1/\ell$ immediately leads to the required identities between integrals of polynomial functions of v (coming from expansion of Ψ (r) in series of $1/\ell$) and functionals of the S.D. (coming from $F(\ell)$).This method is less general than the previous standard one, but it is also more practical. Moreover it is applicable in all the previously mentioned problems, where a Ricatti representation is possible.

We shall state the first and second trace identity for Schrödinger equation in angular momentum. One gets:

$$\int_0^{+\infty} r\, v(r)\, dr = -4 \sum_{K=1}^{N_B} \ell_K + 2 \int_{-\infty}^{+\infty} \frac{\ln D(\tau)}{\pi}\, d\tau \tag{2.17}$$

$$\int_0^{+\infty} r^3 \left((v+1)^2 - 1 \right) dr = \frac{16}{3} \sum_{K=1}^{N_B} \ell_K^3 + 8 \int_{-\infty}^{+\infty} \frac{\ln D(\tau)}{\pi}\, \tau^2 d\tau \tag{2.18}$$

The trace identities, in all the I.S.T. problems which we shall use, have the generic shape:

$$\int_0^{+\infty} r^n\, \overset{*}{\mathcal{P}}(v, \partial v, \dots)\, dr = \sum_{K=1}^{N_B} P_n(\ell_K) + \int_{-\infty}^{+\infty} \frac{\ln D(\tau)}{\pi}\, P_n'(i\tau)\, d\tau \tag{2.19}$$

where . n is an odd integer = 2 k+1

. $\mathcal{P}^*$ is a polynomial in the fields and their derivatives, of order k+1

. P_n is a polynomial of order n.

There remains now to show that this change of variable is consistent, and that there is a one-to-one correspondence between the potential and the scattering data (or at least that one can reconstruct the potential by starting from the scattering data).

Step 3 : The inverse scattering transform

Introducing the integral Kernel K(r,r') such that:

$$\varphi(v \neq 0) = \varphi_0(v = 0) + \int_0^r K(r,r')\,\varphi_0(r')\,dr' \tag{2.20}$$

one can show that this Kernel obeys an integral equation of the Gel'fand-Levitan-Marchenko type:

$$\Omega(r,r') + K(r,r') + \int_r^{+\infty} \frac{ds}{s^2}\, K(r,s)\,\Omega(r',s) = 0 \tag{2.21a}$$

$$K(r,s) = 0 \quad \text{when} \quad r \geqslant s \tag{2.21b}$$

where Ω is completely determined by the scattering data:

$$\Omega(r,r') = \int_0^{+\infty} d\tau \,.\, \frac{\tau^2 \,\mathrm{sh}\,\pi\tau}{2\pi} \,.\, \varphi_0(r,i\tau).\varphi_0(r',i\tau).\left\{\frac{1}{(D(\tau))^2} - 1\right\}$$

$$+ \sum_{K=1}^{N_B} c_K\, \varphi_0(r,\ell_K)\,\varphi_0(r',\ell_K) \tag{2.22}$$

where . c_K , ℓ_K , $D(\tau)$ are the scattering data

. φ_0 are free ($v = 0$) solutions.

It can be shown [17, 31] that if the S.D. verify "consistency conditions" (ℓ_K real, c_K real, $D(\tau) \ll 1/\tau^2$ when $\tau \to +\infty$), equation (2.21) has a unique Kernel-solution K. Moreover, the potential v can be obtained from:

$$\sigma(r) = \frac{1}{r} \quad d_r \quad \frac{K(r, r)}{r} \tag{2.23}$$

The same sort of expression can be derived for Dirac equation in angular momentum. In this case, the Kernels Ω and K are 2x2 matrices, the free solutions φ_0 are no more scalar, but 2-dimensional spinors and the discrete scattering data can remain complex (as it is also the case for Dirac equation in energy, see [1, 6]). Note that unicity of the solution of GLM equation for Dirac equation with energy as spectral parameter is not necessary true [16]. This is not too important as long as we are only concerned with finding saddle-points of the non-local actions: any solution of the GLM equation will generate a "set of potentials" having the requested S.D. but not necessarily continuous, and may be singular ("bursting solitons"). These should be no problem if we stick to continuous non-singular potentials. If we wish to use the S.D. as new variables in the functional integral, this would be critical indeed, but such a use is prohibited by the fact that only spherically symmetric potentials can be used.

Finally one can write a local form of Gel'fand-Levitan-Marchenko equation. This enables one to get the functional derivatives of the potential with respect to the scattering data, and reciprocally. In the case of Schrödinger equation in angular momentum, we get:

$$\delta\sigma(r) = \frac{1}{r} \quad d_r \left\{ \frac{1}{r} \left[-\frac{4}{\pi} \int_0^{+\infty} \frac{\tau \, \mathrm{sh}\,\pi\tau}{(D(\tau))^3} \left(\varphi(r, i\tau)\right)^2 \delta D(\tau) \, d\tau \right.\right.$$
$$\left.\left. + 2 \sum_{K=1}^{N_B} \varphi_K^2(r) \, \delta c_K + 4 \sum_{K=1}^{N_B} c_K \, \varphi_K(r) \, \frac{d\varphi}{d\ell_K}(r, \ell_K) \, \delta\ell_K \right] \right\}$$

Inversely, we get:

$$\delta\ell_K = -\frac{c_K}{2\,\ell_K} \int_0^{+\infty} \varphi_K^2(r) \, \frac{\delta\sigma(r)}{r^2} \, dr \tag{2.25}$$

$$\delta D(\tau) = \frac{1}{2} \int_0^{+\infty} dr \;\; \delta\sigma(r) \;\; \varphi(r, i\tau) \;\; \mathrm{Re}\left(f(r, i\tau)\, e^{i\,\delta(\tau)} \right) \tag{2.26}$$

where
$$\delta(\tau) \equiv \mathrm{Arg} \; \frac{F(i\tau)}{F_0(i\tau)} \tag{2.27}$$

These functional derivatives are especially useful when one wants to compute non-standard trace identities, i.e. relations between local functionals of the fields and local functionals of the S.D. which are not present in the set of identities obtained at step 2. We shall meet such identities in the next chapters, for Non-linear sigma model and Chiral-Gross-Neveu model.

These derivatives are also useful when one has not been able to rewrite the effective action fully as a local functional of the S.D., which is the general case. One can however make safe predictions on the behaviour of saddle-points of the action by using these functional derivatives. We shall see it in the next chapter.

III. BOSONIC EFFECTIVE ACTIONS: $(\vec{\Phi}^2)^2$ IN 1/N [2]

As we have seen on equation (1.3), the (renormalized) effective action that generates 1/N expansion of $(\vec{\Phi}^2)^2$ models in any number of euclidean dimensions $\nu \leqslant 4$ is a clear example of a 1-loop effective action. One immediately sees that the relevant inverse scattering transform involves the previously studied Schrödinger equation with angular momentum as spectral parameter. We have supposed of course that the dominant saddle-points of this action have a maximum (i.e. spherical) symmetry, enabling us to apply the whole scheme of I.S.T. to this effective action.

One gets the following expressions of the effective action in terms of the scattering data:

Dimension 2

$$S_{eff} = \frac{4}{\pi^2} \displaystyle{\not}\int_0^{+\infty} \displaystyle{\not}\int_0^{+\infty} \frac{\tau^2+\tau'^2}{(\tau^2-\tau'^2)^2} \ln D(\tau) \ln D(\tau')\, d\tau\, d\tau' + \ln D(0)$$

$$- \frac{8}{\pi} \int_0^{+\infty} \sum_{K=1}^{N_B} \frac{\ell_K}{\tau^2+\ell_K^2} \ln D(\tau)\,.\, d\tau \quad + 2 \sum_{K=1}^{N_B} \ln\left(4\ell_K^2 \cdot \sin\pi\ell_K\right)$$

$$+ \sum_{1\leq K\neq L\leq N_B} \ln \left(\frac{\ell_K+\ell_L}{\ell_K-\ell_L}\right)^2 - 2 \sum_{K=1}^{N_B} \ln c_K \pm i\pi \sum_{K=1}^{N_B}\Big\{1$$

$$+ 2\, E(\ell_K)\Big\} - \frac{\pi}{4g_B} \int_0^{+\infty} r\, v^2(r)\, dr \qquad (3.1)$$

(E will always stand for "integer part").

A remark must be done on expression (3.1). This effective action is not linear in ln $D(\tau)$, neither does it exhibit the separation between continuum and discrete spectrum that one should expect from (2.19), (2.12) and (2.11). This comes from a non-standard trace identity, expressing $Q = \int_0^{+\infty} \ln r\; v(r)dr$ as a functional of the S.D.. This also accounts for the C_K-dependence of the effective action, which is not expected from (2.11-2.19). It was possible to write a trace identity for Q by derivating with respect to the S.D., using (2.24), and integrating back the expressions thus obtained. Note that, in the language of the theory of integrable models, and of "action-angle variables", this quantity is not a conserved charge, since it contains the normalization coefficient, usually an "angle" variable, C_K. This is a bit puzzling, since trace identities are usually associated to conserved quantities (action variables) [32]. One must however take care of the fact that action-angle variables have not been defined in the Inverse Scattering method in angular momentum, so that this question might not be relevant here.

A second remark is that the term $\int_0^{+\infty} r\; v^2(r)dr$ that breaks separability of action cannot be expressed as a local functional of the S.D.. The method of derivating-integrating with respect to the scattering data is inefficient here.

The case when $g \to \pm\infty$ corresponds to non-linear sigma model with O(N) symmetry [4]. In this case the effective action becomes separable in terms of the fields. This is directly linked to the integrability features of the model [12].

In all other cases the effective action is not separable in term of the S.D.. It is however possible to predict the behaviour of the first instanton using the functional derivatives obtained in the previous section. Analytical and numerical predictions agree very well, giving the following results:

a) $g_B \to 0^+$: v-instanton becomes equal to the (φ^4-instanton) squared. The effective action reads $S_c = -\frac{2.98}{g_B} + \theta(\ln g_B) \pm i\pi$. This potential has one single eigenvalue ℓ_1, that goes to 0 when g goes to zero.

b) $g_B \to +\infty$: This limit, that generates the non-linear sigma model, gives particularly interesting results. The instanton becomes singular; its single eigenvalue ℓ_1 goes to 0.818; the corresponding normalization constant C_1 goes to $+\infty$ as (g_B); the phase of Jost function $\delta(\tau)$ (see (2.27)) goes to zero; $D(\tau) \sim 1/\tau^2$ (which is singular for this S.D.); finally the action S_c behaves like ln $(g_B) + \theta(1) \pm i\pi$, and therefore $|S_c| \to +\infty$.

The conclusions that one can draw are very interesting: it is known that instantons dominate the large-order behaviour of perturbative expansion as long as no renormalization effect (renormalons) spoils this domination; one has [7]:

$$A_K = \frac{K!}{(S_c)^K} \cdot K^{\alpha} \cdot \mathbb{C} \cdot \left[1 + \theta(1/K)\right] \tag{3.2}$$

where
- A is any physical quantity expanded perturbatively as $A = \sum A_K g^K$, g (here 1/N) being the parameter of expansion.
- Sc is the action (classical) of the instanton and determines the Borel summability of the expression [7].
- α is a number, depending on the nature of A itself, and of the symmetries of the instanton (zero modes).
- $\mathbb{C}$ is the functional determinant of fluctuations.
- 1/K expansion of A_K can be computed perturbatively [20].

We see therefore that the limit when the coupling constant g goes to $+\infty$, where one reaches the non-linear, $O(N)$-symmetric, sigma model, corresponds to an expansion in powers of 1/N that becomes less and less divergent. Moreover we know, thanks to the properties of factorizability, the exact expression for the S-matrix [21] and the form factors [22] of this model. Their expansion in powers of 1/N is merely an entire series with a finite radius of convergence. This could indicate that not only on- shell quantities, but also off- shell quantities, on which no information is available until now, could exhibit a convergent expansion in (1/N). This behaviour should be linked to the integrability of the O(N) non-linear sigma model, that could lead (owing to the existence of an infinite number of symmetries of these models) to drastic cancellations of graphs, finally giving a convergent perturbative expansion. We shall see again such a phenomenon for Gross-Neveu and Chiral Gross-Neveu model.

In the case of CP_N model, in two dimensions, we have succeeded [19] in writing the effective action that generates 1/N expansion, in terms of scattering data for the auxiliary linear problem :

$$\left(- d_r^2 + \frac{(\ell - r\theta)^2}{r^2} - \frac{1}{4r^2} + m^2 + \sigma \right) \varphi(r) = 0$$

One obtains a similar form as in the case of non-linear sigma model, but the eigenvalues and continuum spectral function D( ) are no more real. Anyhow, one again finds an absence of instantons for this effective action, suggesting a possible convergence of the 1/N expansion. However the CP model is <u>not</u> quantum-mechanically integrable. This does not spoil the separability of effective action at leading order in 1/N, but this forbids getting exact expressions for the S-matrix; therefore no safe quantitative conclusion can actually be drawn on the 1/N expansion.

<u>Dimension 3</u>

We get:

$$S_{eff} = 2 \int_0^{+\infty} - \tau . \tanh(\pi\tau) . \ln D(\tau) . d\tau - 2\pi \sum_{K=1}^{N_B} \int_0^{\ell_K} x \tan \pi x \, dx$$

$$\pm i\pi \mathcal{N} + \int_0^{+\infty} r^2 v(r) \left(1 - \frac{\pi\sigma(r)}{2 g_B} \right) dr \tag{3.3a}$$

$$\text{where : } \quad \mathcal{N} = \sum_{K=1}^{N_B} \left(\sum_{n=0}^{E(\ell_K - 1/2)} (2n+1) \right) \tag{3.3b}$$

This expression can never be recast as a completely closed function of the scattering data, due to the $\int_0^{+\infty} P(\sigma(r)) \, dr$ term that is not a trace identity, whether standard or non-standard. However, as it was the case in the 2-D case, and will also be the case in 4-dimensional φ^4, the imaginary part depends only on the number and location of the zeroes, and can be determined exactly, which is very important, as we have seen before, in the study of Borel-summability of 1/N expansion.

We can nevertheless obtain results concerning the dominant (lowest-action) instanton, whether by analytical or numerical study.

* $g_B \to 0^+$

First instanton of 1/N expansion is again the square of the classical instanton in 3 dimensions [23]. It has one single eigenvalue $\ell_1 \to 1/2$. Its action reads:

$$S_c = \frac{-9.44}{g_B} + O(\ln g_B) \pm i\pi \tag{3.4}$$

* $g_B \to +\infty$

This corresponds, by renormalization group analysis, to the fixed point of Heisenberg spin model [24], and quantitative results obtained in this case could help one to compute exactly in three dimensions the critical exponents of this model. We find that the first instanton has a limit eigenvalue $\ell_1 = 1$.

The action of the instanton has in this case a finite limit:

$$S_c = 0.94 - \frac{115}{g_B} \pm i\pi \tag{3.5}$$

We find that the 1/N series is always Borel-summable in the whole range of coupling constant g. These results are confirmed by rigorous studies of the 1/N expansion in two [25] and three [26] dimensions. They are linked to the existence of an imaginary part in the instanton action, giving the 1/N series an oscillating sign at large orders: this is known to allow Borel summability for a series $\sim \frac{K!}{a^K}$.

Dimension 4, $m^2 = 0$

Inverse scattering transform with energy as a spectral parameter [30, 33] is most appropriate to treat this case. Moreover the effective action of a spherically symmetric configuration v becomes completely separable in term of the scattering data. This remarkable feature enables to study completely the saddle-points of S_{eff}. One has:

$$S_{eff} = \sum_{j=1}^{N_B} \left\{ \frac{32\pi^2}{g_R} K_j^3 + \int_0^{K_j} x^2 \left(2\Psi(1+x) + \pi \cot \pi x - \frac{1}{x} \right) dx \pm i\pi \left(\sum_{n=1}^{E(K_j)} n^2 \right) \right\} + \frac{1}{4\pi} \int_{-\infty}^{+\infty} \tau^2 \ln |F(\tau)|^2 \left(\mathrm{Re}\, \Psi(i\tau) + \frac{96\pi^2}{g_R} \right) d\tau \tag{3.6}$$

where . $\{K_j\}$ are the eigenvalues of the potential $\mathcal{V}(x) \equiv r^2 v(r)$, $x \equiv \ln r$.

. $F(\tau)$ is the reflection coefficient.

. g_R is the running coupling constant. $\Psi(x) \equiv \frac{d}{dx} \ln \Gamma(x)$

The instantons are obtained by solving the gap equations, which read:

$$\frac{\delta S}{\delta F(\tau)} \Longrightarrow F(\tau) = 0. \tag{3.7a}$$

$$\frac{\delta S}{\delta \kappa_j} = 0 \Longrightarrow \frac{96\pi^2}{g_R} = 2\Psi(1+x_j) + \frac{1}{x_j} - \pi \cot \pi \kappa_j \quad (3.7b)$$

Equation (3.7a) means that any instanton is a reflectionless potential. It will be possible, in this case, to obtain the potential v (once the eigenvalues given by (3.7b) are imposed), as a Bargmann potential [27]. We shall not give the expression of such potentials here. It can be found in [2] or [27].

It must be stated that in the case when $\frac{96\pi^2}{g_R} > 2\gamma$, the eigenvalue equation has a solution ℓ_j smaller than 1, and therefore the corresponding instanton has a real (and indeed positive) action. 1/N expansion is not Borel-summable in this case. This is linked to the existence of an instability of the perturbative vacuum $\langle\phi^2\rangle = 0$: the ($\ell_j < 1$) instanton describes desintegration of this vacuum by tunnel effect towards configurations of large fields $\langle\phi^2\rangle \to \infty$ [8]. This potential can be obtained by analytical I.S.T.. It reads:

$$v(r) = -\frac{8\ell_1^2}{r^2}\left(\left(\frac{r}{r_0}\right)^{\ell_1} + \left(\frac{r_0}{r}\right)^{\ell_1}\right)^{-2} \quad (3.8)$$

where r_0 is an arbitrary scale factor which was introduced to obtain Schrödinger one-dimensional equation from massless radial Schrödinger equation as ($r = r_0 e^x$) [2]. This arbitrariness of r_0 reflects the scale invariance of the action of four-dimensional massless .

Let us now turn to massive Φ^4 model.

Dimension 4, $m^2 \neq 0$

The I.S.T. in angular momentum can again be used here, and leads to the following result:

$$S_{eH} = \frac{2}{\pi} \int_0^{+\infty} \tau^2 \left(\ln 2 - \frac{1}{2} + \mathrm{Re}\, \Psi(i\tau)\right) \ln D(\tau)\, d\tau$$

$$+ \sum_{K=1}^{N_B} \left\{ \frac{2}{3}\left(\ln 2 - \frac{1}{2}\right) \ell_K^3 \pm i\pi \sum_{\ell=1}^{E(\ell_K)^2} \ell^2 + \int_0^{\ell_K} x^2 \left(\right.\right.$$

$$2\psi(1+x)+\pi\cot\pi x-\frac{1}{x}\Big)\,dx\Big\}-\frac{1}{8}\int_0^{+\infty}r^3\ln r\,\big(2$$

$$+\sigma(r)\big)\,\sigma(r)\,dr+\frac{1}{16}\left(\frac{96\pi^2}{g_R}-1\right)\cdot\int_0^{+\infty}r^3\,v^2(r)\,dr \tag{3.9}$$

The running coupling constant g has to be kept positive here. Otherwise we should find tachyons in the 1/N expansion of ϕ^4, see [28].

Anyhow this expression cannot be easily studied by numerical or analytic devices. However it is possible to obtain indications on the instantons when g $\to 0^+$. One finds then (analytically or numerically) a positive real-action instanton, indicating that the 1/N theory is unstable (non-perturbatively) also for the massive case. The instanton is a deep and narrow potential, with depth behaving as g_R^{-1} and range behaving as $g_R^{1/2}$. It behaves as the square of the large-scale limit of massless ϕ^4-instanton, or equivalently like the massive ϕ^4 "pseudo-instanton" (see [7]). One finds again a link between 1/N instanton when $g\to 0$ and ϕ^4-classical instanton; but here this link is more subtle, since no instanton exists (rigorously speaking) for massive ϕ^4-model. Anyhow the main result of our studies is that, at least for a small coupling constant, the 1/N expansion of ϕ^4 (massive) cannot be done in a consistent, stable and "non-tachyonic" way. This suggests that ϕ^4 in four dimensions should actually be trivial. One even gets a result for the life-time of the ϕ^4-vacuum in 1/N expansion as:

$$T^{-1}=\left\{\mathbb{C}\exp-N.\left(\frac{330}{g_R}+\theta(1)\right)\right\}\cdot\left\{1+\theta(1/N)\right\} \tag{3.10}$$

This result should be compared with the one in [29], where a similar "vacuum unstability" is shown to exist in a different framework for $(\vec{\Phi}^2)^2$.

IV. TWO-DIMENSIONAL QUARTIC-COUPLED FERMIONIC MODELS [3]

We shall apply the same methods to study possible instantons of the (fermionic) effective action of 2-D quartic coupled fermionic models which was derived in (1.9). It is clear that the inverse scattering problem relevant for the functional determinant of ($\frac{\not{\partial}+m+v+i\gamma_5 w}{\not{\partial}+m}$) is linked to the linear operator introduced in (2.2). This holds, once one has assumed that the dominant configurations in the functional integral (and especially the leading instantons, if any) have the largest symmetry (i.e. spherical symmetry in two dimensions). We recall that the associated linear problem reads then:

$$\begin{bmatrix} -m-v-iw & d_r+\frac{J+1/2}{r} \\ d_r-\frac{J-1/2}{r} & -m-v+iw \end{bmatrix} \Psi = 0 \tag{4.1}$$

Here J is the full angular momentum (radial + spin) in two dimensions; it will play the role of spectral parameter in the I.S.T. scheme. We shall not give in detail the derivation of the useful identities; actually the procedure is exactly the same as in the case of radial Schrödinger equation, except that one has to deal with matrix - like operators and spinorial eigen functions Ψ. See [3] for complete explanations.

We shall now give the expression of the effective action of a general quartic-coupled model of 2-D fermions (excluding vector-like couplings ($\bar{\Psi}\gamma_\mu\Psi$)²). Note that such couplings could be incorporated in the general study: once the hypothesis of spherically-symmetric auxiliary fields has been done, one can show that the corresponding vector field $A_\mu(x)$ will be absorbed in v and w by a gauge transformation of the operator (4.1): therefore the most general situation is indeed the one which is studied here). The action reads:

$$S_{eff} = -\frac{1}{4\pi^2} \int\!\!\!\!-\!\!\int\!\!\!\!-_{-\infty}^{+\infty} \frac{d\tau\, d\tau'}{(\tau-\tau')^2} \ln D(\tau) \ln D(\tau') - \frac{1}{\pi}\int_{-\infty}^{+\infty} \sum_{K=1}^{N_B} \cdot$$

$$\frac{\text{Re sgn } J_K}{(J_K - i\tau)} \cdot \ln D(\tau)\, d\tau \quad + \sum_{K=1}^{N_B} \frac{\ln \cos \pi J_K}{2\pi}$$

$$- 2 \sum_{K \neq K'} \text{sgn}\,(\text{Re } J_K)\ \text{sgn}(\text{Re } J_{K'})\ \ln\,(J_K - J_{K'}) \ - \sum_{K=1}^{N_B} \ln c_K$$

$$+ \ \frac{1}{2 g_R} \int_0^{+\infty} r\, w^2(r)\, dr \qquad (4.2)$$

where . J_K are the eigenvalues (complex generically)

. C_K are the corresponding normalization coefficients (complex)

. $D(\tau)$ is the continuum spectrum

. $W(r) \equiv 0$ for Gross-Neveu model

. $1/g_R \equiv 0$ for Chiral Gross-Neveu model.

A general numerical study of the extrema of this action is possible,but shall not be given here. Two interesting cases appear, namely the two integrable models G-N [13, 14] and CGN [13, 14]. In both cases, the effective action can be fully expressed in term of the suitable scattering data; and in both cases, no instanton appears owing to the ($\ln c_K$) dependence of S_{eff}. Moreover, one can check analytically that a "limit instanton" exists, when $1/g_R \to 0$, i.e. the integrable chiral Gross-Neveu model. The action of this instanton behaves like $+(\ln g_R)$; therefore, if the 1/N series is indeed dominated by the instanton contribution, and if the spherically symmetric configurations are the most relevant for the study of saddle-points, then the 1/N expansion becomes less divergent when one goes towards CGN parameters ($1/g_R \to 0$). Since the S-matrix [21] and form factors [22] are known, and since their 1/N expansion is a convergent series, we can conjecture that 1/N expansion of any quantity for CGN model is convergent. This should (as it was the case previously for NLσ-model) be linked to the integrability of the model, that probably induces drastic cancellations in the Feynman graphs of 1/N expansion, owing to the existence of infinitely many conserved charges.

V. FOUR-DIMENSIONAL YUKAWA + Φ^4- COUPLED THEORY [6]

We shall finally consider the model introduced in (1.10), with a Yukawa interaction between a fermionic field and a scalar, plus a self-interaction $(\Phi^2)^2$ of the scalar field, necessary in four dimensions to get a renormalizable theory. Integrating over the fermions, and renormalizing the obtained action leads to the following non-local functional (after rescaling the fields v and w):

$$S_{eff} = \ln \det \frac{\not\partial + m + v + i\gamma_5 w}{\not\partial + m} + \left(\frac{\Gamma(1-\nu/2)}{(4\pi)^{\nu/2}} \cdot 2^{\nu/2} - \frac{1}{\lambda_R^2} \right)$$

$$\cdot\, m^{\nu-4} \left\{ \int_{-\infty}^{+\infty} d^\nu x \left((\nabla v)^2 + (\nabla w)^2 + 6m^2v^2 + 2m^2w^2 + 4m^3v^2 + 4mv^3 \right.\right.$$

$$\left.\left. + 4mvw^2 + v^4 + w^4 + 2v^2w^2 \right) \right\} + \frac{m^{\nu-4}}{\lambda_R^2} \int_{-\infty}^{+\infty} d^\nu x \cdot \frac{1}{8\pi^2} \left(K_3 v^2 \right.$$

$$\left. + K_4 w^2 + 32\pi^2 m^3 v + K_5 (v^3 + vw^2) + K_6 (v^2 + w^2)^2 \right) \quad (5.1)$$

One must now again assume that the instantons of S_{eff}, that dominates the λ_R -perturbative expansion (K_3 ... K_6 remaining fixed parameters), have a maximum, i.e. 4-dimensional spherical symmetry. Then it is possible to split into partial waves the functional determinant, and to reduce our problem to the computation of fixed [O(4)-angular-momenta] determinants. One finds that the 4-dimensional spectral problem actually reduces to the previously studied two-dimensional problem due to the assumption of spherical symmetry. It is then straight forward to get the expression (as completely as possible in term of S.D.) of the effective action:

$$S_{eff} = \int_{-\infty}^{+\infty} \frac{d\tau}{2\pi} \ln D(\tau) \left\{ \psi\left(\tfrac{1}{2} + i\tau\right) + \psi\left(\tfrac{1}{2} - i\tau\right) - 2\psi\left(\tfrac{1}{2}\right) \right\} \left(\tau^2 \right.$$

$$\left. - \frac{1}{4} \right) + \sum_{K=1}^{N_B} \operatorname{sgn} \operatorname{Re}(J_K) \int_0^{J_K} dx \left(\frac{1}{4} - x^2 \right) \left(\psi\left(\tfrac{1}{2} - x\right) + \psi\left(\tfrac{1}{2} + x\right) \right.$$

$$\left. - 2\psi\left(\tfrac{1}{2}\right) \right) + \left(\frac{8\pi^2}{\lambda_R^2} + \ln 4 - \gamma \right) \left\{ \int_{-\infty}^{+\infty} d\tau \left(\tau^2 + \frac{1}{4} \right) \ln D(\tau) \right. +$$

$$+ \sum_{K=1}^{N_B} \operatorname{sgn} \operatorname{Re} (J_K) \left(\frac{J_K}{4} - \frac{J_K^3}{3} \right) \Big\} + \frac{1}{2} \int_{-\infty}^{+\infty} \ln D(\tau)\, d\tau + \frac{1}{2} \sum_{K=1}^{N_B} J_K \cdot$$

$$\operatorname{sgn} \operatorname{Re} (J_K) - \frac{1}{4} \int_0^{+\infty} r^3\, dr\, \ln (mr) \left\{ w'^2 + v'^2 + \left((v+m)^2 + w^2 \right)^2 - m^4 \right\} + \mathfrak{X} \tag{5.2}$$

where $\mathfrak{X}$ stands for the third polynomial term in (5.1). $\mathfrak{X}$ cannot be expressed in term of the S.D. since it is not a trace identity of the auxiliary linear problem. This will prevent us from finding analytical instantons in the general case.

However there are two sets of values of the parameters in the effective action that allow for a complete analytical resolution of the saddle-point equation. First case amounts to setting all parameters in $\mathfrak{X}$ equal to zero. The model becomes massless at tree level in 1/N, and the coupling constants of ϕ^4 and $\Psi (\phi_1 + i \gamma_5 \cdot \phi_2)$ are linked by $g_R = \lambda_R^2$. One finds in this case an infinite number of instantons, corresponding to N-bound-states reflectionless potentials (v, w). One of these solutions is simply the standard ϕ^4 instanton, with negative real action; the other solutions, however, have a real positive action S_{eff}, thus giving subleading non-Borel-summable contributions to the large orders in g_R. It seems that introduction of fermion coupling to the scalars (ϕ_1 , ϕ_2) spoils the Borel-summable behaviour of the perturbative expansion of pure Φ^4 theory ($m^2 = 0$).

The second solvable case is Φ^4, without fermions, and massless at tree level. One gets back the instanton of Φ^4, of course, but it is interesting to note that both 1/N generating action and classical action of massless $(\Phi^2)^2$ can be expressed in a closed form as a function of suitably chosen scattering data. This could be linked to deeper properties of the theory.

1. J. Schwinger, Proc. Nat. Acad. Sci. US 37, 452 (1951); G. Jona-Lasinio, Nuov. Cim. 34, 1790 (1964)
2. J. Avan, H.de Vega, Phys. Rev. D 29, 2891, 2904 (1984)
3. J. Avan, H. de Vega, Comm. Math. Phys. (1985), to be published
4. H. de Vega, Phys. Lett. B 98, 280 (1981)
5. A. d'Adda, M. Luscher, P. di Vecchia, Nucl. Phys. B 146, 63 (1978)
6. J. Avan H. de Vega, PAR-LPTHE 85 - 14
7. See J. Zinn-Justin, Phys. Reports 70, 109 (1981); and Les Houches Lectures, Recent advances in quantum field theory, edited by R. Stora and J.B. Zuber, North Holland 1983
8. a) S. Coleman, Phys. Rev. D 15, 2929 (1977)
b) C.Callan, S. Coleman, Phys. Rev. D 16, 1762 (1977)
9. a) M.S. Birman, M.G. Krein, Sov. Phys. Doklady 3, 740 (1962)
b) V.S. Buslaev, Topics. Math. Physics V.1, p. 69 (1968), Consultants Bureau,N.Y.
10. S. Coleman, V. Glaser, A. Martin, Comm. Math. Phys. 58, 211 (1978)
11. H. de Vega, F.Schaposnik, Phys. Rev. D 26, 2814 (1982)
12. See for review: H. Eichenherr, Tvarminne Lectures 1981, ed. by J. Hietarinta and K. Montonen, Springer; Lectures in Physics, vol. 191
13. A. Neveu, H. Papanicolaou, Comm. Math. Phys. 58, 31 (1978)
14. D.J. Gross, A. Neveu, Phys. Rev. D 10, 3235 (1974)
15. L. Faddeev, Amer. Math. Soc. (2), 65, 139 (1964)
16. M. Ablowitz, D. Kaup, A. Newell, H. Segur, Stud. Appl. Math. Vol. III, 4, 249 (1974)
17. H. de Vega, Comm. Math. Phys. 81, 313 (1981)
18. R. Weiss, G. Scharf, Helv. Phys. Acta, 44, 910 (1971)
19. J. Avan, PAR-LPTHE 85-48
20. See R. Seznec, Thèse de 3ème cycle, Univ. Paris VI, 1978 (unpublished)
21. A.B. Zamolodchikov, Al.B. Zamolodchikov, Ann. Phys. 80, 253 (1979)
22. M. Karowski, P. Weisz, Nucl. Phys. B 139, 455 (1978)
23. E. Brézin, G. Parisi, Journ. Stat. Phys. 19, 269 (1978)
24. E. Brézin, J. Zinn-Justin, Phys. Rev. B 14, 3110 (1976)
25. T. Spencer, Comm. Math. Phys. 79, 273 (1980); S.Breen, Comm. Math. Phys. 92, 179 (1981)
26. J. Magnen, V. Rivasseau, Preprint CPhT 652-02-85; C. Billionet, P. Renouard, Comm. Math. Phys. 84, 257 (1982)
27. I. Kay, H. Moses, J. Appl. Phys. 27, 1503 (1956)
28. L. Abbot, J. Kang, H. Schnitzer, Phys. Rev. D 13, 2212 (1976)
29. W. Bardeen, M. Moshe, Phys. Rev. D 28, 1372 (1983)
30. J. Avan, Nucl. Phys. B 237, 159 (1984); J. Avan, H. de Vega, Nucl. Phys. B 224, 61 (1983)
31. L. Faddeev, J. Math. Phys. 4, 72 (1963)
32. L. Faddeev, V.A. Zakharov, Funct. Annal. Appl. 5, 280 (1971)
33. H. de Vega, Tvärminne Lectures 1981, edited by J. Hietarinta and K. Montonen, Springer Lectures in Physics, vol. 191.

GENERAL STRUCTURE AND PROPERTIES OF THE INTEGRABLE NONLINEAR EVOLUTION EQUATIONS IN 1+1 AND 2+1 DIMENSIONS

B.G. Konopelchenko*

Laboratoire de Physique Théorique et Hautes Energies,

Université Pierre et Marie Curie, Paris

ABSTRACT

General form of the integrable equations in 1+1 and 2+1 dimensions and their group - theoretical and Hamiltonian properties are considered. General theory of recursion operators is discussed.

---

* Permanent address : Institute of Nuclear Physics, Novosibirsk-90, 630090, USSR.

I. INTRODUCTION

A study of the structure and properties of nonlinear partial differential equations is one of the most important problems of mathematical physics. There was a great progress in this field during the last ten years mainly due to the discovery [1] and development of the inverse scattering transform (IST) method. The IST method considerably extend the class of differential equations which can be investigated in details (see e.g. [2-4]). The differential equations to which the IST method is applicable possess a number of the remarkable properties: soliton type solutions, infinite sets of the conservation laws, infinite symmetry groups, Backlund transformations, complete integrability etc. The equations integrable by IST method have also the pronounced recursion structure.

Group theoretical structure of the integrable equations and these integrable equations themselves can be analyzed by different methods. Here we consider only one of them which was initiated by Ablowitz, Kaup, Newell and Segur [5]. They proposed a very convenient and beautiful method for the description of nonlinear equations integrable by the second order matrix spectral problem. The main quantity which is used is this approach (AKNS approach) is so-called recursion operator. Recursion operator is calculated directly from the spectral problem.

The advantage of AKNS-approach in comparison with the other versions of IST method consists in the following: it allows

1) to find the general form of nonlinear equations connected with given spectral problem in a simple and convenient form,

2) to calcultate the infinitedimensional group of general Backlund transformations and infinitedimensional symmetry group for these equations,

3) to investigate the Hamiltonian structure simultaneously of the whole class of the equations integrable by the given spectral problem.

The attractive features of the AKNS method lead to its intensive development in the last years. The main activity was to generalize this method to different spectral problems. Here we present a short review of the recent results in this field.

It turns out that the general results of the AKNS method such as form of the integrable equations, general Backlund transformations, symmetry groups are of the same form for the different spectral problems. Only the explicit form of the operators and potentials which are contained in these formulas depends on the concrete form of the spectral problem. So we firstly present the general results and then enumerate the spectral problem. We consider the cases of one and two-dimensional spectral problems separately since the corresponding results are essentially differ.

We consider also the problem of calculation of the recursion operator directly from given nonlinear equation. It is shown that formal (in general, singular) recursion operator exists for any Hamiltonian equation. Regular recursion operators exist only for equations in onedimensional space. The regularity of formal recursion operator is connected with the absence of the singularities of the small denominator types.

II. INTEGRABLE EQUATIONS IN 1+1 DIMENSIONS

1. The most fundamental result of the generalized AKNS-method is a construction of the infinitedimensional group of transformations of the potential of the spectral problem - so called Backlund-Calogero (BC) group. The action: $P \longrightarrow P'$ of this BC group on the manifold of the potentials $P(x,t)$ (where $P(x,t)$ is a appropriate set of the independent potentials) is given by the formula

$$\sum_k B_k(\Lambda^+, t)\,(\mathcal{H}_k P' - \mathcal{M}_k P) = 0 \qquad (1)$$

Here Λ^+ is an integro-differential operator which depends only on P and P' - the so-called recursion operator, $\mathcal{H}_k$ and $\mathcal{M}_k$ are certain integro-differential operators which depend only on P and P'. Functions $B_k(\Lambda^+, t)$ are arbitrary entire functions on Λ^+, i.e. $B_k(\Lambda^+, t) = \sum_n B_{kn}(t)(\Lambda^+)^n$. So the transformation (1) contains only the potentials P, P' and the arbitrary functions $B_{kn}(t)$. Different functions $B_{kn}(t)$ correspond to the different transformations (1). BC transformations (1) act in a simple linear manner on the manifold of the scattering data. Using this fact one can show that the transformations (1) form the infinitedimensional abelian group.

The infinitedimensional abelian BC group of transformations (1) plays a fundamental role in the analysis of nonlinear systems connected with given spectral problem and their group-theoretical properties. BC group contains different subgroups. Let us consider its one-parameter subgroup given by

$$B_k = \exp - \int_t^{t'} ds\, \Omega_k(\Lambda^+, s) \tag{2}$$

where

$\Omega_k(\Lambda^+,t)$ are some (in general, arbitrary) functions entire on Λ^+. The transformations of the potential corresponding to (2) are

$$\sum_k \exp\left\{ - \int_t^{t'} ds\, \Omega_k(\Lambda^+, s)\right\} \left[\mathcal{H}_k P(s) - \mathcal{M}_k P(t) \right] = 0 \tag{3}$$

where in the operators Λ^+, $\mathcal{H}_k$ and $\mathcal{M}_k$ one must put $P'(x,t) = P(x,s)$.

For fixed functions $\Omega_k(\Lambda^+,t)$ one-parameter group of the transformations (3) determine a flow Y_Ω : $P(x,t) \to P(x,t')$, in other words, an evolution system. This evolution system is also described by certain nonlinear evolution equation. Indeed, let us consider the infinitesimal displacement in time t: $t \to t' = t+\varepsilon$ where $\varepsilon \to 0$. In this case we obtain from (3) the following equation

$$\frac{\partial P(x,t)}{\partial t} - \sum_k \Omega_k(L^+,t)\, Z_k P = 0 \tag{4}$$

where $L^+ \equiv \Lambda^+ \big|_{P'=P}$, $Z_k \equiv (\mathcal{H}_k - \mathcal{M}_k)\big|_{P'=P}$.

Thus the consideration of the infinitesimal transformation (3) leads to the nonlinear evolution equation (4). So the evolution equation (4) give the flow Y_Ω : $P(x,t) \to P(x,t')$ in the infinitesimal form. The relation (3) which does not contain the derivative $\frac{\partial P}{\partial t}$ is an "integrated" form of the evolution equation (4). For given spectral problem the choice of concrete functions Ω_k give us concrete equation of the form (4).

The nonlinear evolution equations (4) are just the equations integrable by IST method with the help of given spectral problem.

The infinitesimal group of transformations (1) is the group of general Backlund transformations for the equations (4). A subgroup of BC transformations (1) with $\partial B_k / \partial t = 0$ is the group of auto-Bakclund transformations (B group) for the equations (4): the transformations from B-group convert the solutions of definite equation of the form (4) into the solutions of the same equation. Some well-known Backlund transformations (BT) are the particular cases of these general BT (1).

B-group contains as a subgroup the infinitedimensional abelian symmetry group of the equations (4). Symmetry transformations $p \to p'$ are the transformations (1) with $B_k = \exp f_k(\Lambda^+)$, i.e. the transformations

$$\sum_k \exp f_k(\Lambda^+) \left(\mathcal{H}_k P' - \mathcal{M}_k P \right) = 0 \tag{5}$$

where $f_k(\Lambda^+)$ are arbitrary entire functions. In the infinitesimal form $P \to P' = \quad + \delta P$ symmetry transformations are

$$\delta P(x,t) = \sum_k f_k(L^+) Z_k P \tag{6}$$

The infinitedimensional abelian symmetry group can be also considered as the infinite-parameter abelian Lie group. Indeed, let us expand entire functions $f_k(L^+)$ in the power series: $f_k(L^+) = \sum_{m=0}^{\infty} f_{km} (L^+)^m$

As a result the symmetry transformations (6) is rewritten as $\delta P(x,t) = = \sum_k \sum_{m=0}^{\infty} f_{km} (L^+)^m Z_k P$, i.e. as the superposition of the infinite number of one-parameter symmetry transformations $\delta_{(k,m)} P = f_{km} (L_+)^m Z_k P$ where expansion coefficients f_{km} $(-\infty < f_{km} < +\infty)$ play a role of the transformation parameters. For the quantities $\tilde{\delta}_{(k,m)} P \equiv (f_{km})^{-1} \delta_{(k,m)} P$ we have a simple recursion formula

$$\tilde{\delta}_{(k,m+1)} P = L^+ \tilde{\delta}_{(k,m)} P \qquad (m = 0, 1, 2, \ldots) \tag{7}$$

Let us emphasize that for definite spectral problem B-group of auto Backlund transformations and symmetry group are universal ones: B-group and symmetry group are those for any equation of the form (4).

Lastly, the transformations (1) with time-dependent functions B_k ($\partial B_k / \partial t \neq 0$) are generalized Backlund transformations for the class of the equations (4): such transformations convert the solutions of given equation (4) into the solutions of the other equations of the form (4) (of course, at the fixed spectral problem).

Thus we see that BC-group of the transformations (1) contains complete information on the general group- theoretical properties of the equations integrable by given spectral problem and these integrable equations themselves.

Note that for the first time the role of transformations

of the form (1), (3) (for one simplest spectral problem) was demonstrated by Calogero [6] (see also [4]).

The generalized AKNS-method which we discuss here allows us to prove that all the equations (4) are Hamiltonian systems. Namely one can show that for the fixed spectral problem the equations (4) can be represented in the Hamiltonian form

$$\frac{\partial P(x,t)}{\partial t} = \{ P(x,t), \mathcal{H}_{-n} \}_n \quad (n \in \mathbb{Z}) \tag{8}$$

with respect to the infinite set of Hamiltonians $\mathcal{H}_{-n}$ (n = 0, ± 1, $\pm 2, \ldots$) and infinite set of Poisson brackets $\{\ ,\ \}_n$ where

$$\{F, \mathcal{H}\}_n \equiv \int_{-\infty}^{+\infty} \operatorname{tr} \left\{ \frac{\delta F}{\delta P^T(y,t)} (L^+)^n \Omega^{-1} \frac{\delta \mathcal{H}}{\delta P^T} \right\} \tag{9}$$

Here $\delta/\delta P$ denotes a variational derivative, P^T is a transpose matrix P and Ω^{-1} is the certain matrix differential operator which depends on P (in general case).

The existence of the infinite set of Hamiltonian structures is the characteristic feature of the equations integrable by IST method. This fact was firstly noted in [7,8].

From (8) and (9) it follows that the symmetry transformations are related with the integrals of motion $C_m^{(k)}$ by the formulas

$$\delta_{(k,m)} P(x,t) \equiv f_{km} (L^+)^m Z_k P = - f_{km} \{ P(x,t), C_{m-q+1}^{(k)} \}_q \tag{10}$$

$$= f_{km} (L^+)^q \Omega^{-1} \frac{\delta C_{m-q+1}^{(k)}}{\delta P^T} \quad (m = 0, 1, 2, \ldots;\ q = 0, \pm 1, \pm 2, \ldots)$$

Therefore the symmetry transformations are canonical transformations.

2. Now let us enumerate the spectral problems to which the generalized AKNS method has been applied after the paper [5].

The first problem is

$$\frac{\partial \Psi}{\partial x} = i \lambda A \Psi + i P(x,t) \Psi \tag{11}$$

where P(x,t) is N x N matrix and N is any integer. The case of diagonal matrix A and $P(x,t) \underset{|x| \to \infty}{\longrightarrow} 0$ has been considered in [9-12]: the index k takes the values k = 1, 2, ..., N; then $\mathcal{H}_k \cdot = H_k \cdot$, $\mathcal{M}_k \cdot = \cdot H_k$ where $(H_k)_{il} = \delta_{ik} \delta_{il}$ (i, k, l = 1, ..., N). Recursion operator Λ^+ acts as follows [10,11]

$$\Lambda^+ \varphi = -i \frac{\partial \varphi_A}{\partial x} + \left(\varphi_A(x) P'(x) - P(x) \varphi_A(x) \right)_F -$$

$$- i \int_{-\infty}^{x} dy \left(\varphi_A(y) P'(y) - P(y) \varphi_A(y) \right)_D P'(x) + i P(x) \int_{-\infty}^{x} dy \left(\varphi_A(y) P'(y) - P(y) \varphi_A(y) \right)_D$$

where $[A, \varphi_A] \equiv \varphi$, $(\varphi_D)_{ik} \equiv \varphi_{ii} \delta_{ik}$, $\varphi_F = \varphi - \varphi_D$. The operator $\Omega^{-1} \cdot = [A, \cdot]$.

A very important property of the operator Λ adjoint to Λ^+ is

$$\Lambda \, \Phi_F^F(x, \lambda) = \lambda \, \Phi_P^F(x, \lambda) \tag{12}$$

where $\varphi_{ml}^{ik}(x,\lambda) \equiv (F^+)'_{mk} (F^+)^{-1}_{il}$ (i, k, n, l = 1, ..., N) and F^+ and $F^{+\prime}$ are the solutions of (11) corresponding to the potentials P(x,t) and P'(x,t) with the asymptotics $F^+(F^{+\prime})_{x \to +\infty} \doteq \exp(i \lambda A x)$.

For N = 2 and $A = \begin{pmatrix} 1 & 0 \\ 0 & -1 \end{pmatrix}$ the equations (4) are the class of the equations discovered by AKNS [5]. Among these equations are well-known modified Korteweg-de Vries (mKdv) equation, nonlinear Schroedinger (NLS) equation, sine-Gordon equation. For $N > 2$ the class of the equations (4) contains in particular the multicomponent and matrix mKdv equation, NLS equation and nonabelian sine-Gordon equations.

Generalized AKNS method has been also applied to various generalizations of the problem (11), namely to the problem (11) α) with an arbitrary semisimple matrix A [13], β) with Z_2-graded valued potential P [14], γ) with $\lim_{|x| \to \infty} P(x,t) \neq 0$ [15], δ) when potential P belongs to any classic Lie algebra SL(N,C), SO(N,C) and Sp(2N,C) [16] and ε) in the case of Z_N-type reductions [17] and the reduction $P = -P^T$ [11].

The second spectral problem which was considered by AKNS method is a special quadratic on spectral parameter problem

$$\frac{\partial \Psi}{\partial x} = (\alpha \lambda^2 + 2\beta\lambda) \begin{pmatrix} I_N & 0 \\ 0 & -I_M \end{pmatrix} \Psi + (\alpha\lambda + \beta) \begin{pmatrix} 0 & Q(x,t) \\ R(x,t) & 0 \end{pmatrix} \Psi \tag{13}$$

where α and β are arbitrary constants, $I_N(I_M)$ is identical N x N (M x M) matrix, Q and R are rectangular N x M and M x N matrices. The simplest case $\alpha = 1$, $\beta = 0$, N = M = 1 was considered in [18]. General case has been studied in [19] and for Z_2-graded valued matrices Q and R (supermatrices) in [20].

The general polynomial spectral problem

$$\frac{\partial \Psi}{\partial x} = \sum_{n=1}^{N} \lambda^n P^{(n)}(x,t) \Psi \tag{14}$$

has been considered in the AKNS method framework in [19], but the explicit form of the recursion operator has not been found. A method of the calculation of the recursion operators explicity for polynomial problems was proposed in [21] for the problem

$$\frac{\partial^2 \Psi}{\partial x^2} = \sum_{k} \lambda^k u^{(k)}(x,t) \Psi \tag{15}$$

where $U^{(k)}$ are scalar functions. With the use of this method the different special types of problem (14) has been considered (see [22]).

At last, the generalized AKNS method has been also developed for the general matrix Gelfand-Dikij problem

$$\frac{\partial^N \Psi}{\partial x^N} + V_{N-1}(x,t) \frac{\partial^{N-1} \Psi}{\partial x^{N-1}} + \dots + V_1(x,t) \frac{\partial \Psi}{\partial x} + V_0(x,t) \Psi = \lambda^N \Psi \tag{16}$$

where V_0, V_1, ..., V_{N-1} are matrices of any order [23]. The equations (4) for the problem (16) admit also a gauge-invariant formulation [24] in which a well-known Miura transformations is nothing but the gauge transformation. In more details the transformation properties of the integrable equations in 1+1 dimensions are discussed in [25].

We see that in the AKNS method we start with the spectral problem, then calculate the recursion operator with the property of the type (12) and obtain all general results (1)-(10). The crucial point for the effective applicability of AKNS method for given spectral problem is the possibility to calculate the recursion operator in the explicit form. It seems that in addition to the problems (11)-(16) one can do it for some other one-dimensional problems too.

Let us note that some particular results analogous to those of this section (in particular, the recursion operators) have been obtained by the other techniques for some other spectral problems (see e.g. [26-28]).

III. INTEGRABLE EQUATIONS IN 2+1 DIMENSIONS

Now we discuss the generalization of the AKNS method for two-dimensional spectral problems. Such a generalization is not straight-forward because the two-dimensional spectral problems are essentially differ from one-dimensional ones. Up to now the two-dimensional version of the generalized AKNS method has been given only for two problems, namely for the problem [29]

$$\frac{\partial \psi}{\partial t} + A \frac{\partial \psi}{\partial y} + P(x,y,t)\psi = 0 \tag{17}$$

where P is N x N matrix, A is an arbitrary semisimple constant matrix, and for the problem [30]

$$\frac{\partial^N \psi}{\partial x^N} + V_{N-1}(x,y,t)\frac{\partial^{N-1}\psi}{\partial x^{N-1}} + \cdots + V_0(x,y,t)\psi + \frac{\partial \psi}{\partial y} = 0 \tag{18}$$

where $V_0(x,y,t)$, ..., $V_{N-1}(x,y,t)$ are matrices of arbitrary order.

The problems (17) and (18) are natural generalization of the spectral problems (11), (16) on two dimensions.

The most important feature of the two-dimensional problems (17) and (18) is that instead of the recursion operator Λ with the property (12) there exist the infinite set of the operators $\Lambda_{(m)}$ such that [29, 30]

$$\varphi^F_{(m)F}(x,y) = \Lambda_{(m)}\varphi^F_{(0)F} \tag{19}$$

where

$$\varphi^{ik}_{(m)m\ell}(x,y) \equiv \frac{\partial^m(\hat{F}^+)'_{mk}}{\partial y^m}(\check{F}^+)_{i\ell} \quad (i,k,m,\ell = 1,\ldots,N)$$

and $\check{F}^+$ is a solution of the adjoint problems (17) or (18) (for example the problem $\frac{\partial \check{F}}{\partial x} + \frac{\partial \check{F}}{\partial y}A - \check{F}P = 0$). Let us emphasize that $\Lambda_{(m)} \neq (\Lambda_{(1)})^m$.

A sense of the two-dimensional recursion operators becomes more transparent if one consider a bilocal quantity

$$\varphi^{ik}_{m\ell}(x,\tilde{y},y,\tilde{\lambda},\lambda) \equiv \left(\hat{F}^{+\prime}(x,\tilde{y},\tilde{\lambda})\right)_{mk}\left(\check{F}^+(x,y,\lambda)\right)_{i\ell} \tag{20}$$

which is transformed under the adjoint representation [31]. One has

$$\Phi_F^F(x,\tilde{y},y) = \sum_{n=0}^{\infty} (\tilde{y}-y)^n \frac{\Lambda_{(n)}}{n!} \Phi_F^F(x,y,y) \tag{21}$$

The use of the bilocal quantity (20) essentially simplifies the calculations of recursion operators and general Backlund transformations [32].

Using the recursion operators $\Lambda_{(n)}$ one can construct the infinitedimensional Backlund-Calogero group for the two-dimensional problems (17) and (18). The transformations $P \rightarrow P'$ of this group are of the form

$$\sum_{k=0}^{N} \sum_{n=0}^{\infty} b_{kn}(t) \sum_{m=0}^{N} \Lambda^{+}_{(m)} \left(\mathcal{H}_{(n,m,k)} P' - \mathcal{M}_{(n,m,k)} P\right) = 0 \tag{22}$$

where $b_{kn}(t)$ are arbitrary functions and $\mathcal{H}$, $\mathcal{M}$ are certain matrix integro-differential operators. For example, for the problem (17) $\mathcal{H}_{(n,m,k)} = \frac{n!}{m!\,(n-m)!} \frac{\partial^{n-m}}{\partial y^{n-m}} H_k \cdot$ and $\mathcal{M}_{(n,m,k)} \cdot = \delta_{nm} \cdot H_k \cdot$

The general form of the integrable equations is

$$\frac{\partial P(x,y,t)}{\partial t} + \sum_{k=0}^{N} \sum_{n=0}^{\infty} \sum_{m=0}^{n} \omega_{kn}(t) L^{+}_{(m)} \mathcal{Z}_{(n,m,k)} P = 0 \tag{23}$$

where $\omega_{kn}(t)$ are arbitrary functions. For the problem (18) P is a column with N components V_0, V_1, ..., V_{N-1}. Among the equations (23) are well-known Kadomtsev-Petviashvili (KP) equation and the equation of the resonantly interacting waves in two spatial dimensions as well the new modyfied KP equation, two-dimensional Gardner (combined KdV-mKdV) equation, two-dimensional generalizations of Sawada-Kotera and Kaup-Kupershmidt equations [31, 32].

As in the one-dimensional case the transformations (22) with time-independent b_{kn} form the infinitedimensional group B of auto Backlund transformations for the equations (23). This group contains also the symmetry transformations for the equations (23) which in the infinitesimal form are

$$\delta P(x,y,t) = \sum_{k=0}^{N} \sum_{n=0}^{\infty} \sum_{m=0}^{n} f_{kn} L^{+}_{(m)} \mathcal{Z}_{(n,m,k)} P \tag{24}$$

where f_{kn} are arbitrary constants. Infinitedimensional symmetry group (24) can be considered as the infinite-parameter Lie group with parameters f_{kn}. But in this case for the quantities

$\tilde{\delta}_{(k,m)}P \equiv \sum_{m=0}^{m} L^{+}_{(m)} \mathcal{L}_{(m,m,k)} P$ the property of the type (7) is absent.

Equations (23) and symmetry transformations (24) which contain the terms with $n > 1$ are nonlocal ones as a rule while in the one-dimensional case $\frac{\partial P}{\partial y} = 0$ these equations and transformations are local ones. In general, in the one-dimensional limit $\frac{\partial P}{\partial y} = 0$ all the equations and transformations (such as (19)-(24)) connected with the problems (17) and (18) are reduced to the corresponding equations and transformations for the one-dimensional problems (11) and (16).

Note that equations (23) and symmetry transformations (24) for KP equation are equivalent to those obtained in [33] by τ-scheme.

Recently it was shown [34] that the generalized AKNS method is effective also for the two-dimensional problem

$$\frac{\partial^N \psi}{\partial x^N} + V_{N-1}(x,y,t)\frac{\partial^{N-1}\psi}{\partial x^{N-1}} + \cdots + V_0(x,y,t)\psi + \rho(x,y,t)\frac{\partial \psi}{\partial y} = 0 \quad (25)$$

Among the equations integrable by (25) are the two-dimensional generalization of Harry-Dym equation.

IV. RECURSION OPERATORS FOR NONLINEAR HAMILTONIAN EQUATIONS

Since a recursion operator has an interesting and useful properties it is important have a method for calculation of recursion operator directly from nonlinear equation.

Firstly such an approach was formulated in [35]. Then this idea has been developed in different ways in [36-38].

Here we present short reviews of an approach proposed recently in [39].

Let us consider the nonlinear evolution system in d spatial dimension which is described by n fields $U^1(x,t)$, ..., $U^n(x,t)$ where $x = (x_1, \ldots, x_d)$ and let this system is Hamiltonian one, i.e. is representable in the form

$$\int dx' \, \Omega^{\alpha\beta}(x,x') \frac{\partial U^\beta(x',t)}{\partial t} = \nabla_x^\alpha H \qquad \alpha = 1,\ldots,n \quad (26)$$

where $\nabla_x^\alpha \equiv \dfrac{\delta}{\delta U^\alpha(x,t)}$ ($\alpha = 1, \ldots, n$), Hamiltonian H is certain functional on U^1, ..., U^n and $\Omega^{\alpha\beta}(x,x')$ is a kernel of nondegenerate linear operator which satisfies the well known closeness and skewsymmetry conditions.

<u>Definitions</u>. Operator L_N is refered as H - weak recursion operator if any it s possible power convert gradient H into gradients: $\int dx'(L^n_H)^{\alpha\beta}(x,x')\ \nabla^\beta_{x'} H = \nabla^\alpha_x H$ (n = 1, 2, 3, ...). Operator L_Ω is called Ω- weak recursion operator if any it positive power convert closed symplectic forms Ω into closed symplectic forms Ω_m: $\int dx'(L^n_\Omega)^{\alpha\beta}(x,x')\ \Omega^{\beta\gamma}(x',x") = \Omega^{\alpha\gamma}_m(x,x")$, n = 1, 2, 3, ... Operator L is refered as the strong recursion operator if it is both H - and Ω - weak recursion operator.

Recursion operator allows recurrently multiple given equation (26) without leaving the class of Hamiltonian equations. If equation (26) admit H-weak recursion operator we can construct starting from this equation the following infinite family of equations

$$\int dx'\, \Omega^{\alpha\beta}(x,x')\, \frac{\partial U^\beta(x',t)}{\partial t} = \int dx''\, (\varphi(L_H))^{\alpha\gamma}(x,x'')\ \nabla^\gamma_{x''} H$$

where $\varphi(L_H)$ is an arbitrary entire function. If equation (26) possess Ω-weak recursion operator L_Ω one can construct family of equations $\int dx'(f(L_\Omega)\Omega)^{\alpha\beta}(x,x')\ \partial_t U^\beta(x',t) = \nabla^\alpha_x H$ where $f(L_\Omega)$ is an arbitrary entire function and

$$(f(L_\Omega)\Omega)^{\alpha\beta}(x,x') \equiv \int dx''\, (f(L_\Omega))^{\alpha\gamma}(x,x'')\ \Omega^{\gamma\beta}(x'',x')$$

Strong recursion operator generates starting from (26) the following infinite family of evolution equations

$$\frac{\partial U^\alpha(x,t)}{\partial t} = \int dx'\, (\Omega^{-1}\varphi(L))^{\alpha\beta}(x,x')\ \nabla^\beta_{x'} H \tag{27}$$

where $\varphi(L)$ is an arbitrary meromorphic function. Each equation (27) possess the infinite set of integrals of motion in involution and each from these equations is Hamiltonian one with respect to the infinite family of Poisson brackets $\{\ ,\ \}$ of the form

$$\{F, H\}_f = \int dx\, dx'\ \nabla^\alpha_x F\, (\Omega^{-1} f(L))^{\alpha\beta}(x,x')\ \nabla^\beta_{x'} H \tag{28}$$

where f is an arbitrary meromorphic function.

The equations which define the weak and strong recursion operators are of the form

$$\int d\tilde{x} \{ (\nabla^\beta_{x'} L^{\alpha\delta}(x,\tilde{x}) - \nabla^\alpha_x L^{\beta\delta}(x',\tilde{x}))\, L^{\delta\gamma}(\tilde{x},x'') +$$
$$+ L^{\alpha\delta}(x,\tilde{x})\, \nabla^\delta_{\tilde{x}} L^{\delta\gamma}(x',x'') - L^{\beta\delta}(x',\tilde{x})\, \nabla^\delta_{\tilde{x}} L(x,x'') \} = 0 \tag{29}$$

$$\int dx'' \{ \nabla^{\beta}_{x'} L^{\alpha\rho}(x,x'') \nabla^{\rho}_{x''} H + L^{\alpha\rho}(x,x'') \nabla^{\beta}_{x'} \nabla^{\rho}_{x''} H$$

$$- \nabla^{\alpha}_{x} L^{\beta\rho}(x',x'') \nabla^{\rho}_{x''} H - L^{\beta\rho}(x',x'') \nabla^{\alpha}_{x} \nabla^{\rho}_{x''} H \} = 0 \qquad (30)$$

$$\nabla^{\gamma}_{x''} (L\Omega)^{\alpha\beta}(x,x') + \nabla^{\beta}_{x'} (L\Omega)^{\gamma\alpha}(x'',x) + \nabla^{\alpha}_{x} (L\Omega)^{\beta\gamma}(x',x'') = 0 \qquad (31)$$

$$(L\Omega)^{\alpha\beta}(x,x') = -(L\Omega)^{\beta\alpha}(x',x) \qquad (32)$$

Equation (30) is equivalent to the condition $L\nabla H = \nabla H_1$. Equations (31) and (32) mean the closeness and skewsymmetry of the form $(L\Omega)^{\alpha\beta}(x,x')$. The quadratic on L equation (29) we will refer as Nijenhuis equation.

<u>Theorem</u>. If operator L satisfies to equations (29) and (30) then it is a H-weak recursion operator. If operator L satisfies equations (29), (31), (32), it is a Ω-weak recursion operator. If operator L satisfies the whole system of equations (29)-(32) then it is the strong recursion operator.

In virtue of formal relation $\Omega^{-1} L = L^{+} \Omega^{-1}$ the expressions (28) and (9) for Poisson brackets equivalent, in essence. Note also that all the recursion operators L, which are calculated starting with the spectral problem, are strong recursion operators.

Let us consider the class of nonlinear translationary invariant systems which have the hamiltonians and symplectic forms of the form

$$H = \sum_{m=2}^{\infty} \int dp_2 \dots dp_m \, \delta(p_2 + \dots + p_m) H^{\gamma_2 \dots \gamma_m}_{(m) p_2 \dots p_m} a^{\gamma_2}_{p_2} \dots a^{\gamma_m}_{p_m} \qquad (33)$$

$$\Omega^{\alpha\beta}_{pq} \equiv (2\pi)^{-d} \int dx \, dx' \, \Omega^{\alpha\beta}(x,x') \exp[-ipx - iqx'] =$$

$$= \sum_{m=0}^{\infty} \int dp_1 \dots dp_m \, \delta(p + q - p_1 - \dots - p_m) \Omega^{\alpha\beta(\gamma_1 \dots \gamma_m)}_{(m) pq (p_1 \dots p_m)} a^{\gamma_1}_{p_1} \dots a^{\gamma_m}_{p_m} \qquad (34)$$

where $a^{\alpha}_{p} = (2\pi)^{-\frac{d}{2}} \int dx \, U^{\alpha}(x,t) \exp(-ipx)$. For the systems of such type it is quite natural to look for the recursion operator as the "entire" function on a^{γ}_{p} too:

$$L^{\alpha\beta}_{pq} = (2\pi)^{-d}\int dx\,dx'\, L^{\alpha\beta}(x,x')\exp[-ipx-iqx'] =$$
$$= \sum_{m=0}^{\infty}\int dp_1 \ldots dp_m\, L_{(m)\,pq\,(p_1\ldots p_m)}^{\alpha\beta\,(\gamma_1\ldots\gamma_m)}\, a^{\gamma_1}_{p_1}\ldots a^{\gamma_m}_{p_m} \tag{35}$$

Transition in (29)-(32) to the Fourier representation and use of (33)-(35) gives a system of equations for $L_{(m)\,pq\,(p_1\ldots p_m)}^{\alpha\beta\,(\gamma_1\ldots\gamma_m)}$. This system is a complete system of alge raic functional equations for the calculation of all $L_{(m)\,pq\,(p_1\ldots p_m)}^{\alpha\beta\,(\gamma_1\ldots\gamma_m)}$ which define the recursion operator L.

From this system of equations one can obtain the expressions for functions $L_{(m)\,pq\,(p_1\ldots p_m)}^{\alpha\beta\,(\gamma_1\ldots\gamma_m)}$ through the functions $H_{(m)}(p_2,\ldots,p_m)$, i.e. the Hamiltonian H. For example, in the simplest case n = 1, we have

$$L_{(1)\,pq\,(p+q)} = 3\,\frac{\varphi(p) - \varphi(q)}{f_q\,[\omega(p+q)-\omega(p)-\omega(q)]}\,H_{(3)\,-p,-q,\,p+q} \tag{36}$$

where $\varphi(p) = L_{(0),\,p,-p} = L_{(0)\,-p,\,p}$, $\Omega_{pq} = f_p\,\delta(p+q)$, $= -f_{-q}$ and $\omega(p)$ $V_{(2)\,p,-p}/f_p$ is a dispersion law. Using these expressions for $L_{(m)}(\ldots)$ one can show that any Hamiltonian equation has a formal recursion operator L_{pq}.

The existence of a formal strong recursion operator for any Hamiltonian system of the form (26), (33-34) becomes obvious if one takes into account the following three circumstances. The first one is: any system of equations (29)-(32) is invariant under the general transformations of "coordinates" U(x,t) and, therefore, the existence of the recursion operator for this system is independent of the choice of variables U(x,t). Secondly, any nonlinear system with a Hamiltonian of the form (33) can be linearized by a suitable canonical transformation [40]. The third point is: any linear equation $\frac{\partial b_p}{\partial t} = \omega(p)\,b_p$ with odd function $\omega(p)$ possesses the recursion operator of the form $L_{pq} = \varphi(p)\,\delta(p+q)$ where $\varphi(p)$ is an arbitrary even function.

Indeed, let we have the dynamical system with Hamiltonian (33). Let us linearize this system (i.e. reduce the Hamiltonian to the form $H = \int dp_1\,dp_2\,\delta(p_1+p_2)\,H_{(2)\,p_1p_2}\,b_{p_1}b_{p_2})$ by the canonical transformation $a_p \to b_p$:

$$a_p = b_p + \sum_{m=2}^{\infty}\int dp_2\ldots dp_m\,\delta(p-p_2-\ldots-p_m)\,b_{p_2}\ldots b_{p_m} \tag{37}$$

Using the condition of the canonical character of the transformation (37) for $\Omega_{pq} = f_p \, \delta(p+q)$, i.e.

$$\{a_p, a_q\}_b = \int dk \, \frac{\delta a_p}{\delta b_k} \, \frac{1}{f_k} \, \frac{\delta a_q}{\delta b_{-k}} = \frac{1}{f_p} \, \delta(p+q) = \{a_p, a_q\}_a$$

one can express the functions $R_{(m)p}(p_2 \cdots p_m)$ through the functions $H_{(m)}(p_2, \ldots, p_m)$. The linear equation $\partial b_p / \partial t = \omega(p) \, b_p$, which appears after this canonical transformation, possesses the recursion operator $L^{(linear)}_{pq} = \varphi(p) \, \delta(p+q)$ where $\varphi(p)$ is an arbitrary even function. Let us now perform the inverse canonical transformation $b_p \to a_p$ into the initial nonlinear system. The recursion operator is transformed under this transformation as follows:

$$L^{(linear)}_{pq} \qquad L_{pq} = \int dk_1 \, dk_2 \, S_{-p,-k_1} \, L^{(linear)}_{k_1 k_2} \, \tilde{S}_{k_2 q} = \tag{38}$$
$$= \int dk \, S_{-p,k} \, \varphi(k) \, \tilde{S}_{k,q}$$

where $S_{pq} \overset{def}{=} \delta b_q / \delta a_p$, $\tilde{S}_{pq} = \delta a_q / \delta b_p$. In particular, formula (38) gives the relation (36).

Thus, any Hamiltonian equation possess formal recursion operator. In the case of one field $U(x,t)$ the expression of this formal recursion operator through Hamiltonian is given by the formula (38).

In general case this formal recursion operator is a singular one due to the denominators of the forms $\omega(p+q) - \omega(p) - \omega(q)$, $\omega(p+q-k) + \omega(k) - \omega(p) - \omega(q)$, etc. The regular recursion operator, i.e. the operator for which such singularities are absent, exists only for the onedimensional space [39].

The simplest solution of equations (29)-(32) in the onedimensional space is

$$L_{pq} = (\alpha + \beta p^2) \, \delta(p+q) + \gamma \, \frac{p-q}{p} \, a_{p+q} \tag{39}$$

where α, β, γ are arbitrary constants. For $\alpha = 0$, $\beta = -1$ and $\gamma = -2$ the operator (39) is the recursion operator for KdV equation in the momentum representation.

Essentially different situation take place for $d \geq 2$. Nonlinear evolution equations in two and higher dimensional spaces have no regular recursion operators [39]. In particular, the Kadomtsev-Petviashvili equation has no regular recursion operator. So, the regular recursion operator is a purely onedimensional phenomenon.

References

1. C.S. Gardner, J.M. Green, M.D. Kruskal, R.. Miura, Phys. Rev. Lett., 19, 1095 (1967)
2. V.E. Zakharov, S.V. Manakov, S.P. Novikov, L.P. Pitaevski, Soliton Theory. The method of the Inverse Problem, Nauka, Moscow (1980)
3. M.J. Ablowitz, H. Segur, Solitons and the Inverse Scattering Transform, SIAM Philadelphia, 1981
4. F. Calogero, A. Degasperis, Spectral transform and Solitons, I. North-Holland P.C., 1982
5. M.J. Ablowitz, D.J. Kaup, A.C. Newell, H. Segur, Stud. Appl. Math., 53, 249 (1974)
6. F. Calogero, Lett. Nuovo Cimento, 14, 537 (1975)
7. F. Magri, J. Math. Phys., 19, 1156 (1978); Lecture Notes in Physics, 120, 233 (1980)
8. P.P. Kulish, A.G. Reiman, Notes of LOMI scientific seminars, 77, 134 (1978)
9. A.C. Newell, Proc. Roy. Soc. (London), A365, 283 (1979)
10. B.G. Konopelchenko, Phys. Lett., 75A, 447 (1980); preprint INP 79-82 (1979)
11. B.G. Konopelchenko,, J. Phys. A: Math. Gen., 14, 1237 (1981)
12. V.S. Gerdjikov, P.P. Kulish, Physica 3D, 549 (1981)
13. B.G. Konopelchenko, Phys. Lett., 79A, 39 (1980)
14. B.G. Konopelchenko, Phys. Lett., 95B, 83 (1980)
15. B.G. Konopelchenko, Phys. Lett., 108B, 26 (1982)
16. B.G. Konopelchenko, V.G. Mokhnachev, J. Phys. A: Math. Gen., 14, 1849 (1981)
17. B.G. Konopelchenko, Funct. Analysis and its Appl., 16, N3, 63 (1982); Lett. Math. Phys., 6, 309 (1982); Physica D (1985); preprint INP 80-223 (1980)
18. V.S. Gerdjikov, M.I. Ivanov, P.P. Kulish, Teor. Math. Phys., 44, 342 (1980)
19. B.G. Konopelchenko, J. Phys. A: Math. Gen., 14, 3125 (1981)
20. B.G. Konopelchenko, I.B. Formusatic, J. Phys. A: Math. Gen., 15, 2017 (1982)
21. L.M. Alonso, J. Math. Phys., 21, 2342 (1980)
22. V.S. Gerdjikov, M.I. Ivanov, JINR reprint E2-82-545, Dubna (1982)
23. V.G. Dubrovsky, B.G. Konopelchenko, Fortschrt. Physik, 32, 25 (1984); preprint INP 82-09 (1982)
24. B.G. Konopelchenko, Phys. Lett., 92A, 323 (1982)
B.G. Konopelchenko, V.G. Dubrovsky, Annals of Physics (NY), 156, 265 (1984); preprint INP 83-26 (1983)
V.G. Dubrovsky, B.G. Konopelchenko, preprint INP 83-57 (1983)
25. B.G. Konopelchenko, Fortschrt. Physik., 31, 253 (1983); Phys. Lett., 100B, 254 (1981)
26. M. Adler, Inventions Math., 50, 219 (1979)
27. M. Bruschi, O. Ragnisco, D. Levi, J. Math. Phys., 22, 2463 (1981)
M. Bruschi, D. Levi, O. Ragnisco, Phys. Lett., 88A, 379 (1982)
28. M. Boiti, Tu Gui-zhang, Nuovo Cim., 71B, 253 (1982)
M. Boiti, C. Laddomada, F. Pempinelli, G.Z. Tu, J. Math. Phys., 24, 2035 (1983)
29. B.G. Konopelchenko, Commun. Math. Phys., 87, 105 (1982); Phys. Lett., 86A, 346 (1981)
30. B.G. Konopelchenko, Commun. Math. Phys., 88, 531 (1983)
31. B.G. Konopelchenko, Phys. Lett., 93A, 379 (1983)
32. B.G. Konopelchenko, V.G. Dubrovsky, preprint INP N 83-115 (1983); to be published in Physica D (1985)
33. W. Oevel, B. Fuchssteiner, Phys. Lett., 88A, 323 (1982)
34. V.G. Dubrovsky, B.G. Konopelchenko, to be published; preprint INP 84-50 (1984)
35. P.J. Olver, J. Math. Phys., 18, 1212 (1977)
36. I.M. Gelfand, I.Ya. Dorphman, Funct. Anal. and its Appl., 13, N4 (1979); 14, N3 (1980)
37. B.Fuchssteiner, Prog. Theor. Phys., 65, 861 (1981); 68, 1082 (1982)

B.Fuchssteiner, A.S. Fokas, Physica 4D, 47 (1981)
38. H.H. Chen, Y.C. Lee, C.S. Liu, Physica Scripta, 20, 490 (1979)
39. V.E. Zakharov, B.G. Konopelchenko, Commun. Math. Phys., 94, 483 (1984); preprint INP N 83-151 (1983)
40. V.E. Zakharov, Izvestiya VUZov. Radiofizika, 17, 431 (1974).

HIERARCHIES OF POISSON BRACKETS FOR ELEMENTS OF THE SCATTERING MATRICES

B.G. KONOPELCHENKO*

*Laboratoire de Physique Théorique et Hautes Energies*** - Paris

ABSTRACT

The infinite families of Poisson brackets $\{S_{i_1k_1}(\lambda_1), S_{i_2k_2}(\lambda_2)\}_n$ (n = 0,1,2,...) between the elements of scattering matrices are calculated for the linear NxN matrix spectral problem and differential spectral problem of an arbitrary order.

---

* Permanent address : Institute of Nuclear Physics, Novosibirsk 90, 630090 USSR

** Postal Address : Université Pierre et Marie Curie
Tour 16 - 1er étage
4, place Jussieu 75230 PARIS CEDEX 05
- FRANCE -

I. INTRODUCTION

One of the most important problems for the nonlinear equations, solvable by the inverse scattering transform method, is the investigation of their complete integrability [1,2] . Calculation of the Poisson brackets between the elements of a scattering matrix is a main step in the construction of action-angle type variables. This calculation has been performed for some concrete equations for the simplest Poisson brackets [1-5] . On the other hand the whole infinite families of Hamiltonian structures are connected with the integrable equations [6,7] . So the problem of calculation of the families of Poisson brackets for the scattering matrix elements arises. This problem is also closely connected with the theory of the classical r-matrix which plays an important role in the Hamiltonian treatment of the integrable equations [8-11] .

In the present paper we consider two general spectral problems, namely NxN matrix spectral problem

$$\frac{\partial \psi}{\partial x} = i\lambda A\psi + iP(x)\psi \tag{1.1}$$

and differential spectral problem of an arbitrary order

$$\frac{\partial^N \psi}{\partial x^N} + V_{N-2}(x)\frac{\partial^{N-2}\psi}{\partial x^{N-2}} + \dots + V_1(x)\frac{\partial \psi}{\partial x} + V_0(x)\psi = \lambda^N \psi \ . \tag{1.2}$$

It is shown that the hierarchies of the Poisson brackets between the elements of the scattering matrices for the spectral problems (1.1) and (1.2) are of the form

$$\begin{aligned} &\{S_{i_1k_1}(\lambda_1), S_{i_2k_2}(\lambda_2)\}_n = \\ &= r^{(n)+}_{i_1i_2m_1m_2}(\lambda_1,\lambda_2) S_{m_1k_1}(\lambda_1) S_{m_2k_2}(\lambda_2) - \\ &- S_{i_1m_1}(\lambda_1) S_{i_2m_2}(\lambda_2) r^{(n)-}_{m_1m_2k_1k_2}(\lambda_1,\lambda_2) + t^{(n)}_{i_1k_1i_2k_2}(\lambda_1,\lambda_2) S_{i_1k_1}(\lambda_1) S_{i_2k_2}(\lambda_2) \ . \end{aligned} \tag{1.3}$$

The explicit expressions for $r^{(n)\pm}_{i_1 i_2 m_1 m_2}$ and $t^{(n)}_{i_1 k_1 i_2 k_2}$ are found.

Matrices $r^{(n)+}$ and $r^{(n)-}$. are the usual classical r -matrices [8,9] . Matrices $t^{(n)}$ are of the projection type. From the point of view of the theory of the classical r-matrix, it is important that the third term in (1.3) is essentially distinguished from the first two terms. The existence of this $t^{(n)}$-type term is the main feature of the higher Poisson brackets $\{ , \}_n$.

II. MATRIX SPECTRAL PROBLEM

We will consider the problem (1.1) with diagonal matrix A : $A_{ik} = \delta_{ik} a_i$ (i,K = 1,..., N), $a_i \neq a_K$, all a_i are real constants and P(x,t) is an NxN matrix such that $P_{ii} = 0$ (i = 1,..., N) and $P(x) \xrightarrow[|x| \to \infty]{} 0$.

The infinite family of Poisson brakets $\{ , \}_n$ which correspond to the problem (1.1) looks like [12,13]

$$\{\mathcal{F}, \mathcal{H}\}_n = -\frac{1}{4}\int_{-\infty}^{+\infty} dx\, tr\left(\frac{\delta \mathcal{F}}{\delta P^T} (L_A^+)^n \left[A, \frac{\delta \mathcal{H}}{\delta P^T}\right]\right) - (\mathcal{F} \leftrightarrow \mathcal{H}) \tag{2.1}$$

where n = 0,1,2,3,..., $\mathcal{F}$ and $\mathcal{H}$ are arbitrary functionals and operator L_A^+ acts as follows $L_A^+ Q = - L^+ Q_A$ where

$$L^+ \cdot = i\frac{\partial}{\partial x} + [P(x), \cdot]_F + i\left[P(x), \int_{-\infty}^{x} dy [P(y), \cdot]_D\right] \tag{2.2}$$

and

$$[A, Q_A] \stackrel{def}{=} Q \ , \quad (Q_D)_{ik} \stackrel{def}{=} \delta_{ik} Q_{ii} \ , \quad Q_F \stackrel{def}{=} Q - Q_D \ .$$

Scattering matrix $S(\lambda)$ for the problem (1.1) is defined in the usual manner as $F^+(x,\lambda) = F^-(x,\lambda) S(\lambda)$ where $F^+(x,\lambda)$, $F^-(x,\lambda)$ are the solutions of the problem (1.1) such that $F^{\pm}(x,\lambda) \xrightarrow[x\to\pm\infty]{} \exp i\lambda Ax$.

An important quantity is the tensor product $\varphi^{in}_{kl}(x,\lambda) \stackrel{def}{=} F^{+}_{kn}(x,\lambda)\left(F^{-}(x,\lambda)\right)^{-1}_{il}$ $(i,K,n,l = 1,\ldots, N)$. This quantity satisfies the following relations [12,14]

$$\frac{\delta S_{in}(\lambda)}{\delta P_{kl}(x)} = -i\,\varphi^{in}_{lk}(x,\lambda) \qquad (i,k,l,n = 1,\ldots,N) \tag{2.3}$$

and

$$\begin{aligned} L^{+}\,\varphi^{in}_{F}(\lambda) &= -\lambda\left[A, \varphi^{in}_{F}(x,\lambda)\right] - \left[P(x), \varphi^{in}_{D}(\lambda, x=-\infty)\right], \\ L\,\varphi^{in}_{F}(\lambda) &= \lambda\left[A, \varphi^{in}_{F}(x,\lambda)\right] + \left[P(x), \varphi^{in}_{D}(\lambda, x=+\infty)\right] \end{aligned} \tag{2.4}$$

where L is the operator adjoint to L^{+}.

The relations (2.3) and (2.4) are basic ones for the calculation of $\{S_{i_1k_1}(\lambda_1), S_{i_2k_2}(\lambda_2)\}_n$. Using (2.3) one has

$$\{S_{i_1k_1}(\lambda_1), S_{i_2k_2}(\lambda_2)\}_n =$$

$$= \frac{1}{4}\int_{-\infty}^{+\infty} dx\, tr\left(\varphi^{i_1k_1}_{F}(x,\lambda_1)\left(L^{+}_{A}\right)^{n}\left[A, \varphi^{i_2k_2}_{F}(x,\lambda_2)\right]\right) - (1\leftrightarrow 2). \tag{2.5}$$

The use of (2.4) gives $\left(\mathcal{P}^{(n)}_{i_1k_1i_2k_2}(\lambda_1,\lambda_2) \stackrel{def}{=} \frac{1}{2}\int_{-\infty}^{+\infty} dx\, tr\left(\varphi^{i_1k_1}_{F}(x,\lambda_1)\left(L^{+}_{A}\right)^{n}\left[A, \varphi^{i_2k_2}_{F}(\lambda_2)\right]\right)\right)$

$$\mathcal{P}^{(n)}_{i_1k_1i_2k_2}(\lambda_1,\lambda_2) = \lambda_2\,\mathcal{P}^{(n-1)}_{i_1k_1i_2k_2}(\lambda_1,\lambda_2) + K^{(n-1)}_{i_1k_1i_2k_2}(\lambda_1,\lambda_2) \tag{2.6}$$

and

$$K^{(n)}_{i_1k_1i_2k_2}(\lambda_1,\lambda_2) = \lambda_1\,K^{(n-1)}_{i_1k_1i_2k_2}(\lambda_1,\lambda_2) \tag{2.7}$$

where

$$K^{(n)}_{i_1k_1i_2k_2}(\lambda_1,\lambda_2)=\frac{1}{2}\int_{-\infty}^{+\infty}dx\,tr\left(\varphi_F^{i_1k_1}(\lambda_1)\left(L_A^{+}\right)^n\left[P,\ \varphi_0^{i_2k_2}(\lambda_2,\ x=-\infty)\right]\right).$$

As a results one gets

$$\mathcal{P}^{(n)}_{i_1k_1i_2k_2}(\lambda_1,\lambda_2)=\lambda_2^n\,\mathcal{P}^{(0)}_{i_1k_1i_2k_2}(\lambda_1,\lambda_2)+$$
$$+\left(\lambda_2^{n-1}+\lambda_2^{n-2}\lambda_1+\dots+\lambda_2\lambda_1^{n-2}+\lambda_1^{n-1}\right)K^{(0)}_{i_1k_1i_2k_2}(\lambda_1,\lambda_2)\,. \tag{2.8}$$

Quantity $\mathcal{P}^{(0)}$ can be calculated by the trick proposed in the cases N = 2,3 in [3,4]. One obtains for real λ_1 and λ_2

$$\mathcal{P}^{(0)}_{i_1k_1i_2k_2}(\lambda_1,\lambda_2)=\Gamma^{(0)+}_{i_1i_2m_1m_2}(\lambda_1,\lambda_2)\,S_{m_1k_1}(\lambda_1)\,S_{m_2k_2}(\lambda_2)-$$
$$-\,S'_{i_1m_1}(\lambda_1)\,S'_{i_2m_2}(\lambda_2)\,\Gamma^{(0)-}_{m_1m_2k_1k_2}(\lambda_1,\lambda_2) \tag{2.9}$$

where

$$\Gamma^{(0)\pm}_{m_1m_2k_1k_2}(\lambda_1,\lambda_2)=-\frac{i}{2}\,\delta_{m_1k_1}\,\delta_{m_2k_2}\left(\frac{\delta_{k_1k_2}}{\lambda_1-\lambda_2}\pm\right.$$
$$\left.\pm\,i\pi\,\mathrm{sgn}(a_{k_1}-a_{k_2})\,\delta(\lambda_1-\lambda_2)\right) \tag{2.10}$$

where $\mathrm{sgn}(a)=\begin{cases}1, & a>1\\ 0, & a=0\\ -1, & a<1\end{cases}$.

The quantity $K^{(0)}$ is easily calculated by the use of the equation $\frac{\partial\varphi}{\partial x}=i\lambda[A,\varphi]+i[P,\varphi]$. Namely one has

$$K^{(0)}_{i_1k_1i_2k_2}(\lambda_1,\lambda_2)-(1\leftrightarrow 2)=$$
$$=\frac{i}{2}\,S_{i_1k_1}(\lambda_1)\,S'_{i_2k_2}(\lambda_2)\left(\delta_{k_1i_2}-\delta_{i_1k_2}\right). \tag{2.11}$$

Taking into account the relations (2.8)-(2.11) we finally obtain [15] the formula (1.3) where

$$\Gamma^{(n)\pm}_{ikml}(\lambda_1,\lambda_2)=-\frac{i}{4}(\lambda_1^n+\lambda_2^n)\delta_{km}\delta_{il}\left(\frac{\delta_{ml}}{\lambda_1-\lambda_2}\pm i\pi\,\mathrm{sgn}(a_m-a_l)\,\delta(\lambda_1-\lambda_2)\right)$$

(2.12)

and

$$t^{(n)}_{i_1k_1i_2k_2}(\lambda_1,\lambda_2)=\frac{i}{4}\,\frac{\lambda_1^n-\lambda_2^n}{\lambda_1-\lambda_2}\,(\delta_{i_2k_1}-\delta_{i_1k_2})\,. \qquad (2.13)$$

There is no summation over repeated indices in the third term in (1.3).

We emphasize that for the problem (1.1) the classical r-matrices $\Gamma^{(n)+}(\lambda_1,\lambda_2)$ and matrices $t^{(n)}(\lambda_1,\lambda_2)$ are bounded for arbitrary real λ_1 and λ_2.

III. GENERAL DIFFERENTIAL SPECTRAL PROBLEM

We will consider the problem (1.2) with the scalar potentials $V_K(x)$ ($K = 0,1,\ldots, N-2$) such that $V_K(x) \xrightarrow[|x|\to\infty]{} 0$. The infinite family of Poisson brackets is of the form [16-18]

$$\{\mathcal{F},\mathcal{H}\}_n=\frac{1}{2N}\int_{-\infty}^{+\infty}dx\,\mathrm{tr}\left(\frac{\delta\mathcal{F}}{\delta V^T}(L_+^+)^n\,\mathcal{Y}\,\frac{\delta\mathcal{H}}{\delta V}\right)-(\mathcal{F}\leftrightarrow\mathcal{H}) \qquad (3.1)$$

where $V\equiv(V_0, V_1,\ldots, V_{N-2})^T$. Here $\mathcal{Y}$ is the Hamiltonian operator which was calculated by Gelfand and Dikij [16] and L_+^+ is the so-called recursion operator. For the scattering matrix S and elements of the tensor product of the solutions of (1.2) we have the relations [17,18]

$$\frac{\delta S_{ik}(\lambda)}{\delta V_{l-1}(x)}=i\,\varphi^{ik}_{lN}(x,\lambda)\equiv i\,\chi^{ik}_{l}(x,\lambda) \qquad (3.2)$$

$$(l=1,\ldots,N-1)$$

and

$$L_\pm\,\chi^{in}(\lambda)=\lambda^N\chi^{in}(x,\lambda)-\tilde{G}_\pm\,\chi^{in}(\lambda,x=\pm\infty) \qquad (3.3)$$

where L_+ and L_- are operators adjoint to $\overset{+}{L}$. They are distinguished only by the form of operator inverse to $\frac{\partial}{\partial x}$: $\partial_+^{-1} = -\int_x^{\infty}$, $\partial_-^{-1} = \int_{-\infty}^{x}$.

From (3.3) is follows that

$$\lambda^N \mathcal{Y} \chi^{in}(\lambda) = \mathcal{Y} L_- \chi^{in}(\lambda) \overset{def}{=} \Gamma \chi^{in}(\lambda). \quad (3.4)$$

The method of calculation of $\{S_{i_1 m_1}(\lambda_1), S_{i_2 m_2}(\lambda_2)\}_n$ is similar to those given for matrix spectral problem. Using (3.2) one gets

$$\{S_{i_1 k_1}(\lambda_1), S_{i_2 k_2}(\lambda_2)\}_n = -\frac{1}{2N}\int_{-\infty}^{+\infty} dx\, tr\left(\chi_T^{i_1 k_1}(\lambda_1)\left(\overset{+}{L}_+\right)^n \mathcal{Y}\, \chi^{i_2 k_2}(\lambda_2)\right) - \quad (1 \leftrightarrow 2) =$$

$$\equiv \mathcal{P}^{(n)}_{i_1 k_1, i_2 k_2}(\lambda_1, \lambda_2) - \quad (1 \leftrightarrow 2).$$

Taking into account (3.3) one gets

$$\mathcal{P}^{(n)}_{i_1 k_1, i_2 k_2}(\lambda_1, \lambda_2) = \lambda_2^N \mathcal{P}^{(n-1)}_{i_1 k_1, i_2 k_2}(\lambda_1, \lambda_2) + K^{(n-1)}_{i_1 k_1, i_2 k_2}(\lambda_1, \lambda_2), \quad (3.5)$$

$$K^{(n)}_{i_1 k_1, i_2 k_2}(\lambda_1, \lambda_2) = \lambda_1^N \; K^{(n-1)}_{i_1 k_1, i_2 k_2}(\lambda_1, \lambda_2) \quad (3.6)$$

where

$$K^{(n)}_{i_1 k_1, i_2 k_2}(\lambda_1, \lambda_2) = -\frac{1}{2N}\int_{-\infty}^{+\infty} dx\, tr\left(\chi_T^{i_1 k_1}(\lambda_1)\left(\overset{+}{L}_+\right)^n \left(\overset{+}{L}_+ \mathcal{Y} - \mathcal{Y} L_-\right) \chi^{i_2 k_2}(\lambda_2)\right).$$

Recurrent relations (3.5) and (3.6) gives

$$\mathcal{P}^{(n)}_{i_1 k_1, i_2 k_2}(\lambda_1, \lambda_2) = \left(\lambda_2^N\right)^n \mathcal{P}^{(0)}_{i_1 k_1, i_2 k_2}(\lambda_1, \lambda_2) +$$
$$+ \left(\left(\lambda_2^N\right)^{n-1} + \left(\lambda_2^N\right)^{n-2}\lambda_1^N + \dots + \lambda_2^N\left(\lambda_1^N\right)^{n-2} + \left(\lambda_1^N\right)^{n-1}\right) K^{(0)}_{i_1 k_1, i_2 k_2}(\lambda_1, \lambda_2) \quad (3.7)$$

where

$$\mathcal{P}^{(0)}_{i_1k_1i_2k_2}(\lambda_1,\lambda_2) = -\frac{1}{2N}\int_{-\infty}^{+\infty} dx\, tr\left(\mathfrak{X}_T^{i_1k_1}(\lambda_1)\, \mathcal{Y}\, \mathfrak{X}^{i_2k_2}(\lambda_2)\right) , \tag{3.8}$$

$$K^{(0)}_{i_1k_1i_2k_2}(\lambda_1,\lambda_2) = -\frac{1}{2N}\int_{-\infty}^{+\infty} dx\, tr\left(\mathfrak{X}_T^{i_1k_1}(\lambda_1)\left(L_+^{+}\mathcal{Y} - \mathcal{Y}L_-\right) \mathfrak{X}^{i_2k_2}(\lambda_2)\right) . \tag{3.9}$$

Taking into account (3.7) we have

$$\begin{aligned} &\{ S_{i_1k_1}(\lambda_1), S_{i_2k_2}(\lambda_2)\}_n = \\ &= (\lambda_2^N)^n \mathcal{P}^{(0)}_{i_1k_1i_2k_2}(\lambda_1,\lambda_2) - (\lambda_1^N)^n \mathcal{P}^{(0)}_{i_2k_2i_1k_1}(\lambda_2,\lambda_1) + \\ &+ \frac{(\lambda_2^N)^n - (\lambda_1^N)^n}{\lambda_2^N - \lambda_1^N}\left(K^{(0)}_{i_1k_1i_2k_2}(\lambda_1,\lambda_2) - K^{(0)}_{i_2k_2i_1k_1}(\lambda_2,\lambda_1)\right) . \end{aligned} \tag{3.10}$$

In order to calculate $\mathcal{P}^{(0)}$ one can use the following relation

$$(\lambda_1^N - \lambda_2^N)\mathcal{P}^{(0)}_{i_1k_1i_2k_2}(\lambda_1,\lambda_2) = \frac{1}{2N}\int_{-\infty}^{+\infty} dx\, tr\left(\mathfrak{X}_T^{i_1k_1}(\lambda_1)\left(\Gamma - \lambda_1^N \mathcal{Y}\right) \mathfrak{X}^{i_2k_2}(\lambda_2)\right) . \tag{3.11}$$

Taking into account that the first Hamiltonian structure $\mathcal{Y}$ and the second Hamiltonian structure Γ are purely differential operators one can transform the integrand in r.h.s. of (3.11) into the total derivative. As a result we obtain the formula (1.3) where [19]

$$\begin{aligned} \Gamma^{(n)+}_{i_1i_2m_1m_2}(\lambda_1,\lambda_2) = -\frac{1}{\lambda_1^N - \lambda_2^N}\Big\{ &(\lambda_2^N)^n \Delta\left(\bar{\bar{\mathfrak{X}}}^{i_1m_1}(\lambda_1), \bar{\bar{\mathfrak{X}}}^{i_2m_2}(\lambda_2)\right) + \\ &+ (\lambda_1^N)^n \Delta\left(\bar{\bar{\mathfrak{X}}}^{i_2m_2}(\lambda_2), \bar{\bar{\mathfrak{X}}}^{i_1m_1}(\lambda_1)\right)\Big\}\Big|_{x\to-\infty} \end{aligned} \tag{3.12a}$$

$$\Gamma^{(n)-}_{m_1 m_2 k_1 k_2}(\lambda_1,\lambda_2) = -\frac{1}{\lambda_1^N-\lambda_2^N}\Big\{(\lambda_2^n)^N\,\Delta\big(\overset{++}{\chi}{}^{m_1k_1}(\lambda_1),\overset{++}{\chi}{}^{m_2k_2}(\lambda_2)\big) +$$

$$+(\lambda_1^N)^n\,\Delta\big(\overset{++}{\chi}{}^{m_2k_2}(\lambda_2),\overset{++}{\chi}{}^{m_1k_1}(\lambda_1)\big)\Big\}\Big|_{x\to+\infty} \qquad (3.12b)$$

where $\overset{--}{\chi}{}^{ik}_{e} \overset{\text{def}}{=} F^-_{ek}(F^-)^{-1}_{iN}$, $\overset{++}{\chi}{}^{ik}_{e} \overset{\text{def}}{=} F^+_{ek}(F^+)^{-1}_{iN}$,

$$F^{\pm}_{ik} \xrightarrow[x\to\pm\infty]{} \frac{1}{\sqrt{N}}(\lambda q^{k-1})^{i-1}\exp(iq^{k-1}\lambda x) \ , \quad q=\exp\frac{2\pi i}{N}$$

and operator Δ acts as follows $\left(\partial\equiv\frac{\partial}{\partial x}\right)$

$$\Delta(\chi^{i_1k_1}(\lambda_1),\chi^{i_2k_2}(\lambda_2)) = \frac{1}{2N}\sum_{p,q=1}^{N-1}\sum_{m=0}^{2N-p-q}\big((i\partial)^m\chi_p^{i_1k_1}(\lambda_1)\big)\gamma_{pq}(-i\partial)^{2N-p-q-m}\chi_q^{i_2k_2}(\lambda_2) -$$

$$-\frac{\lambda_1^N}{2N}\sum_{p,q=1}^{N-1}\sum_{m=0}^{N-p-q}\big((i\partial)^m\chi_p^{i_1k_1}(\lambda_1)\big)\,j_{pq}(-i\partial)^{N-p-q-m}\chi_q^{i_2k_2}(\lambda_2) \qquad (3.13)$$

where

$$\gamma_{pq} = \sum_{e=1}^{N} C^{p-1}_{N-e}\Big[C^{q-e}_{N}\theta(q-e) - \frac{1}{N}C^{e}_{N}C^{q-1}_{N}\Big](-1)^{N-q-e}(1-\delta_{e1})\theta(N+e-p-q) \qquad (3.14)$$

$$j_{pq} = \Big[C^{p-1}_{N-q} + (-1)^{N-p-q}C^{q-1}_{N-p}\Big]\theta(N-p-q) \qquad (3.15)$$

and $C^n_m = \frac{m!}{(m-n)!\,n!}$, $\theta(q)=\begin{cases}1, & q\geq 0\\ 0, & q<0.\end{cases}$

Taking into account the asymptotic behaviour of $\overset{-}{\chi}$ and $\overset{++}{\chi}$ one can rewrite the formulas (3.12) as follows

$$\Gamma^{(n)+}_{i_1 i_2 m_1 m_2}(\lambda_1,\lambda_2) = \Pi^{(n)+}_{i_1 i_2 m_1 m_2}(\lambda_1,\lambda_2) \times$$

$$\times \frac{1}{\lambda_1^N-\lambda_2^N} \exp\left\{ i\left[(q^{m_1-1}-q^{i_1-1})\lambda_1+(q^{m_2-1}-q^{i_2-1})\lambda_2\right]x\right\}\Big|_{x\to-\infty}, \qquad (3.16)$$

$$\Gamma^{(n)-}_{m_1 m_2 k_1 k_2}(\lambda_1,\lambda_2) = \Pi^{(n)-}_{m_1 m_2 k_1 k_2}(\lambda_1,\lambda_2) \times$$

$$\times \frac{1}{\lambda_1^N-\lambda_2^N} \exp\left\{ i\left[(q^{k_1-1}-q^{m_1-1})\lambda_1+(q^{k_2-1}-q^{m_2-1})\lambda_2\right]x\right\}\Big|_{x\to+\infty} \qquad (3.17)$$

where $\Pi^{(n)^+}$ and $\Pi^{(n)^-}$ are polynomials on λ_1, λ_2 which can be easily calculated with the use of (3.12), (3.13) and (3.14).

It is obvious from (3.16) and (3.17) that elements of the classical r-matrices are not well defined for all values of λ_1 and λ_2. In order the matrix $\Gamma^{(n)+}_{i_1 i_2 m_1 m_2}(\lambda_1,\lambda_2)$ be bounded, λ_1 and λ_2 should satisfy the inequality

$$\mathrm{Im}\left[(q^{i_1-1}-q^{m_1-1})\lambda_1+(q^{i_2-1}-q^{m_2-1})\lambda_2\right]\geqslant 0. \qquad (3.18)$$

For $\Gamma^{(n)-}_{m_1 m_2 k_1 k_2}(\lambda_1,\lambda_2)$ the condition of boundnessness is

$$\mathrm{Im}\left[(q^{k_1-1}-q^{m_1-1})\lambda_1+(q^{k_2-1}-q^{m_2-1})\lambda_2\right]\geqslant 0. \qquad (3.19)$$

In order that the whole r.h.s. of (1.3) be bounded it is necessary that the inequalities (3.18), (3.19) should take place for any m_1 and m_2 (m_1, m_2 = 1,..., N).

For the spectral problem (1.2) only the elements $S_{1K}(\lambda)$ (K = 1,..., N) of the scattering matrix are independent. So it is sufficient to consider the Poisson brackets $\{S_{1K_1}(\lambda_1),\ S_{1K_2}(\lambda_2)\}_{(n)}$.

For such Poisson brackets the system of inequalities (3.18) and (3.19) with m_1, m_2 = 1,..., N is equivalent to the following system of inequalities

$$\mathrm{Im}\left[\left(1-q^{m_1-1}\right)\lambda_1\right] \geqslant 0 , \tag{3.20a}$$

$$\mathrm{Im}\left[\left(q^{k_1-1}-q^{m_1-1}\right)\lambda_1\right] \geqslant 0 \tag{3.20b}$$

$(m_1 = 1, \ldots, N)$

and

$$\mathrm{Im}\left[\left(1-q^{m_2-1}\right)\lambda_2\right] \geqslant 0 , \tag{3.21a}$$

$$\mathrm{Im}\left[\left(q^{k_2-1}-q^{m_2-1}\right)\lambda_2\right] \geqslant 0 . \tag{3.21b}$$

$(m_2 = 1, \ldots, N)$

It is not difficult to show that system of inequalities (3.20) gives $\left(q = \exp\frac{2\pi i}{N}\right)$

$$\frac{\pi}{2}+\frac{\pi}{N} \geqslant \arg\lambda_1 \geqslant \frac{\pi}{2}-\frac{\pi}{N} , \tag{3.22a}$$

$$\frac{\pi}{2}+\frac{\pi}{N} \geqslant \arg(q^{k_1-1}\lambda_1) \geqslant \frac{\pi}{2}-\frac{\pi}{N} . \tag{3.22b}$$

Analogously from (3.21) one has

$$\frac{\pi}{2}+\frac{\pi}{N} \geqslant \arg\lambda_2 \geqslant \frac{\pi}{2}-\frac{\pi}{N} \tag{3.23a}$$

$$\frac{\pi}{2}+\frac{\pi}{N} \geqslant \arg(q^{k_2-1}\lambda_2) \geqslant \frac{\pi}{2}-\frac{\pi}{N} . \tag{3.23b}$$

Thus the Poisson brackets $\{S_{1K1}(\lambda_1), S_{1K2}(\lambda_2)\}_{(n)}$ are bounded only if λ_1 and λ_2 belong to certain regions which are determined by the inequalities (3.22) and (3.23).

The situation is quite different for different N. For N = 2 system (3.22) gives a well known result : $\pi \geq \arg\lambda_1 \geq 0$ for $K_1 = 1$ and $\mathrm{Im}\,\lambda_1 = 0$ for $K_1 = 2$. Similarily for λ_2. For N = 3 one has $q = \exp\frac{2\pi i}{3}$. As a result system (3.22) gives : for $K_1=1$, $\frac{5\pi}{6} \geq \arg\lambda_1 \geq \frac{\pi}{6}$; for $K_1 = 2$, $\arg\lambda_1 = \frac{\pi}{6}$ and $K_1 = 3$, $\arg\lambda_1 = \frac{5\pi}{6}$ So for N = 3 the boundnessness condition fix $\arg\lambda$ for K = 2,3. For $N \geq 4$ one can show that the inequalities (3.22a) and (3.22b) are consistent only for K = 1, K = 2 and K = N. So in the case $N \geq 4$ the Poisson brackets $\{S_{1K_1}(\lambda_1), S_{1K_2}(\lambda_2)\}_n$ are bounded only for K_1, K_2 = 1,2,N.

The circumstance that the Poisson brackets $\{s_{i1K1}(\lambda_1), s_{i2K2}(\lambda_2)\}_n$ are bounded only for certain special values of $\arg\lambda_1$ and $\arg\lambda_2$ is the main feature of the spectral problem (1.2) for $N \geq 3$.

For λ_1 and λ_2 which belong to the regions of boundnessness (3.22), (3.23) one can calculate all elements of $r^{(n)+}$ and $r^{(n)-}$. These elements contain both pole type terms $\left(\sim \frac{1}{\lambda_1^N - \lambda_2^N}\right)$ and delta-function type terms. The explicit expressions for $r^{(n)\pm}_{iklm}(\lambda_1, \lambda_2)$ are given in the paper [19].

One can also shown that the elements of matrix $t^{(n)}$ are bounded just in the same regions as $r^{(n)}$. For λ_1 and λ_2 which belong to these regions one can calculate $t^{(n)}$ explicitly. Namely, one has [19]

$$t^{(n)}_{i_1k_1i_2k_2}(\lambda_1, \lambda_2) = \frac{1}{2}\,\frac{(\lambda_2^N)^n - (\lambda_1^N)^n}{\lambda_2^N - \lambda_1^N}\times$$

$$\times\sum_{K=1}^{N-1}\Big\{\lambda_1^N \mathcal{D}_{N-K,i_2}(\lambda_2)\mathcal{D}^{-1}_{i_2N}(\lambda_2)\left[\mathcal{D}_{Ki_1}(\lambda_1)\mathcal{D}^{-1}_{i_1N}(\lambda_1) - \mathcal{D}_{Kk_1}(\lambda_1)\mathcal{D}^{-1}_{k_1N}(\lambda_1)\right] - \tag{3.24}$$

$$-\lambda_2^N \mathcal{D}_{N-K,i_1}(\lambda_1)\mathcal{D}^{-1}_{i_1N}(\lambda_1)\left[\mathcal{D}_{Ki_2}(\lambda_2)\mathcal{D}^{-1}_{i_2N}(\lambda_2) - \mathcal{D}_{Kk_2}(\lambda_2)\mathcal{D}^{-1}_{k_2N}(\lambda_2)\right]\Big\}$$

where

$$\mathcal{D}_{nK} = \frac{1}{\sqrt{N}}\left(\lambda q^{K-1}\right)^{n-1} .$$

The knowledge of $r^{(n)}$ and $t^{(n)}$ gives us the possibility to calculate the action-angle type variables. For N = 3 see [19].

The similar problem with boundnessness of the classical r-matrices $r^{(n)}$ arise also for other spectral problems, for instance for the matrix problem (1.1) in the case $\lim_{|x|\to\infty} P(x) \neq 0$. This problem will be considered elsewhere.

REFERENCES

1 V.E. Zakharov, S.V. Manakov, S.P. Novikov, L.P. Pitaevski, Theory of solitions. Method of Inverse Problem, Moscow, Nanka, 1980.

2 L.D. Faddeev, in "Solitons", p. 339, Topics in Current Physics, vol. 17 (1980), Eds. R. Bullough, P. Caudrey.

3 V.E. Zakharov, S.V. Manakov, Teor. Mat. Fyz., 19, 332 (1974).

4 S.V. Manakov, Teor. Mat. Fyz., 28, 172 (1976).

5 L.A. Takhtajan, L.D. Faddeev, Teor. Mat. Fyz., 21, 160 (1974).

6 F. Magri, J. Math. Phys. 19, 1156 (1978).

7 P.P. Kulish, A.G. Reiman, Notes of LOMI Scientific seminars, 77, 134 (1978).

8 E.K. Sklyanin, Notes of LOMI Scientific Seminars, 95, 55 (1980).

9 P.P. Kulish, E.K. Sklyanin, Lecture Notes in Physics, 151, 61 (1982).

10 A.A. Belavin, V.G. Drinfeld, Func. Anal. and its Appl., 16, 1 (1982).

11 O. Babelon, H.J. de Vega, C.M. Maillet, Nucl. Phys. B200 (FS4), 266 (1982).

12 B.G. Konopelchenko, J. Phys. A : Math. Gen., 14, 1237 (1981).

13 V.S. Gerdjikov, P.P. Kulish, Physica 3D, 549 (1981).

14 B.G. Konopelchenko, Forts. Phys. 31, 253 (1983).

15 B.G. Konopelchenko, V.G. Dubrovsky, Lett. Math. Phys., 8, 273 (1984).

16 I.M. Gelfand, L.A. Dikij, Funct. Anal. and its Appl. 10, (1), 18 (1976) ; 10, (4), 13 (1976) ; 11 (2), 11 (1977) ; 12 (2), 8 (1978).

17 V.G. Dubrovsky, B.G. Konopelchenko, Forts. Phys., 32, 25 (1984).

18 B.G. Konopelchenko, V.G. Dubrovsky, Ann. Phys., (NY) 156, 265 (1984).

19 B.G. Konopelchenko, V.G. Dubrovsky, Comm. Math. Phys., (to be published).

Multidimensional Inverse Scattering and Nonlinear Equations

Adrian I. Nachman
Dept. of Mathematics, University of Rochester
Rochester, NY 14627/USA

Is it possible to duplicate in higher dimensions the great success of the one-dimensional inverse scattering transform (IST) in the exact solution of integrable nonlinear equations? The question immediately splits into two subquestions: i) is it possible to develop a genuinely multidimensional inverse scattering transform? and ii) are there corresponding nonlinear equations solvable by IST? This lecture presented recent joint work with Mark Ablowitz addressing mostly the analytical problem i). We have introduced a general method for solving a wide class of multidimensional inverse scattering problems and treating the overdeterminacy issues involved. In [1], [2] we have applied this method to the classical inverse scattering problems for:

a) the time-dependent Schrödinger operator in $\mathbb{R} \times \mathbb{R}^n$

$$i \frac{\partial}{\partial t} - \Delta_x + v(t,x)$$

b) the time-independent Schrödinger operator in $\mathbb{R}^n$

$$-\Delta + v(x)$$

c) hyperbolic systems of the form

$$\frac{\partial}{\partial x_0} - \sum_{\ell=1}^{n} J_\ell \frac{\partial}{\partial x_\ell} - \mathbf{Q}(x_0,x),$$

with J_ℓ constant real diagonal $m \times m$ matrices and $Q = (\mathbf{Q}^{ij})$ an $m \times m$ off-diagonal matrix-valued potential.

Question (i) thus appears to have a satisfactory affirmative answer. As for (ii), we note that the IST associated to c) allows us to solve the Cauchy problem for the multidimensional N-wave resonant interaction equations:

$$\frac{\partial \Omega^{ij}}{\partial t} - a_{ij} \frac{\partial \Omega^{ij}}{\partial x_0} + \sum_{\ell=1}^{n} (B_\ell^i - a_{ij} J_\ell^i) \frac{\partial \Omega^{ij}}{\partial x_\ell} + \sum_{\nu=1}^{m} (a_{i\nu} - a_{\nu j}) \Omega^{i\nu} \Omega^{\nu j} = 0,$$

with $a_{ij} = (B_\ell^j - B_\ell^i)/(J_\ell^j - J_\ell^i)$, $1 \le \ell \le n$. This provides an answer to ii) but it isn't a very satisfactory one: the question of finding such equations which are genuinely 3+1 dimensional and have soliton solutions is still open. However our study of (i) does point out explicitly the obstructions arising (only) in several variables, which a solution of (ii) must overcome.

To describe some of our results, we briefly recall the definition of the scattering operator associated with c). When Q is absent, given $f:\mathbb{R}^n \to \mathbb{R}^m$, the solution of the Cauchy problem

$$\frac{\partial u}{\partial x_0} = \sum_{\ell=1}^{n} J_\ell \frac{\partial u}{\partial x_\ell}, \quad u(0,x) = f(x)$$

is given explicitly by $u(x_0,x) = f(x + x_0 J)$ (this is shorthand notation for the vector valued function with components

$$u^i(x_0,x) = f^i(x_1 + x_0 J_1^i, \ldots, x_n + x_0 J_n^i), \quad 1 \le i \le m).$$

When Q is a reasonable potential decaying sufficiently rapidly for large $|x|$, given any square integrable f there is a unique solution of

$$\frac{\partial u}{\partial x_0} = \sum_{\ell=1}^{n} J_\ell \frac{\partial u}{\partial x_\ell} + Qu \quad \text{with} \quad u(x_0,x) \sim f(x + x_0 J) \quad \text{as} \quad x_0 \to -\infty \ ;$$

furthermore there is a unique square integrable g such that $u(x_0,x) \sim g(x + x_0 J)$ when $x_0 \to \infty$. The scattering operator is defined by $g = Sf$. A more concrete expression for S can be obtained starting from the generalized eigenfunctions $\phi(x_0,x,k)$ solutions of the integral equation $\phi(x_0,x,k) = e^{i(x+x_0 J)\cdot k} + \iint G(x_0-y_0,x-y)Q(y_0,y)\phi(y_0,y,k)\,dy_0 dy$, with k a real vector in $\mathbb{R}^n$ and G the retarded Green's function $G(x_0,x) = \frac{\theta(x_0)}{(2\pi)^n}\int e^{i(x+x_0 J)\cdot\xi}\,d\xi$; $\theta(x_0) = 1$ when $x_0 > 0$, 0 when $x_0 < 0$.

From the behaviour of ϕ for large x_0 we then find that on the Fourier transform side S is given by:

$$\widehat{Sf}(\xi) = \hat{f}(\xi) + \frac{1}{(2\pi)^n}\int_{\mathbb{R}^n} S(\xi,k)\hat{f}(k)\,dk,$$

with $S(\xi,k) = \iint e^{-i(x+x_0 J)\cdot\xi}\,Q(x_0,x)\phi(x_0,x,k)\,dx_0 dx.$

A naive variable count ($S(\xi,k)$ is a function of 2n variables while Q is a function of n+1 variables) already indicates that (for n>1) an arbitrary function $S(\xi,k)$ is not likely to correspond to any local potential Q and the problem is to characterize those which do. A characterization of admissible $S(\xi,k)$ is crucial not only for inverse scattering questions of stability, reconstruction from partial data, etc., but also for the application of IST to nonlinear equations: any evolution of Q corresponds to an evolution of S through the manifold of admissible data. Once the characterization conditions are known, the question becomes whether there are linear (or at least explicitly solvable) evolutions of $S(\xi,k)$ compatible with these conditions, corresponding to interesting (possibly non-local) equations for Q. Our method for understanding the nonlinear overdetermined inverse scattering problem is to relate it to the linear overdetermined Cauchy-Riemann equations in several complex variables. It turns out to be technically simpler to do this first for the system

$$\frac{\partial}{\partial x_0} + \sigma \sum_{\ell=1}^{n} J_\ell \frac{\partial}{\partial x_\ell} - Q(x_0,x) \quad \text{where} \quad \sigma = \sigma_R + i\sigma_I, \ \sigma_I \neq 0, \ \text{is a complex}$$

parameter. The $\bar{\partial}$ method yields a useful generalized scattering transform for this non-hyperbolic system. The compatibility conditions for $\bar{\partial}$ lead to the characterization conditions on the scattering transform; furthermore, the Bochner-Martinelli formula translates into a linear integral reconstruction equation. Let us mention in passing that this generalized IST allows us to solve a boundary value problem for the system

$$\frac{\partial Q^{ij}}{\partial t} = \frac{1}{\sigma} a_{ij} \frac{\partial \Omega^{ij}}{\partial x_0} + \sum (a_{ij} J^i_\ell - B^i_\ell) \frac{\partial \Omega^{ij}}{\partial x_\ell} + \frac{1}{\sigma} \sum (a_{i\nu} - a_{\nu j}) Q^{i\nu} Q^{\nu j}.$$

(If $Q(t,x_0,x)$ is a well-behaved solution in the half-space $t \geq 0$ then we can write, via IST, a formula relating the values of Ω at some $t>0$ to those at $t=0$; for $\sigma=i$, one can think of this as a nonlinear analogue of the Cauchy integral formula for the upper half plane.) While the above equation may not be physically significant, it serves as a prototype of how one may be able to solve some nonlinear elliptic boundary problems using IST.

The hyperbolic case $\sigma_I = 0$ can be obtained by a limiting argument; the upshot is a new scattering transform $T(\xi,k,k_I)$ depending on 3n real parameters. The important feature of $T(\xi,k,k_I)$ is that its characterization is explicit. Thus it is a good substitute for S in the study of nonlinear equations. Its usefulness for classical inverse scattering comes from the fact that there is a relatively simple linear integral equation to find $T(\xi,k,k_I)$, given $S(\xi,k)$. The interested reader is referred to [2] for this equation, the explicit characterization of T and further details and references on all of the above.

[1] A.I. Nachman and M.J. Ablowitz, A Multidimensional Inverse Scattering Method, Studies in Applied Mathematics 71: 243-250 (1984).

[2] A.I. Nachman and M.J. Ablowitz, Multidimensional Scattering for First Order Systems, Studies in Applied Mathematics 71: 251-262 (1984).

AN SL(3)-SYMMETRICAL F-GORDON EQUATION: $Z_{\alpha\beta} = \frac{1}{3}(e^{Z}-e^{-2Z})$

B. Gaffet
Service d'Astrophysique, Centre d'Etudes Nucléaires de Saclay
91191 Gif sur Yvette Cedex, FRANCE

Abstract

An equation originally derived from non-relativistic ideal gasdynamics turns out to be reducible to a Lorentz invariant nonlinear version of the Klein-Gordon equation. We present its interacting soliton solutions, which are here constructed by means of a Bäcklund transformation, starting from the "vacuum".

I - Introduction

Among the class of soliton bearing equations, one of the most well-known is the Sine-Gordon equation

$$Z_{\alpha\beta} = \sin Z \tag{1.1}$$

which first appeared in the literature with the work of Bäcklund(1) on pseudospherical surfaces. Equation (1.1) is manifestly Lorentz invariant in 1+1 dimensions, and has been proposed by various authors(2-7) as a model for elementary particles, replacing the Klein-Gordon equation. At the same time it was realized (Perring and Skyrme(8); see also Seeger et al.(9)) that the equation admits soliton solutions, which may be viewed as classical analogues of the elementary particle; thus the Sine-Gordon equation has excited considerable interest, and it has been the subject of numerous studies.

In particular, people have attempted to obtain generalizations and to extend the soliton property to higher dimensionality spaces(10). An obvious way of adding dimensions would seem to generalize Bäcklund's theory to N-dimensional pseudospherical surfaces; this probably does not work however, as noted by Chinea(11); this author also remarked that successful generalizations were more likely to come from group theory. It is well-known (see in particular ref. 11-13) that SG is based on an SL(2) structure: the next step would appear to be to look for a similar equation endowed with SL(3) symmetry, as suggested by Hermann[12]; one of the conclusions of the present work is that a symmetry higher than SL(2) may require generalization of the concept of Bäcklund transformation.

From a practical viewpoint, one may note that SG is a member of the more general class of so-called F-Gordon equations:

$$Z_{\alpha\beta} = F(Z) \tag{1.2}$$

which are manifestly Lorentz invariant as well; thus an obvious alternative approach would be to determine the class of functions F for which eq.(1.2) possesses soliton solutions, or Bäcklund transformations, or an infinite set of conservation laws. This problem has been examined by Dodd and Bullough[14,15] and others[13]; their result is that FG has Bäcklund transformations if and only if F satisfies the equation:

$$F''(Z) = k^2 F \quad (k \text{ constant}) \tag{1.3}$$

which is essentially the Sine-Gordon equation. They have also considered the case where F satisfies a general second-order equation with constant coefficients, i.e.:

$$F = A\, e^{aZ} + B\, e^{bZ} \tag{1.4}$$

to which our proposed generalization belongs : they remark that eq.(1.4) is a necessary condition for the existence of polynomial conserved densities of rank greater than two; still, they find that FG has no auto-Bäcklund transformation in that case.

The apparent discrepancy between Dodd and Bullough's negative results and those presented here appears to be resolved as follows:

Let us first recall the general definition of Bäcklund transformations (B.T.) as a pair of ordinary differential equations (o.d.e.) of the first order for the new solution Z', with coefficients depending on the "original" solution Z and its derivatives. These equations usually turn out to be simple Ricatti equations, and in such cases the B.T. can be reformulated in terms of a pair of linear, second-order o.d.e.'s for the unknown Z': that property is, without doubt, at the heart of the SL(2) symmetry underlying the Sine-Gordon equation. On the other hand, the soliton bearing equation proposed here, which is of the type (1.4), is found to be invariant under B.T.'s of a generalized type, which are constituted by a pair of linear o.d.e.'s of the third order; whence the SL(3) symmetry of the equation.

It is interesting to note that the new equation was originally derived in the context of non-relativistic ideal gasdynamics(16,17,33); it represents the exact system of the Euler equations, assuming a particular power-law entropy distribution. The generalized B.T. that we derive has its roots in a symmetry property of 1-D gasdynamics first remarked by Stanyukovich(18,19) and by Martin and Ludford(20) (see also ref. 21-24).

II - A manifestly SL(3)-symmetrical formulation, resulting from a generalized Bäcklund transformation

In this section we summarize some recent results which have already been reported elsewhere (ref. 16,17,25).

1) The Generalized Bäcklund Transformation

We shall start with the pair of equations:

$$\begin{cases} \Omega X_{\alpha\alpha} = \Omega_\alpha X_\alpha + \dfrac{\nu^3}{\sqrt{3}} X_\beta \\ \\ \Omega X_{\beta\beta} = \Omega_\beta X_\beta + \dfrac{1}{\nu^3\sqrt{3}} X_\alpha \end{cases} \tag{2.1}$$

where ν is an arbitrary constant, and the reducible factor $\sqrt{3}$ has been introduced here for convenience. With the notation:

$$\Phi \equiv X_{\alpha\beta}/\Omega \ .$$

we obtain by differentiation:

$$\begin{cases} \Omega^2 \ \Phi_\alpha = X_\alpha \left(\Omega_{\alpha\beta} - \dfrac{\Omega_\alpha \Omega_\beta}{\Omega} + \dfrac{1}{3\Omega} \right) \\ \\ \Omega^2 \ \Phi_\beta = X_\beta \left(\Omega_{\alpha\beta} - \dfrac{\Omega_\alpha \Omega_\beta}{\Omega} + \dfrac{1}{3\Omega} \right) \end{cases} \tag{2.2}$$

and hence the relation between exact differentials:

$$\Omega d\Phi = \left[(\ln\Omega)_{\alpha\beta} + \frac{1}{3\Omega^2} \right] dX \tag{2.3}$$

Thus Φ is an (a priori arbitrary) function of X only, and we may write in place of (2.2):

$$(\ln\Omega)_{\alpha\beta} + \frac{1}{3\Omega^2} = \Omega\ \Phi'(X) \tag{2.4}$$

Choosing:

$$\Phi(X) \equiv X/3 \tag{2.5}$$

eq.(2.4) becomes a partial differential equation (p.d.e.) for the single unknown Ω:

$$(\ln\Omega)_{\alpha\beta} = \frac{(\Omega^3 - 1)}{3\Omega^2} \tag{2.6}$$

- or, equivalently, with $Z \equiv \ln\Omega$:

$$Z_{\alpha\beta} = \frac{1}{3}\,(e^{Z} - e^{-2Z}) \tag{2.7}$$

It must be pointed out that Ω usually is a better choice of unknown than Z; in particular, the Painlevé expansion given in § II.6 holds for Ω, not Z.

In conclusion, equations (2.1), with the additional constraint:

$$X_{\alpha\beta} = \frac{1}{3}\,\Omega\, X \tag{2.8}$$

constitute a differential transformation of the F-Gordon equation (2.7) into another - since the equations satisfied by X differ from (2.7): they can be derived very easily by substituting $\Omega = 3X_{\alpha\beta}/X$ in eq.(2.1), and the result is a pair of <u>third-order</u>, <u>cubic homogeneous</u> equations in X:

$$\begin{cases} X_\alpha X_{\alpha\alpha\beta} - X_{\alpha\beta}\left(X_{\alpha\alpha} + \dfrac{X_\alpha^{\,2}}{X}\right) + \dfrac{\nu^3}{3\sqrt{3}}\, XX_\beta = 0 \\[2ex] X_\beta X_{\alpha\beta\beta} - X_{\alpha\beta}\left(X_{\beta\beta} + \dfrac{X_\beta^{\,2}}{X}\right) + \dfrac{1}{3\nu^3\sqrt{3}}\, XX_\alpha = 0 \end{cases} \tag{2.9}$$

We remark that these equations <u>are invariant under the very simple transformation</u>:

$$X' = 1/X \tag{2.10}$$

and, by virtue of eq.(2.8), the resulting transformation formula for Ω reads:

$$\Omega + \Omega' = \frac{6\ X_\alpha X_\beta}{X^2} \tag{2.11}$$

The above equation, with X defined by equations (2.1),(2.8), constitutes our generalized auto-Bäcklund transformation of the p.d.e. (2.6); as we shall see, it leads to N-soliton solutions, starting from the "vacuum" solution, Z = 0 (Ω = 1).

2) O.d.e. Form of the Bäcklund Transformation

It is possible to rewrite the B.T. formulae (2.1) in a form which merely involves ordinary differential equations in X: differentiating (2.1), one obtains:

$$\begin{aligned} X_{\alpha\alpha\alpha} - \frac{\Omega_{\alpha\alpha}}{\Omega} X_\alpha &= \frac{\nu^3}{3\sqrt{3}} X \\ X_{\beta\beta\beta} - \frac{\Omega_{\beta\beta}}{\Omega} X_\beta &= \frac{1}{3\nu^3\sqrt{3}} X \end{aligned} \tag{2.12}$$

which is a pair of bilinear o.d.e.'s, second-order in Ω, third-order in X.

It is worth pointing out here that a single 1st-order p.d.e. (but not a pair!) may also be derived for X, in the following way:

Introducing the notation: $\xi \equiv (\ln X)_\alpha$, $\eta \equiv (\ln X)_\beta$,

$$\begin{cases} S \equiv (\Omega + \Omega') = 6\ \xi\eta \\ P \equiv \Omega\ \Omega' \end{cases} \qquad \begin{cases} L \equiv (\Omega'\ \Omega_\alpha - \Omega\ \Omega'_\alpha) \\ M \quad (\Omega'\ \Omega_\beta - \Omega\ \Omega'_\beta) \end{cases}$$

differentiating eq.(2.11) with respect to α, and taking account of eq.(2.1), (2.8), we find:

$$\frac{(\Omega_\alpha + \Omega'_\alpha)}{6} = \frac{\Omega_\alpha}{\Omega}\xi\eta + \frac{\Omega}{3}\xi - 2\ \xi^2\eta + \frac{\nu^3}{\Omega\sqrt{3}}\eta^2$$

We substitute $\eta = S/6\xi$ in order to eliminate all β-derivatives, and get:

$$2P\xi^3 - L\xi^2 - \frac{\nu^3}{6\sqrt{3}} S^2 = 0$$

which is an o.d.e. relating the functions Ω, Ω' and X. We may also choose to eliminate ξ in favor of η, in which case the above equation becomes:

$$PS - 3L\eta - 6\sqrt{3}\,\nu^3\,\eta^3 = 0\ .$$

Exchanging the roles of α and β we similarly obtain:

$$2P\eta^3 - M\eta^2 - \frac{S^2}{6\nu^3\sqrt{3}} = 0\ .$$

The elimination of η between the last two equations produces an algebraic relation between P,S,L,M, i.e. a (strongly nonlinear) first-order p.d.e. for the unknown function Ω'.

It is also worth recalling that eq.(2.8) itself is a linear, second-order p.d.e. for Ω'.

3) An SL(3)-symmetrical Formulation

It is well-known that the solutions of a 3rd-order linear o.d.e. form a vector space of dimension 3; we accordingly may rewrite eq.(2.12) in vector form, by merely replacing the symbol X by $\vec{X}$:

$$\vec{X}_{\alpha\alpha\alpha} - \frac{\Omega_{\alpha\alpha}}{\Omega}\vec{X}_\alpha = \frac{\nu^3}{3\sqrt{3}}\vec{X}$$

$$\vec{X}_{\beta\beta\beta} - \frac{\Omega_{\beta\beta}}{\Omega}\vec{X}_\beta = \frac{1}{3\nu^3\sqrt{3}}\vec{X} \qquad (2.13)$$

and the same holds of the equations (2.1):

$$\text{(a)}\quad \Omega\vec{X}_{\alpha\alpha} = \Omega_\alpha\vec{X}_\alpha + \frac{\nu^3}{\sqrt{3}}\vec{X}_\beta$$

$$\text{(b)}\quad \Omega\vec{X}_{\beta\beta} = \Omega_\beta\vec{X}_\beta + \frac{1}{\nu^3\sqrt{3}}X_\alpha \qquad (2.14)$$

Equations (2.13) indicate that the triple product $(\vec{X},\vec{X}_\alpha,\vec{X}_{\alpha\alpha})$ is independent of α; and from eq.(2.14) together with (2.8), it does not depend on β either; therefore it is a constant; and the same is true of the triple product $(\vec{X},\vec{X}_\beta,\vec{X}_{\beta\beta})$. Furthermore, the triple product formed with the 1st-order derivatives:

$$(\vec{X},\vec{X}_\alpha,\vec{X}_\beta)$$

is easily shown from eq.(2.14) to be proportional to Ω:

$$\Omega(\vec{X},\vec{X}_\alpha,\vec{X}_{\alpha\alpha}) = \frac{\nu^3}{\sqrt{3}}(\vec{X},\vec{X}_\alpha,\vec{X}_\beta)$$

$$\Omega(\vec{X},\vec{X}_\beta,\vec{X}_{\beta\beta}) = \frac{1}{\nu^3\sqrt{3}}(\vec{X},\vec{X}_\alpha,\vec{X}_\beta) \qquad (2.15)$$

We may then choose the normalization contraints as follows:

$$(\vec{X},\vec{X}_\alpha,\vec{X}_{\alpha\alpha}) = \nu^3/3\sqrt{3}$$

$$(\vec{X},\vec{X}_\beta,\vec{X}_{\beta\beta}) = 3/\nu^3\sqrt{3} \qquad (2.16)$$

and obtain, from (2.15):

$$(\vec{X},\vec{X}_\alpha,\vec{X}_\beta) = \Omega/3 \qquad (2.17)$$

In this way the original equation for Ω (eq.(2.6)) has been transformed into a single 3-D vector equation for $\vec{X}$:

$$\vec{X}_{\alpha\beta} = (\vec{X},\vec{X}_\alpha,\vec{X}_\beta)\vec{X} \qquad (2.18)$$

which is manifestly SL(3)-invariant.

Unlike eq.(2.6), the system (2.18) is of the 4th-order; the fact that it can be reduced to second-order is a consequence of its invariance under arbitrary gauge transformations of the characteristic coordinates α, β, of the type:

$$\alpha' = f(\alpha), \qquad \beta' = g(\beta) ,$$

together with the gauge-invariant properties:

$$\partial_\beta(\vec{X}, \vec{X}_\alpha, \vec{X}_{\alpha\alpha}) = 0$$

$$\partial_\alpha(\vec{X}, \vec{X}_\beta, \vec{X}_{\beta\beta}) = 0 \qquad (2.19)$$

which obviously result from (2.16) but may also be derived from (2.18). The above equations (2.19) establish the quantities $(\vec{X}, \vec{X}_\alpha, \vec{X}_{\alpha\alpha})$, $(\vec{X}, \vec{X}_\beta, \vec{X}_{\beta\beta})$ - or, rather, closely related quantities - as Riemann invariants, whose existence ensures reducibility of the order of the system by two units. The presence of Riemann invariants for the equation (2.18) was first noted in ref. 26.

4) A Duality Transformation

From $\vec{X}$, a new vector $\vec{U}$, which will turn out to possess the same properties as $\vec{X}$ itself, may be constructed:

$$\vec{U} = \frac{3\sqrt{3}}{\nu^3} \vec{X}_\alpha \wedge \vec{X}_{\alpha\alpha} \tag{2.20}$$

First we note that the scalar product $\vec{U}.\vec{X}$ is unity, owing to the normalization constraint (2.16):

$$\vec{U}.\vec{X} = 1 \tag{2.21}$$

From eq.(2.13) the following relation is easily derived:

$$\vec{U}_\alpha = - \vec{X} \wedge \vec{X}_\alpha \tag{2.22}$$

That a similar relation holds vs variable β is almost as easily demonstrated:

$$\vec{U}_\beta = \frac{3\sqrt{3}}{\nu^3} \left\{ \vec{X}_{\alpha\beta} \wedge \vec{X}_{\alpha\alpha} + \vec{X}_\alpha \wedge \partial_\alpha (\vec{X}_{\alpha\beta}) \right\}$$

and, substituting eq.(2.8) for $\vec{X}_{\alpha\beta}$, eq.(2.14) for $\vec{X}_{\alpha\alpha}$, we find:

$$\vec{U}_\beta = + \vec{X} \wedge \vec{X}_\beta \tag{2.23}$$

The system of equations (2.21-23) is complete, and is equivalent to eq.(2.18). It is straightforward to show that it is also equivalent to the following system, obtained by exchanging the roles of $\vec{X}$ and $\vec{U}$ (see ref. 17):

$$\left\{ \begin{aligned} \vec{X}_\alpha &= + \vec{U} \wedge \vec{U}_\alpha \\ \vec{X}_\beta &= - \vec{U} \wedge \vec{U}_\beta \end{aligned} \right. \tag{2.24}$$

together with: $\vec{U}.\vec{X} = 1$ (eq.2.21). The operation of exchanging $\vec{X}$ and $\vec{U}$ may be called the duality transformation.

That the system is effectively of the 4th-order may be seen from the fact that a subsystem of 4 equations involving 4 unknowns only (e.g. U_1, U_2, X_1, X_3) may be isolated.

The two dual vectors $\vec{X}$ and $\vec{U}$ may serve to construct new potentials, or conservation laws, such as (ref. 17):

$$d\,A_{ij} = X_i(\vec{X} \wedge d\vec{X})_j + U_i(\vec{U} \wedge d\vec{U})_j \tag{2.25}$$

where $(\vec{X} \wedge \vec{Y})_i$ denotes the i component of the cross-product.

5) On the Applicability of the Inverse Scattering Transform

We show in the present article that the F-Gordon equation (2.7) admits soliton solutions (§ III), and possesses the Painlevé property as defined by Weiss, Tabor and Carnevale[27] for partial differential equations (§ II.6). According to the well-known Ablowitz-Ramani-Segur conjecture[28,29], the Inverse Scattering Transform (IST) method ought to be applicable.

We observe that the parameter ν, which arises as a consequence of Lorentz invariance, plays the role of eigenvalue in eq.(2.12), or in its vector form (2.13) as well, in the same way as the spectral parameter does in the case of Lax pairs[30,31].

As already noted the 3rd-order nature of our problem is not accidental, and reflects both the underlying SL(3) symmetry and (equivalently?) the fact that our B.T. is of a generalized type, involving higher-order equations; whereas in the Sine-Gordon case, both the SL(2) symmetry and the presence of B.T.'s of ordinary type naturally result in the emergence of linear equations which are of second-order, such as, in particular, the Lax equation.

In view of the above considerations, the possibility of adapting the IST method to SL(3)-symmetrical cases should be considered as doubtful, unless one resorts to a third-order scattering problem (instead of the simple Schrödinger equation which is used for solving the KdV and SG equations)[33].

6) The Painlevé Expansion

Recently, Weiss, Tabor and Carnevale[27] have proposed a general definition of the Painlevé property which is applicable to p.d.e.'s;

and they have shown that the main equations exhibiting solitons, including the Sine-Gordon equation, satisfy this new Painlevé criterion. They also point out that, in various cases, a Bäcklund Transformation can be derived in the form of a truncated Painlevé expansion, i.e. one which contains only a finite number of terms; they do not present such a B.T. for the Sine-Gordon case though.

The Painlevé expansion of eq.(2.6) may be written in the form (see ref. 25):

$$\Omega = \frac{6\varphi_\alpha\varphi_\beta}{\varphi^2} - \frac{6\varphi_{\alpha\beta}}{\varphi} + u_o + \sum_{j=1}^{+\infty} u_j \varphi^j \tag{2.26}$$

where the functions φ and u_o are arbitrary; there are thus two resonances as required, located at j = -2 and 0. To within that order, eq.(2.26) does not differ from the expansion derived by Weiss et al. for the SG case, and also coincides with that of the simple Liouville equation:

$$z_{\alpha\beta} = e^z (\equiv \Omega) \tag{2.27}$$

as well.

It is remarkable that our proposed B.T. (eq.(2.11)) indeed assumes the form of a truncated Painlevé expansion in the manner suggested by Weiss et al.: with X playing the role of φ, and the original solution Ω that of u_o, eq.(2.26) reads:

$$\Omega' = \frac{6\, X_\alpha\, X_\beta}{X^2} - \frac{6\, X_{\alpha\beta}}{X} + \Omega \tag{2.28}$$

which, taking account of eq.(2.8), does coincide with the B.T. formula (2.11).

Another remarkable feature of our Bäcklund transformation is that it is strictly speaking, a reciprocal transformation, i.e. the transformation squared is the identity, as is evident from the form of eqs. (2.10), (2.11); we will further comment on this point in § III.

III - Construction of Interacting Soliton Pairs, by Means of the Bäcklund Transformation

The Bäcklund transformation (2.11) is used in the present section to derive 1-soliton and 2-soliton solutions, starting from the vacuum:

$Z = 0 \quad ; \quad \Omega = 1$.

1) 1-Soliton

The potential X associated with the vacuum is determined by a pair of constant coefficient third-order equations (eqs.(2.12)):

$$\begin{cases} X_{\alpha\alpha\alpha} = \dfrac{i\lambda^3}{3\sqrt{3}} X \\ \\ X_{\beta\beta\beta} = \dfrac{1}{3i\sqrt{3}\lambda^3} X \end{cases} \tag{3.1}$$

where the constant λ (which may be complex) is arbitrary; there are three independent solutions:

$$X_k = \exp \frac{1}{i\sqrt{3}} (\lambda u_k \alpha - \beta/\lambda u_k) \qquad (k = 1,2,3) \tag{3.2}$$

where $u_1 = 1$, $u_2 = e^{+2i\pi/3}$, $u_3 = e^{-2i\pi/3}$ are the three cubic roots of unity.

Substituting any of these three elementary solutions X_k in eq.(2.11) merely gives back the vacuum solution: $\Omega' \equiv 1$; in order to obtain new solutions it is clear that we must introduce a potential X that is not separable vs coordinates α, β; we choose the (real) linear combination:

$$Y = \frac{1}{2} (X_2 + X_3) = \exp \left[\frac{i}{2\sqrt{3}} (\lambda\alpha - \beta/\lambda) \right] \cosh(\varphi/2) \tag{3.3}$$

where:

$$\varphi \equiv (\lambda\alpha + \beta/\lambda) \tag{3.4}$$

will be interpreted as the phase.

This yields the new solution:

$$\Omega' = 1 - \frac{3}{(1 + \cosh \varphi)} \tag{3.5}$$

which may also be expressed rationally in terms of a variable z:

$$z \equiv e^{\varphi} \equiv \exp(\lambda\alpha + \beta/\lambda) \qquad (3.6)$$

namely:

$$\Omega' = 1 - \frac{6z}{(z+1)^2} \qquad (3.7)$$

Furthermore there exist a family of phase-shifted solutions, which may be deduced from (3.7) by an arbitrary rescaling of z:

$$\Omega = 1 - \frac{6mz}{(z+m)^2} \qquad (3.8)$$

The solution $Z(\varphi) \equiv \ln\Omega$ exponentially decreases to zero as the phase φ tends to $\pm\infty$; it is localized, and the results of the next section (III.2) show that it represents a soliton. The soliton's propagation velocity u is clearly related to the parameter λ by:

$$\lambda = \sqrt{\frac{1-u}{1+u}} \quad ; \quad u = \frac{1-\lambda^2}{1+\lambda^2} \quad .$$

The analytical form of (3.5),(3.7) resembles, but is less simple than, the 1-soliton of the Sinh-Gordon equation: $Z_{\alpha\beta} = 1/2\ (e^Z - e^{-Z})$, $(Z \equiv \ln\Omega)$, which reads:

$$\Omega_{Sh-G} = 1 - \frac{2}{(\cosh\varphi + 1)} \equiv \frac{(z-1)^2}{(z+1)^2} \qquad (3.9)$$

In order to be able to apply the B.T. (2.11) once again, we must first determine the potentials X_k associated with the 1-soliton (3.7). The pair of equations (2.12) is now non-trivial, if still linear:

$$\begin{cases} X_{\alpha\alpha\alpha} + \dfrac{6\lambda^2 z}{(z+1)^2} X_\alpha = \dfrac{\nu^3}{3\sqrt{3}} X \\[2ex] X_{\beta\beta\beta} + \dfrac{6z}{\lambda^2(z+1)^2} X_\beta = \dfrac{1}{3\sqrt{3}\nu^3} X \end{cases} \qquad (3.10)$$

where ν may be chosen arbitrarily. Both equations reduce to the same o.d.e. vs variable z, though with different eigenvalues:

$$\begin{cases} \text{(a)}\ z^3 X'''(z) + 3z^2 X''(z) + \left[1 + \frac{6z}{(z+1)^2}\right] zX'(z) - \frac{\mu^3}{3\sqrt{3}} X = 0 \\ \text{(b)}\ z^3 X'''(z) + 3z^2 X''(z) + \left[1 + \frac{6z}{(z+1)^2}\right] zX'(z) - \frac{1}{3\sqrt{3}\mu^3} X = 0 \end{cases} \tag{3.11}$$

with: $\mu \equiv \nu/\lambda$.

These are found to be of Fuchsian type, i.e. their three singularities at $z = 0, \infty, -1$ are regular (Ince, ref. 32); in addition the three characteristic exponents at $z = -1$ are all integer, and that turns out to be an apparent singularity(25); thus we expect the solution to be integrable in closed form. An analysis of the characteristic exponents shows that any rational solution must have the form:

$$X_k(z) = \frac{P_1(z)}{(z+1)} z^{\mu_k/\sqrt{3}} \tag{3.12}$$

where $\mu_k \equiv \mu u_k$ is one of the cubic roots of μ^3, and $P_1(z)$ is polynomial in z and of the first degree. By identification one easily obtains:

$$P_1(z) \equiv z + \frac{(\mu_k^2 + \sqrt{3}\mu_k + 1)}{(\mu_k^2 - \sqrt{3}\mu_k + 1)} \tag{3.13}$$

The three elementary solutions of the combined equations (3.11a,b) are thus found as:

$$X_k = \frac{P_1(z)}{(z+1)} \exp \frac{1}{\sqrt{3}} (\nu_k\alpha + \beta/\nu_k) \tag{3.14}$$

with $\nu_k \equiv \lambda\mu_k$; since ν is arbitrary we may drop the index k, and the general expression of the elementary solution X associated with the 1-soliton (3.8) becomes, after appropriate rescaling of z:

$$\begin{cases} X = \dfrac{(z + mA)}{(z + m)} \exp \dfrac{1}{\sqrt{3}} (\nu\alpha + \beta/\nu) \\ A = \dfrac{(\nu^2 + \sqrt{3}\lambda\nu + \lambda^2)}{(\nu^2 - \sqrt{3}\lambda\nu + \lambda^2)} \\ z = \exp(\lambda\alpha + \beta/\lambda) \end{cases} \tag{3.15}$$

Clearly, if we now substitute the above expression in the B.T. formula (2.11), $(\ln X)_\alpha$ and $(\ln X)_\beta$, as well as Ω, will be functions of the single variable z only, and the "new" solution Ω' will be a mere phase-shifted 1-soliton, of the general form (3.8): once again, we must choose a non-trivial linear combination of two elementary solutions, X_1 and X_2 say.

In order to obtain the 2-soliton in its center-of-mass frame, we expect that ν should be taken proportional to $1/\lambda$. For our first potential X_1 we shall choose:

$$\nu_1 = iu_3/\lambda = (\sqrt{3}-i)/2\lambda \tag{3.16}$$

with the scaling: $m = 1/A$, in order to have a factor $(z+1)$ at the numerator; that is:

$$m = \frac{\lambda^2 - i\sqrt{3}\, u_3 - u_2/\lambda^2}{\lambda^2 + i\sqrt{3}\, u_3 - u_2/\lambda^2} \equiv \frac{(\lambda - 1/\lambda)}{(\lambda+1/\lambda)} \; \frac{[2\lambda \; -(1-i\sqrt{3})/\lambda]}{[2\lambda \; +(1-i\sqrt{3})/\lambda]} \tag{3.17}$$

For the second potential X_2 we choose:

$$\nu_2 = -\nu_1^* = +iu_2/\lambda \tag{3.18}$$

and keep the same factor $(z+m)$ at the denominator; then the numerator is proportional to: $(z+mn)$, with n given by:

$$n = \frac{\lambda^2 + i\sqrt{3}\, u_2 - u_3/\lambda^2}{\lambda^2 - i\sqrt{3}\, u_2 - u_3/\lambda^2} \tag{3.19}$$

i.e.: $n = m^*$.

From (3.17) it is straightforward to derive an (at least formally) real equation for m: separating the real and imaginary parts we find:

$$\frac{(\lambda+1/\lambda)}{(\lambda-1/\lambda)} (\lambda^2 + 1 + 1/\lambda^2) m = (\lambda^2 - 1/\lambda^2) + i\sqrt{3} \tag{3.20}$$

and, taking the square of the pure imaginary term, we obtain:

$$a_o m^2 - 2a_1 m + 1 = 0 \tag{3.21}$$

the announced result, where the three constants a_o, a_1, and Θ are:

$$\left\{\begin{array}{l} \Theta = (\lambda^2 + 1/\lambda^2) \equiv \dfrac{2(1+u^2)}{(1-u^2)} , \\ a_1 = \dfrac{(\Theta+2)}{(\Theta-1)} , \\ a_o = \dfrac{(\lambda+1/\lambda)^2(\lambda^2+1+1/\lambda^2)}{(\lambda-1/\lambda)^2(\lambda^2-1+1/\lambda^2)} \equiv \dfrac{(\Theta+1)(\Theta+2)}{(\Theta-1)(\Theta-2)} \equiv \dfrac{(1+3/u^2)}{(1+3u^2)} \end{array}\right. \tag{3.22}$$

The parameter $n = m^*$ is clearly the second root of the equation (3.21) (in fact, whether λ is real or not); and accordingly the numerator in X_2 is just: (a_oz+1). Our choice of linear combination is thus:

$$Y = X_1 + X_2 ,$$

$$\left\{\begin{array}{l} X_1 = \dfrac{(z+1)}{(z+m)} \exp \dfrac{+1}{\sqrt{3}} (\nu_1\alpha + \beta/\nu_1) \\ X_2 = \dfrac{(a_oz+1)}{(z+m)} \exp \dfrac{-1}{\sqrt{3}} (\nu_1^*\alpha + \beta/\nu_1^*) \end{array}\right. \tag{3.23}$$

That is:

$$\left\{\begin{array}{l} Y = \dfrac{D(y,z)}{(z+m)} \exp \left\{ \dfrac{i}{2\sqrt{3}} (- \dfrac{\alpha}{\lambda} + \lambda\beta) + \dfrac{1}{2} (\dfrac{\alpha}{\lambda} + \lambda\beta) \right\} , \text{ with:} \\ D(y,z) \equiv 1 + (y+z) + a_oyz \\ y \equiv \exp - (\frac{\alpha}{\lambda} + \lambda\beta) \end{array}\right. \tag{3.24}$$

Applying now the Bäcklund Transformation (2.11) with Y as potential instead of X produces the new solution:

$$\Omega' = 1 - \frac{6N}{D^2} \tag{3.25}$$

with:

$$\left\{\begin{array}{l} N(y,z) \equiv (y+z)-2a_1yz + a_oyz(y+z) \\ D(y,z) \equiv 1 +(y+z) + a_oyz \end{array}\right.$$

The above equation (3.25) represents the interaction of two solitons in their center-of-mass frame, each propagating with velocity $\pm$ u. In the limit $y \to 0$ we recover the 1-soliton formula, $\Omega = 1 - 6z/(z+1)^2$ and in the limit $y \to \infty$, the phase-shifted 1-soliton:

$\Omega = 1 - \dfrac{6a_oz}{(a_oz+1)^2}$; thus ln a_o is the phase-shift resulting from the interaction.

For comparison, the Sinh-Gordon 2-soliton formula reads:

$$\Omega = \frac{[1 - (y+z) + a_o yz]^2}{[1 + (y+z) + a_o yz]^2} \tag{3.26}$$

where the phase-shift parameter is: $a_o = 1/u^2$.

We remark that, in the Sinh-Gordon case, the "original" soliton to which the B.T. is applied has a phase which is just the arithmetic average of the two phases that obtain respectively before and after interaction. In our case we started from a 1-soliton of the form: $\Omega = 1 - \frac{6mz}{(z+m)^2}$, which has <u>complex</u> phase-shift ln m with respect to the $y \to 0$ limit, and a complex conjugate shift ln m* with respect to the other limit ($y \to \infty$).

It is remarkable that (complex) superpositions of elementary solutions X had to be performed in order to obtain new solutions by means of the Bäcklund transformation: in the present case the nonlinear superposition principle is thus most clearly related to the ordinary linear superposition which holds in the vector space $\{\vec{X}\}$.

IV - An SL((N+2) Generalization, in N + 1 Dimensions

It may be shown[17] that the self-dual formulation of § II.4 (eq.2.24) is the characteristic-coordinate form of the Monge-Ampère equation:

$$s^2 - rt = 1/z^4 \tag{4.1}$$

where, following standard notation:

$p = z_x$, $q = z_y$, $r = p_x$, $s = p_y$, $t = q_y$;

and the equivalence results from the following identifications:

$$\begin{cases} x = X_1/X_3 \ , \ y = X_2/X_3 \ , \ z = 1/X_3 \\ p = U_1 \ , \ q = U_2 \end{cases} \tag{4.2}$$

Equation (4.1) admits a natural (N+1)-dimensional generalization:

$$\frac{\partial \vec{p}}{\partial \vec{x}} = f(z) \tag{4.3}$$

where $\vec{x} = (x_1,\ldots x_{N+1})$, $\vec{p} = (p_1,\ldots p_{N+1}) \equiv (z_{x_1},\ldots z_{x_{N+1}})$ and $\partial\vec{p}/\partial\vec{x}$ denotes a Jacobian, which may also be viewed as the determinant constructed with the second-order derivatives:

$$\det \left\{ z_{x_i x_j} \right\} .$$

Eq.(4.3) has an obvious SL(N+1) symmetry, whatever the form of the arbitrary function f is: if a (constant) linear transformation $\mathcal{L}$ is applied on $\vec{x}$, and $dz = \vec{p}.d\vec{x}$ is assumed invariant, the resulting transformation on $\vec{p}$ is $\tilde{\mathcal{L}}^{-1}$ and consequently:

$$\frac{\partial \vec{p}\,'}{\partial \vec{x}\,'} = (\det \mathcal{L})^{-2} \frac{\partial \vec{p}}{\partial \vec{x}} \tag{4.4}$$

which proves the announced result. Now, it can be shown that a higher symmetry SL(N+2) obtains in the particular case where $f(z) \propto 1/z^{N+3}$:

$$\frac{\partial \vec{p}}{\partial \vec{x}} = \frac{1}{z^{N+3}} \tag{4.5}$$

and a fundamental representation of the group is:

$$\begin{cases} X_i = x_i/z & (i \leqslant N+1) \\ X_{N+2} = 1/z \end{cases} \tag{4.6}$$

The point is that eq.(4.5) turns out to be invariant under the transformation:

$$\begin{cases} x'_1 = -1/x_1 \quad ; \quad x'_i = x_i/x_1 & (i > 1) \\ z' = z/x_1 \end{cases} \tag{4.7}$$

which is a mere permutation of variables X_i (up to a sign); that property enlarges the existing SL(N+1) symmetry into the higher-dimensional group SL(N+2).

A preliminary investigation reveals that the self-similar equations associated with eq.(4.5) are integrable algebraically; so that, according to the ARS conjecture(28,29), the equation may turn out to have the Painlevé property for partial differential equations as defined by Weiss et al.(27). There remains to check whether the transformation:

$$X'_i = 1/X_i \tag{4.8}$$

(for some i, arbitrarily chosen), together with the constraint of invariant characteristic surfaces, constitutes a Bäcklund transformation. With a Bäcklund transformation multi-soliton solutions can be derived starting from the "vacuum", which in the present case is represented by Primakoff-type (i.e. Sedov degenerate) self-similar solutions, whereas 1-solitons are Sedov-Taylor type.

Note:

The present work originated in a systematic study of the symmetry properties of 1 D gasdynamics, and the recent discovery (ref. 26) of the presence of a pair of Riemann Invariants for a 3-parameter class of flows; which resulted in the reduction of the Euler's equations to the nonlinear KG equation $Z_{\alpha\beta} = e^{Z}-e^{-2Z}$. The same equation has also been considered earlier by Mikhailov(33,34) in relation with a bi-dimensional Toda lattice; and by Fordy and Gibbons(35), who derived it from an analysis of Lax pairs having a third-order scattering operator. Mikhailov(36) also discussed multisoliton solutions of that equation.

References

(1) L.P. Eisenhart (1960): "Differential Geometry of Curves and Surfaces", New York: Dover.
(2) N. Rosen and H.B. Rosenstock (1952) Phys. Rev. 85, 257.
(3) T.H.R. Skyrme (1958) Proc. Roy. Soc. A247, 260.
(4) T.H.R. Skyrme (1961) Proc. Roy. Soc. A262, 237.
(5) U. Enz (1963) Phys. Rev. 131, 1392.
(6) A.C. Scott (1969) Am. J. of Phys. 37, 52.
(7) J. Rubinstein (1970) J. Math. Phys. 11, 258.
(8) J.K. Perring and T.H.R. Skyrme (1962) Nucl. Phys. 31, 550.
(9) A. Seeger, H. Donth and A. Kochendörfer (1953) Z. Phys. 134, 173.
(10) G. Leibbrandt, R. Morf and Shein-Shion Wang (1980) J. Math. Phys. 21, 1613.
(11) F.J. Chinea (1980) J. Math. Phys. 21, 1588.
(12) R. Hermann (1978), in "Solitons in Action", K. Lonngren and A. Scott eds., Academic Press.
(13) W.F. Shadwick (1978) J. Math. Phys. 19, 2312.
(14) R.K. Dodd and R.K. Bullough (1976) Proc. Roy. Soc. A351, 499; A352, 481.
(15) D.W. McLaughlin and A.C. Scott (1973) J. Math. Phys. 14, 1817.
(16) B. Gaffet (1984) Physica 11D, 287.
(17) B. Gaffet (1984) J. Math. Phys. 25, 245.
(18) K.P. Stanyukovich (1954) Dokl. Akad. Nauk. SSSR 96, 441.
(19) K.P. Stanyukovich (1960) "Unsteady Motion of Continuous Media", Pergamon Press.
(20) M.H. Martin and G.S.S. Ludford (1954) Commun. on Pure and Applied Math. 7, 45.
(21) H. Cabannes (1960) Handbuch der Physik Vol. IX, p. 200. Springer-Verlag.
(22) J.A. Steketee (1972) Qu. of Appl. Math. 30, 167.
(23) J.A. Steketee (1976) J. of Eng. Math. 10, 69.
(24) B. Gaffet (1983) J. Fluid Mech. 134, 179.
(25) B. Gaffet (1985) "A Painlevé nonlinear generalization of the Klein-Gordon equation, from nonrelativistic gasdynamics", subm. J. Math. Phys., Aug. 1985.
(26) B. Gaffet (1985) "A 3-parameter class of 1-D gas flows possessing Riemann invariants", subm. J. Fluid Mech., July 1985.
(27) J. Weiss, M. Tabor and G. Carnevale (1983) J. Math. Phys. 24, 522.
(28) M.J. Ablowitz, A. Ramani and H. Segur (1978) Nuovo Cimento 23, 233.
(29) M.J. Ablowitz, A. Ramani and H. Segur (1980) J. Math. Phys. 21, 715, 1006.
(30) P.D. Lax (1968) Commun. Pure Appl. Math. 21, 467.
(31) A.C. Scott, F.Y.F. Chu and D.W. McLaughlin (1973), Proc. I.E.E.E. 61, 1443.
(32) E.L. Ince (1956) "Ordinary differential equations" Dover, New York.
(33) I was not aware of Mikhailov's[34] work on the equation: $\varphi_{tt} - \varphi_{xx} + 2e^{4\varphi} - 2e^{-2\varphi} = 0$ until the present work has been completed. In particular, Mikhailov shows that the equation is completely integrable and can be solved by a third-order Inverse Scattering Transform. An infinite number of polynomial conserved densities has also been found to exist.
(34) A.V. Mikhailov (1980) JETP Letters 30, 414.
(35) A.P. Fordy and J. Gibbons (1981) J. Math. Phys. 22, 1170.
(36) A.V. Mikhailov (1981) Physica 3D, 73.

THE SOLUTION OF THE CARTAN EQUIVALENCE PROBLEM FOR $\frac{d^2y}{dx^2} = F(x,y,\frac{dy}{dx})$ UNDER THE PSEUDO-GROUP $\bar{x} = \varphi(x)$, $\bar{y} = \psi(x,y)$

by

N. KAMRAN
Centre de Recherches Mathématiques
Université de Montréal
Case Postale 6128, Succursale "A"
Montréal, Québec, H3C 3J7, Canada

and

W.F. SHADWICK
Department of Pure Mathematics
University of Waterloo
Waterloo, Ontario, N2L 3G1, Canada

ABSTRACT

We give a complete solution to the local equivalence problem for $\frac{d^2y}{dx^2} = F(x,y,\frac{dy}{dx})$ under the pseudo-group of coordinate transformations $\bar{x} = \varphi(x)$, $\bar{y} = \psi(x,y)$. Applying Cartan's equivalence method, we obtain an {e}-structure on $J^1(\mathbb{R},\mathbb{R}) \times G$, where G is a certain three-dimensional real Lie group. We show that except for the equivalence class of $\frac{d^2y}{dx^2} = 0$, the G-action can be used to reduce this {e}-structure on $J^1(\mathbb{R},\mathbb{R}) \times G$ to an {e}-structure on a lower-dimensional space $J^1(\mathbb{R},\mathbb{R}) \times G_{(1)}$, where the Lie group $G_{(1)}$ is at most one-dimensional. We then show how the invariants obtained by this procedure can be used to obtain necessary and sufficient conditions for equivalence.

1. INTRODUCTION

This paper is the sequel to a previous article [5] which dealt with the local equivalence problem for

$$\frac{d^2y}{dx^2} = F(x,y,\frac{dy}{dx}), \tag{1.1}$$

under the pseudo-group of coordinate transformations of the form

$$\bar{x} = \varphi(x), \quad \bar{y} = \psi(x,y). \tag{1.2}$$

The approach taken in Reference 5 was motivated by the fact that the invariants provided by Cartan's equivalence method [1] would give a complete solution to the problem. To the equivalence problem given by Eqs. (1.1) and (1.2) we associated a

G-structure on $J^1(\mathbb{R},\mathbb{R})$, where G is a certain three-dimensional real Lie group, and we were led after one prolongation of this G-structure to an {e}-structure on $J^1(\mathbb{R},\mathbb{R}) \times G$ which provided us with three invariants I_1, I_2 and I_3.

These invariants can then be used to produce necessary and sufficient conditions for equivalence without having to integrate any differential equations. If these necessary and sufficient conditions are satisfied, all the maps solving the equivalence problem are then obtained by a procedure which involves at worst integrating a system of differential equations which will by construction be involutive.

However one can, except for the case admitting the largest symmetry group, that is the equivalence class of $\frac{d^2y}{dx^2} = 0$, reduce this {e}-structure on $J^1(\mathbb{R},\mathbb{R}) \times G$ to an {e}-structure on a lower-dimensional space, thereby reducing the number of coordinate functions involved, at the price of generally introducing higher derivatives of the original data. This has been done in Reference 5 for the equivalence classes of the equations defining the six Painlevé transcendents [4] (for which one has $I_1 \neq 0$, $I_2 = I_3 = 0$), where we showed that our {e}-structure on $J^1(\mathbb{R},\mathbb{R}) \times G$ could generically be reduced to an {e}-structure on $J^1(\mathbb{R},\mathbb{R})$ by using the G-action to cast I_1 and further torsion coefficients arising in the reduction process into normal forms, thereby leaving us with invariants which are functions on $J^1(\mathbb{R},\mathbb{R})$. In particular, we used these invariants to give necessary and sufficient conditions of an algebraic nature for an equation of the form (1.1) to be equivalent to the Painlevé I or Painlevé II equations under a transformation of the form (1.2).

The purpose of the present paper is to deal with the remaining cases in the reduction process, so as to give a complete solution to the equivalence problem defined by Eqs. (1.1) and (1.2). More precisely, we show that unless the three invariants I_1, I_2 and I_3 vanish (which precisely corresponds to the most symmetric case, that is the equivalence class of $\frac{d^2y}{dx^2} = 0$, for which the structure equations of our {e}-structure are simply the Maurer-Cartan equations for the affine group in the plane [5]), the G-action can always be used to reduce our {e}-structure on $J^1(\mathbb{R},\mathbb{R}) \times G$ to an {e}-structure on a lower-dimensional space $J^1(\mathbb{R},\mathbb{R}) \times G_{(1)}$, where the reduced group $G_{(1)}$ is at most one-dimensional, thereby producing invariants which live on this lower-dimensional space and give us necessary and sufficient conditions for equivalence.

As an interesting consequence of this reduction process, we prove that an equation of the form (1.1) cannot have a five-parameter(*) group of symmetries of the form (1.2).

In Section 2, we briefly recall from reference 5 the set-up of the equivalence problem and the expression of the structure equations for the {e}-structure on $J^1(\mathbb{R},\mathbb{R}) \times G$. In Section 3, we give an exhaustive discussion of all the branches that occur in the reduction procedure, provide parametric expressions for the invariants that are obtained for each branch and show how these invariants are used to obtain necessary and sufficient conditions for equivalence.

(*) See note added before the acknowledgments.

Finally, we wish to emphasize that although the question of equivalence can be extremely intricate when dealt with in the absence of proper geometric tools, the calculations involved in this paper do not require more than exterior differentiation and basic exterior algebra, a fact that confirms the power of Cartan's equivalence method and suggests that the tests presented here could easily be programmed on a computer for practical uses.

2. THE EQUIVALENCE PROBLEM

We first briefly recall the solution given in Reference 5 to the local equivalence problem for the equation

$$\frac{d^2y}{dx^2} = F(x,y,\frac{dy}{dx}), \tag{2.1}$$

under the first prolongation

$$\begin{aligned} p^1\Phi : J^1(\mathbb{R},\mathbb{R}) &\to J^1(\mathbb{R},\mathbb{R}) \\ (x,y,p) &\to (\varphi(x),\psi(x,y),\frac{\psi_{,x}+p\psi_{,y}}{\varphi_{,x}}) \end{aligned} \tag{2.2a}$$

of point transformations of the form

$$\begin{aligned} \Phi : J^0(\mathbb{R},\mathbb{R}) &\to J^0(\mathbb{R},\mathbb{R}) \\ (x,y) &\to (\varphi(x),\psi(x,y)). \end{aligned} \tag{2.2b}$$

From the fact that Eq. (2.1) and an equation given by

$$\frac{d^2\bar{y}}{d\bar{x}^2} = \bar{F}(\bar{x},\bar{y},\frac{d\bar{y}}{d\bar{x}}), \tag{2.3}$$

will be equivalent under $p^1\Phi$ if and only if

$$(p^1\Phi)^* \begin{pmatrix} d\bar{x} \\ d\bar{y}-\bar{p}d\bar{x} \\ d\bar{p}-\bar{F}d\bar{x} \end{pmatrix} = \begin{pmatrix} A & 0 & 0 \\ 0 & B & 0 \\ 0 & BC & B/A \end{pmatrix} \begin{pmatrix} dx \\ dy-pdx \\ dp-Fdx \end{pmatrix} \tag{2.4a}$$

where A and B are nowhere vanishing functions on $J^1(\mathbb{R},\mathbb{R})$, we obtained, following a procedure given by E. Cartan [1] as described by R. Gardner [3], an {e}-structure on $J^1(\mathbb{R},\mathbb{R}) \times G$, where G is the group of matrices of the form

$$\begin{pmatrix} a & 0 & 0 \\ 0 & b & 0 \\ 0 & bc & b/a \end{pmatrix}, \qquad (2.4b)$$

with $ab \neq 0$.

The structure equations of this {e}-structure are the following

$$d\omega^1 = \alpha\Lambda\omega^1, \qquad (2.5a)$$

$$d\omega^2 = \beta\Lambda\omega^2 + \omega^1\Lambda\omega^3, \qquad (2.5b)$$

$$d\omega^3 = \gamma\Lambda\omega^2 + (\beta-\alpha)\Lambda\omega^3, \qquad (2.5c)$$

$$d\alpha = 2\omega^1\Lambda\gamma, \qquad (2.5d)$$

$$d\beta = \omega^1\Lambda\gamma + I_1\omega^2\Lambda\omega^3 + I_2\omega^1\Lambda\omega^2, \qquad (2.5e)$$

$$d\gamma = \gamma\Lambda\alpha + I_2\omega^1\Lambda\omega^3 + I_3\omega^1\Lambda\omega^2, \qquad (2.5f)$$

where $(\omega^1,\omega^2,\omega^3)$ is a collection of semi-basic 1-forms obtained by lifting the coframe $(dx,dy-pdx,dp-Fdx)$ on $J^1(\mathbb{R},\mathbb{R})$ to the space $J^1(\mathbb{R},\mathbb{R}) \times G$ of G-coframes on $J^1(\mathbb{R},\mathbb{R})$

$$\begin{pmatrix} \omega^1 \\ \omega^2 \\ \omega^3 \end{pmatrix} := \begin{pmatrix} A & 0 & 0 \\ 0 & B & 0 \\ 0 & BC & B/A \end{pmatrix} \begin{pmatrix} dx \\ dy-pdx \\ dp-F(x,y,p)dx \end{pmatrix}, \qquad (2.6a)$$

where I_1, I_2 and I_3 are invariants given by

$$I_1 := \frac{A}{2B^2} F_{,ppp}, \quad I_2 := \frac{1}{2AB}\left(\frac{dF_{,pp}}{dx} - F_{,py}\right), \qquad (2.6b)$$

$$I_3 := -CI_2 + \frac{1}{2A^2B}\left(\frac{dF_{,py}}{dx} + F_{,pp}F_{,y} - F_{,py}F_{,p} - 2F_{,yy}\right), \qquad (2.6c)$$

where (α,β,γ) is a collection of 1-forms congruent mod $\omega^1\omega^2\omega^3$ to right-invariant 1-forms on G

$$\alpha := \frac{dA}{A} - \left(2C + \frac{F_{,p}}{A}\right)\omega^1, \quad \beta := \frac{dB}{B} - C\omega^1 + \frac{F_{,pp}}{2B}\omega^2 \qquad (2.6d)$$

$$\gamma := dC + C\frac{dA}{A} + \left(\frac{F_{,y}}{A^2} - \frac{CF_{,p}}{A} - C^2\right)\omega^1 + \left(\frac{F_{,py}}{2AB} - \frac{CF_{,pp}}{2B}\right)\omega^2 + \frac{F_{,pp}}{2B}\omega^3, \qquad (2.6e)$$

and where in Eqs (2.6b,c) $\frac{d}{dx}$ denotes the total derivative operator

$$\frac{d}{dx} := \frac{\partial}{\partial x} + p\frac{\partial}{\partial y} + F\frac{\partial}{\partial p} . \tag{2.6f}$$

3. THE REDUCTION TO AN {e}-STRUCTURE ON A LOWER-DIMENSIONAL SPACE

The invariants I_1, I_2 and I_3 given by the structure equations (2.5) are functions on $J^1(\mathbb{R},\mathbb{R}) \times G$ which can be used to produce necessary and sufficient conditions for the equivalence of two equations of the form (2.1) under a transformation of the form (2.2). However, we show in the present section that unless the three invariants vanish, the natural G-action can be used to cast I_1, I_2, I_3 and further torsion coefficients arising in the reduction process into normal forms and reduce the {e}-structure on $J^1(\mathbb{R},\mathbb{R}) \times G$ given by Eqs. (2.5) to an {e}-structure on a lower-dimensional space $J^1(\mathbb{R},\mathbb{R}) \times G_{(1)}$, where the Lie group $G_{(1)}$ is at most one-dimensional, thereby reducing the number of coordinate functions involved, albeit at the price of generally introducing higher derivatives of the original data. This procedure will also allow us to identify the equivalence classes of equations admitting symmetry groups as well as the structure of these groups.

The G-action on I_1, I_2 and I_3 can easily be computed by expressing the integrability conditions for Eqs. (2.5). We have

$$(dI_2+I_2(\alpha+\beta))\wedge\omega^1\wedge\omega^2 + (dI_2+I_1(2\beta-\alpha))\wedge\omega^2\wedge\omega^3 = 0, \tag{3.1a}$$

$$(dI_3+I_3(2\alpha+\beta)+\gamma I_2)\wedge\omega^1\wedge\omega^2 + (dI_2+I_2(\alpha+\beta))\wedge\omega^1\wedge\omega^3 = 0, \tag{3.1b}$$

from which it follows that

$$dI_1 + I_1(2\beta-\alpha) \equiv 0 \bmod \omega^1\omega^2\omega^3, \tag{3.2a}$$

$$dI_2 + I_2(\alpha+\beta) \equiv 0 \bmod \omega^1\omega^2\omega^3, \tag{3.2b}$$

$$dI_3 + I_3(2\alpha+\beta) + \gamma I_2 \equiv 0 \bmod \omega^1\omega^2\omega^3. \tag{3.2c}$$

Eqs. (3.2) lead us to consider an exhaustive set of cases in the reduction procedure. These cases are defined by conditions which are invariant under all transformations of the form (2.2a) and which are expressed in terms of the function F(x,y,p) appearing in Eq. (2.1) and its derivatives.

CASE A: $I_1 I_2 \neq 0$, that is

$$F_{,ppp}\left(\frac{dF_{,pp}}{dx} - F_{,py}\right) \neq 0. \tag{3.3a}$$

From Eqs. (3.2) it follows that we can use the G-action to scale I_1 and I_2 to one translate I_3 to zero. The group G has now been reduced to the identity and we obtain easily from Eqs. (3.1)

$$\alpha = k\omega^1 + \ell\omega^2 + m\omega^3, \tag{3.3b}$$

$$\beta = r\omega^1 + s\omega^2 + (k-2r-m)\omega^3, \tag{3.3c}$$

$$\gamma = u\omega^1 + v\omega^2 + (\ell+s)\omega^3, \tag{3.3d}$$

so that from Eqs. (2.5) and (3.3), we obtain the following {e}-structure on $J^1(\mathbb{R},\mathbb{R})$

$$d\omega^1 = \ell\omega^2\wedge\omega^1 + m\omega^3\wedge\omega^1, \tag{3.4a}$$

$$d\omega^2 = r\omega^1\wedge\omega^2 + (k-2r-m)\omega^3\wedge\omega^2 + \omega^1\wedge\omega^3, \tag{3.4b}$$

$$d\omega^3 = u\omega^1\wedge\omega^2 + 2\ell\omega^3\wedge\omega^2 + (r-k)\omega^1\wedge\omega^3. \tag{3.4c}$$

The parametric expressions of the five invariants k,ℓ,m,r,u appearing in Eqs. (3.4) are easily computed using Eqs. (3.6). Indeed, after scaling I_1 and I_2 to one and translating I_3 to zero, we have, using Eqs. (2.6b,c)

$$A = -(2F_{,ppp})^{-1/3}\left(\frac{dF_{,pp}}{dx} - F_{,py}\right)^{2/3}, \tag{3.5a}$$

$$B = -\left[\frac{F_{,ppp}}{4}\left(\frac{dF_{,pp}}{dx} - F_{,py}\right)\right]^{1/3}, \tag{3.5b}$$

$$C = -(2F_{,ppp})^{1/3}\left(\frac{dF_{,pp}}{dx} - F_{,py}\right)^{-5/3}\left(-2F_{,yy} + \frac{dF_{,py}}{dx} - F_{,yp}F_{,p}+F_{,pp}F_{,y}\right), \tag{3.5c}$$

from which it follows, using Eqs. (2.6a,d,e,f) that

$$k = \frac{1}{A^2}\frac{dA}{dx} - 2C - \frac{F_{,p}}{A}, \quad \ell = \frac{1}{AB}(A_{,y}-ACA_{,p}) \tag{3.6a}$$

$$m = \frac{A_{,p}}{B}, \quad r = \frac{1}{AB}\frac{dB}{dx} - C, \tag{3.6b}$$

$$u = \frac{1}{A}\frac{dC}{dx} + \frac{C}{A^2}\frac{dA}{dx} + \frac{F_{,y}}{A^2} - \frac{CF_{,p}}{A} - C^2, \tag{3.6c}$$

where A, B and C are given by Eqs. (3.5a,b,c). To test for the equivalence of two equations of the form

$$\frac{d^2y}{dx^2} = F(x,y,\frac{dy}{dx}), \quad \frac{d^2\bar{y}}{d\bar{x}^2} = \bar{F}(\bar{x},\bar{y},\frac{d\bar{y}}{d\bar{x}}), \tag{3.6d}$$

satisfying the invariant conditions defining Case A, one performs for both equations the group reduction yielding Eqs. (3.4), (3.5) and (3.6). Necessary and sufficient conditions for equivalence are [2] that there be the same number of constant invariants amongst k,ℓ,m,r,u as there are amongst $\bar{k},\bar{\ell},\bar{m},\bar{r},\bar{u}$, that constant invariants with the same label take the same numerical values and that choosing amongst k,ℓ,m,r,u and their covariant derivatives a maximal set of $p \le 3$ functionally independent invariants $\{J_a | 1 \le a \le p\}$ while making the identical choice amongst $\bar{k},\bar{\ell},\bar{m},\bar{r},\bar{u}$ to obtain a set $\{\bar{J}_a | 1 \le a \le p\}$, one should have

$$dJ_a = F_{ai}(J_b)\omega^i, \quad d\bar{J}_a = F_{ai}(\bar{J}_b)\bar{\omega}^i. \tag{3.6e}$$

If these conditions are satisfied, all the equivalences $f : J^1(\mathbb{R},\mathbb{R}) \to J^1(\mathbb{R},\mathbb{R})$ are obtained by solving

$$\bar{J}_a \circ f = J_a, \tag{3.6f}$$

which together with the relations

$$f^*\bar{\omega}^i = \omega^i, \tag{3.6g}$$

form a completely integrable system, the general solution of which depends on 3-p arbitrary constants which parametrize the 3-p dimensional group of symmetries of the equation. It should be remarked that some of the relations between the invariants and their covariant derivatives may be automatically satisfied as a result of Eqs. (3.3) and (3.4). For example the integrability condition for (3.4a) tells us that

$$\ell_{,3} - m_{,2} = \ell(2r-k-m). \tag{3.6h}$$

Such relations therefore do not need to be checked when testing for equivalence as they are direct consequences of Eqs. (3.4). Also, let us remark that since we were able to reduce our original $\{e\}$-structure on $J^1(\mathbb{R},\mathbb{R}) \times G$ to an $\{e\}$-structure on $J^1(\mathbb{R},\mathbb{R})$ there are no equations within Case A admitting a symmetry group of dimension greater than three. Finally, it should be remarked that the equivalence class of $\frac{d^2y}{dx^2} = 0$ is the only class admitting a six-parameter symmetry group, as it is easily shown that Eqs. (2.5) will be the Maurer-Cartan equations of a Lie algebra if and only if $I_1 = I_2 = I_3 = 0$.

CASE B: $I_1 = 0$, $\bar{I}_2 \neq 0$, that is

$$F(x,y,p) = p^2M(x,y) + pN(x,y) + Q(x,y), \tag{3.7a}$$

$$2M_{,x} - N_{,y} =: G(x,y) \neq 0. \tag{3.7b}$$

From Eqs. (3.2b,c) we see that we can scale I_2 to one and translate I_3 to zero, in which case Eqs. (3.1) tell us that

$$\beta = -\alpha + a\omega^1 + f\omega^2, \tag{3.8a}$$

$$\gamma = c\omega^1 + e\omega^2 + f\omega^3. \tag{3.8b}$$

Parametrically, the choice made to scale I_2 to one and translate I_3 to zero amounts an account of Eqs. (2.6) and (3.7) to setting

$$2AB = G, \quad AC = \frac{pG_{,y}+H}{G}, \tag{3.8c}$$

where

$$H(x,y) := -2Q_{,yy} + N_{,yx} - N_{,y}N + 2(QM)_{,y}, \tag{3.8d}$$

in which case the parametric expressions of the functions a, c, e and f are easily computed using Eqs. (2.6d,e,f)

$$a = \frac{1}{A}\,(-2p((\log G)_{,y}+M) + (\log G)_{,x}-N-3\,\frac{H}{G}), \tag{3.9a}$$

$$c = \frac{1}{A^2}\left[\frac{d}{dx}\,(\frac{pG_{,y}+H}{G})+p^2M_{,y}+pN_{,y}+Q_{,y}-(\frac{pG_{,y}+H}{G})(2pM+N)-(\frac{pG_{,y}+H}{G})^2\right], \tag{3.9b}$$

$$e = \frac{2}{G}\left[(\frac{pG_{,y}+H}{G})_{,y} - \frac{1}{2}\,((\frac{pG_{,y}+H}{G})^2)_{,p}\right] + \frac{1}{G}\,(2pM_{,y}+N_{,y}) - 2(\frac{pG_{,y}+H}{G^2})M, \tag{3.9c}$$

$$f = \frac{2A}{G}\,((\log G)_{,y}+M). \tag{3.9d}$$

The structure equations (2.5) become, an account of Eqs. (3.8a,b)

$$d\omega^2 = \alpha\wedge\omega^1, \tag{3.9e}$$

$$d\omega^2 = -\alpha\wedge\omega^2 + a\omega^1\wedge\omega^2 + \omega^1\wedge\omega^3, \tag{3.9f}$$

$$d\omega^3 = -2\alpha\wedge\omega^3 + c\omega^1\wedge\omega^2 + a\omega^1\wedge\omega^3, \tag{3.9g}$$

$$d\alpha = 2e\omega^1\wedge\omega^2 + 2f\omega^1\wedge\omega^3. \tag{3.9h}$$

We thus have an {e}-structure on $J^1(\mathbb{R},\mathbb{R}) \times G_{(1)}$, where $G_{(1)}$ os now one-dimensional, and we have to perform one further reduction in order to obtain an {e}-structure on

$J^1(\mathbb{R},\mathbb{R})$. We do so by computing the action of the one-dimensional reduced group $G_{(1)}$ on the non-constant components a, c, e and f of the structure tensor. Again this is most easily done by expressing the integrability conditions for Eqs. (3.9) which read

$$(da+a\alpha)\wedge\omega^1\wedge\omega^2 + 2f\omega^1\wedge\omega^2\wedge\omega^3 = 0 \tag{3.10a}$$

$$(dc+2c\alpha)\wedge\omega^1\wedge\omega^2 + (da+a\alpha)\wedge\omega^1\wedge\omega^3 + 4e\omega^1\wedge\omega^2\wedge\omega^3 = 0, \tag{3.10b}$$

$$de\wedge\omega^2\wedge\omega^1 + (df+f\alpha)\wedge\omega^3\wedge\omega^1 = 0. \tag{3.11c}$$

We are now led to consider several subcases.

<u>CASE B.I</u>: $a \neq 0$ or $c \neq 0$ or $f \neq 0$. Assume that $a \neq 0$, that is

$$[(\log G)_{,y}+M]^2 + [(\log G)_{,x}-N-3\,\tfrac{H}{G}]^2 \neq 0. \tag{3.11a}$$

It follows from Eq. (3.10a) that we can normalize a to one using the $G_{(1)}$-action, in which case we have

$$a = k\omega^1 + \ell\omega^2 - 2f\omega^3, \tag{3.11b}$$

and the final {e}-structure is given by

$$d\omega^1 = \ell\omega^2\wedge\omega^1 - 2f\omega^3\wedge\omega^1, \tag{3.11c}$$

$$d\omega^2 = (1-k)\omega^1\wedge\omega^2 + 2f\omega^3\wedge\omega^2 + \omega^1\wedge\omega^3, \tag{3.11d}$$

$$d\omega^3 = (1-2k)\omega^1\wedge\omega^3 + c\omega^1\wedge\omega^2 - 2\ell\omega^2\wedge\omega^3. \tag{3.11e}$$

We now have invariants c, f, k and ℓ on $J^1(\mathbb{R},\mathbb{R})$. Their parametric expressions easily computed using Eqs. (3.9), (2.6d) and the fact that we have

$$A = -2p((\log G)_{,y}+M) + (\log G)_{,x} - N - 3\,\frac{H}{G}, \tag{3.11f}$$

once a has been normalized to one. Necessary and sufficient conditions for the equivalence of two equations of the form (3.6d) satisfying the invariant conditions (3.7a,b) and (3.11a) are now obtained by a procedure identical to that described for Case A.

The reduction procedure from an {e}-structure on $J^1(\mathbb{R},\mathbb{R}) \times G_{(1)}$ to an {e}-structure on $J^1(\mathbb{R},\mathbb{R})$ in the case $a = 0$ is similar to the one given when $a \neq 0$ since from Eqs. (3.10) we see that we can use the $G_{(1)}$-action to normalize f to one if it is non-zero or C to $\varepsilon = \frac{|c|}{c}$ if it is non-zero, thereby reducing $G_{(1)}$ to {e}.

CASE B.II: $a = c = f = 0$.
This case is impossible as the conditions $a = c = f = 0$ lead to a contradiction. Indeed, from Eqs. (3.10) we obtain that $e = 0$ so that Eqs. (3.8a,b) reduce to $\beta = -\alpha$ and $\gamma = 0$. Then going back to Eqs. (2.5d,e) we have $d\alpha = 0$ and $d\beta = \omega^1\wedge\omega^2$ which contradicts the fact that $\beta = -\alpha$. Let us remark that the impossibility of Case B.II could also have been inferred from the parametric expressions (3.9), as the conditions $a = c = f = 0$ substituted therein lead to $G \equiv 0$. From the fact that the first reduction led to an {e}-structure on $J^1(\mathbb{R},\mathbb{R}) \times G_{(1)}$ where $G_{(1)}$ is one-dimensional, we conclude that there are no equations within Case B admitting a five-parameter symmetry group. Let us also remark that there are no equations within Case B admitting a four-parameter symmetry group either as it is easily seen that Eqs. (3.9) will be the Maurer-Cartan equations of a Lie algebra if and only if $a = c = f = 0$, a case we just showed to be impossible.

CASE C: $I_2 = 0$, $I_1 I_3 \neq 0$, that is

$$\frac{dF_{,pp}}{dx} - F_{,py} = 0, \tag{3.12a}$$

$$F_{,ppp}\left(-2F_{,yy} + \frac{dF_{,py}}{dx} - F_{,yp}F_{,p} + F_{,pp}F_{,y}\right) \neq 0. \tag{3.12b}$$

We see from Eqs. (3.2a,c) that we can scale I_1 and I_3 to one, in which case Eqs. (3.1) give

$$\alpha = 2r\omega^1 + s\omega^2 - \ell\omega^3, \tag{3.13a}$$

$$\beta = r\omega^1 + k\omega^2 + 2\ell\omega^3. \tag{3.13b}$$

Parametrically, the choice made to scale I_1 and I_3 to one enables us by Eqs. (2.6) to solve for A and B in terms of F and its derivatives

$$A = -(2F_{,ppp})^{-1/5}\left(-2F_{,yy} + \frac{dF_{,py}}{dx} - F_{,yp}F_{,p} + F_{,pp}F_{,y}\right)^{2/5}, \tag{3.14a}$$

$$B = 2^{-3/5}F_{,ppp}^{\;2/5}\left(-2F_{,yy} + \frac{dF_{,py}}{dx} - F_{,yp}F_{,p} + F_{,pp}F_{,y}\right)^{1/5}, \tag{3.14b}$$

in which case the parametric expressions of the functions r, s, k and ℓ are easily computed using Eqs. (2.6d,e,f)

$$r = \frac{1}{2A^2}\frac{dA}{dx} - C - \frac{F_{,p}}{2A} = \frac{1}{AB}\frac{dB}{dx} - C, \tag{3.15a}$$

$$s = \frac{1}{AB}\left(A_{,y} - ACA_{,p}\right), \tag{3.15b}$$

$$k = \frac{1}{B^2}(B_{,y} - ACB_{,p}) + \frac{F_{,pp}}{2B}, \quad \ell = -\frac{A_{,p}}{B} = \frac{A}{2B^2}B_{,p}, \tag{3.15c}$$

where A and B are given by Eqs. (3.14a,b) and where the equalities given above between the different expressions of r and ℓ follow from Eqs. (3.12a) and (3.14a,b). The structure equations (2.5) become, an account of Eqs. (3.13a,b).

$$d\omega^1 = s\omega^2\wedge\omega^1 - \ell\omega^3\wedge\omega^1, \tag{3.16a}$$

$$d\omega^2 = r\omega^1\wedge\omega^2 + 2\ell\omega^3\wedge\omega^2 + \omega^1\wedge\omega^3, \tag{3.16b}$$

$$d\omega^3 = \gamma\wedge\omega^2 - r\omega^1\wedge\omega^3 + (k-s)\omega^2\wedge\omega^3, \tag{3.16c}$$

$$d\gamma = 2r\gamma\wedge\omega^1 + s\gamma\wedge\omega^2 - \ell\gamma\wedge\omega^3 + \omega^1\wedge\omega^2. \tag{3.16d}$$

Again, we have an {e}-structure on $J^1(\mathbb{R},\mathbb{R}) \times G_{(1)}$, where $G_{(1)}$ is one-dimensional, and we have to perform one further reduction in order to obtain an {e}-structure on $J^1(\mathbb{R},\mathbb{R})$. To do so, we first compute $G_{(1)}$'s action on the non-constant components r, s, k and ℓ of the structure tensor. We have

$$(ds-\ell\gamma)\wedge\omega^2\wedge\omega^1 + d\ell\wedge\omega^1\wedge\omega^3 - \ell(k+s)\omega^1\wedge\omega^2\wedge\omega^3, \tag{3.17a}$$

$$(dr+\gamma)\wedge\omega^1\wedge\omega^2 + 2d\ell\wedge\omega^3\wedge\omega^2 + (r\ell-k)\omega^1\wedge\omega^2\wedge\omega^3 = 0, \tag{3.17b}$$

$$(dr+\gamma)\wedge\omega^3\wedge\omega^1 + (d(k-s)+3\ell\gamma)\wedge\omega^2\wedge\omega^3 + rk\omega^1\wedge\omega^2\wedge\omega^3 = 0. \tag{3.17c}$$

It follows from Eqs. (3.17) that we can always use $G_{(1)}$'s action to translate r to zero, in which case we have

$$\gamma = c\omega^1 + e\omega^2 + f\omega^3, \tag{3.18a}$$

and the final {e}-structure is given by

$$d\omega^1 = s\omega^2\wedge\omega^1 - \ell\omega^3\wedge\omega^1, \tag{3.18b}$$

$$d\omega^2 = 2\ell\omega^3\wedge\omega^2 + \omega^1\wedge\omega^3, \tag{3.18c}$$

$$d\omega^3 = c\omega^1\wedge\omega^2 + (k-s-f)\omega^2\wedge\omega^3. \tag{3.18d}$$

We now have invariants k, s, ℓ, c and f on $J^1(\mathbb{R},\mathbb{R})$ whose parametric expressions are easily computed using Eqs. (3.15), (2.6e) and the fact that we have

$$C = \frac{1}{AB}\frac{dB}{dx} = \frac{1}{2A^2}\frac{dA}{dx} - \frac{F_{,p}}{2A}, \tag{3.18e}$$

once r has been translated to zero. Necessary and sufficient conditions for the equivalence of two equations of the form (3.6d) satisfying the invariant conditions (3.12a,b) are now obtained by a procedure similar to the one described for Case A.

From the fact that the first reduction led to an {e}-structure on $J^1(\mathbb{R},\mathbb{R}) \times G_{(1)}$ where $G_{(1)}$ is one-dimensional, we conclude that there are no equations within Case C admitting a five-parameter symmetry group. Let us also remark that we would obtain a contradiction if we were to impose on Eqs. (3.16) that they be the Maurer-Cartan equations of a Lie algebra. It then follows that there are no equations within Case C admitting a four-parameter symmetry group.

<u>CASE D</u>: $I_1 \neq 0$, $I_3 = 0$ ($\Rightarrow I_2 = 0$), that is

$$F_{,ppp} \neq 0, \tag{3.19a}$$

$$\frac{dF_{,pp}}{dx} - F_{,py} = 0, \quad -2F_{,yy} + \frac{dF_{,py}}{dx} - F_{,yp}F_{,p} + F_{,pp}F_{,y} = 0. \tag{3.19b}$$

It follows from Eq. (3.2a) that we can use G's action to scale I_1 to one, in which case Eq. (3.1a) gives

$$2\beta = \alpha + 2r\omega^2 + 2s\omega^3. \tag{3.20}$$

Parametrically, scaling I_1 to one amounts on account of Eq. (3.6b) to setting

$$-\frac{2B^2}{A} = F_{,ppp}. \tag{3.21a}$$

The parametric expressions of the functions r and s are now easily computed using Eq. (2.6d)

$$r = \frac{1}{2B}\left(\frac{F_{,pppy}}{F_{,ppp}} + F_{,pp}\right) - Cs, \quad s = -B\,\frac{F_{,pppp}}{F^2_{,ppp}}. \tag{3.21b}$$

Using Eq. (3.20), the structure equations (2.5) become

$$d\omega^1 = \alpha\wedge\omega^1, \tag{3.22a}$$

$$d\omega^2 = \frac{1}{2}\,\alpha\wedge\omega^2 + \omega^1\wedge\omega^3 + s\omega^3\wedge\omega^2, \tag{3.22b}$$

$$d\omega^3 = \gamma\wedge\omega^2 - \frac{1}{2}\,\alpha\wedge\omega^3 + r\omega^2\wedge\omega^3, \tag{3.22c}$$

$$d\alpha = -2\gamma\wedge\omega^1, \tag{3.22d}$$

$$d\gamma = \gamma\wedge\alpha. \tag{3.22e}$$

We now have an {e}-structure on $J^1(\mathbb{R},\mathbb{R}) \times G_{(1)}$, where the reduced group $G_{(1)}$ is two-dimensional, and we have to perform further reductions in order to obtain an {e}-structure on $J^1(\mathbb{R},\mathbb{R})$. To do so, we first compute $G_{(1)}$'s action on the non-constant components r and s of the structure tensor. We have

$$(ds - \tfrac{1}{2}s\alpha)\wedge\omega^3\wedge\omega^2 + r\omega^1\wedge\omega^3\wedge\omega^2 = 0, \tag{3.23a}$$

$$(dr + \tfrac{1}{2}r\alpha+s\gamma)\wedge\omega^2\wedge\omega^3 = 0, \tag{3.23b}$$

and we see that there are several cases to be considered.

<u>CASE D.I:</u> $s \neq 0$, that is

$$F_{,pppp} \neq 0. \tag{3.24a}$$

It follows from Eqs. (3.23a,b) that we can use $G_{(1)}$'s action to scale s to $\varepsilon = \frac{|s|}{s}$ and translate r to zero, in which case we have

$$\alpha = b\omega^2 + c\omega^3, \quad \gamma = k\omega^2 + m\omega^3, \tag{3.24b}$$

and the final {e}-structure is given by

$$d\omega^1 = b\omega^2\wedge\omega^1 + c\omega^3\wedge\omega^1, \tag{3.24c}$$

$$d\omega^2 = (\tfrac{c}{2} + \varepsilon)\omega^3\wedge\omega^2 + \omega^1\wedge\omega^3, \tag{3.24d}$$

$$d\omega^3 = (m + \tfrac{b}{2})\omega^3\wedge\omega^2. \tag{3.24e}$$

We now have invariants b, c and m on $J^1(\mathbb{R},\mathbb{R})$ whose parametric expressions are easily computed using Eqs. (2.6d,e) and the following expressions for B and C

$$B = -\varepsilon\frac{F^2_{,ppp}}{F_{,pppp}}, \quad C = -\frac{1}{2}\frac{F_{,pppp}}{F^2_{,ppp}}\left(\frac{F_{,pppy}}{F_{,ppp}} + F_{,pp}\right), \tag{3.24f}$$

which follow from Eq. (3.21b) and the choice made to scale s to ε and translate r to zero. Necessary and sufficient conditions for the equivalence of two equations of the form (3.6d) satisfying the invariant conditions (3.24a) and (3.19b) are now obtained by a procedure similar to the one described for Case A.

CASE D.II: $s = 0$, $r \neq 0$, that is

$$F_{,pppp} = 0, \quad \frac{F_{,pppy}}{F_{,ppp}} + F_{,pp} \neq 0. \tag{3.25a}$$

This case is impossible since as a consequence of Eq. (3.23a) we have $s = 0 \Rightarrow r = 0$.

CASE D.III: $s = 0$, $r = 0$, that is

$$F_{,pppp} = 0, \quad \frac{F_{,pppy}}{F_{,ppp}} + F_{,pp} = 0. \tag{3.25b}$$

This case is impossible as well. Indeed from Eq. (3.20) we obtain that $\beta = \frac{1}{2}\alpha$. But from Eqs. (2.5d,e) we have $d\beta = \omega^1 \wedge \gamma + \omega^2 \wedge \omega^3$ and $d\alpha = 2\omega^1 \wedge \gamma$, which contradicts the fact that $d\beta = \frac{1}{2} d\alpha$. Let us note that the impossibility of Cases D.II and D.III could also have been deduced from the parametric expressions (3.21b), (3.19b) and (3.25b); for example it is easily seen that Eq. (3.25b) implies that $F_{,ppp} = 0$ which contradicts the condition $I_1 \neq 0$ expressed by Eq. (3.19a).

Let us remark that imposing on Eqs. (3.22) that they be the Maurer-Cartan equation of a Lie algebra leads to $r = s = 0$, which is impossible. We conclude that there are no equations within Case D admitting a five-parameter symmetry group. Let us finally observe that there are no equations within Case D admitting a four-parameter symmetry group either since we saw that the original {e}-structure on $J^1(\mathbb{R},\mathbb{R}) \times G$ could always be reduced to an {e}-structure on $J^1(\mathbb{R},\mathbb{R})$.

The cases remaining to be considered, that is $I_1 = 0$, $I_2 = 0$, $I_3 \neq 0$ and $I_1 = I_2 = I_3 = 0$ have been treated in detail in Reference [5].

Note: In fact, we have proved that an equation of the form(1.1) cannot have a four-parameter group of symmetries either. This follows from the fact contained in ref. [5] that one can always reduce to an 2 -structure on $J^1(\mathbb{R},\mathbb{R})$ in the case $I_1=I_2=0$, $I_3 \neq 0$.

ACKNOWLEDGMENTS

This work was supported by an NSERC post-doctoral fellowship for the first-named author and by an NSERC operating grant for the second-named author.

This research was done while one of us (N. Kamran) was a visitor in the Mathematics Department at the University of North Carolina at Chapel Hill. Professor R.B. Gardner's helpful comments and kind hospitality are gratefully acknowledged.

REFERENCES

[1] E. Cartan, Ann. Ecole Normale 25, 1908, p. 57 (collected works part II, p. 719).

[2] E. Cartan, Séminaire de Math., exposé D, 11 janvier 1937; Selecta, p. 113 (collected works part II, p. 1311).

[3] R.B. Gardner, Differential Geometric Control Theory, R. Brockett, R. Millman and H. Sussman eds. Progress in Mathematics, Vol. 27, Birkhaüser, Boston 1983.

[4] E.L. Ince, Ordinary Differential Equations, Green and Co. London 1927.

[5] N. Kamran, K. Lamb and W. Shadwick, J. Diff. Geom., 1985 (in press).

Quantum R matrix related to the generalized Toda system: an algebraic approach

Michio Jimbo

Research Institute for Mathematical Sciences
Kyoto University, Kyoto 606, Japan

Abstract. We report on some recent progress concerning the Yang-Baxter equation and related algebraic structures —the q-analogue $\hat{U}(\mathfrak{g})$ of the universal enveloping algebra and the Hecke algebra. The results include i) construction of trigonometric R matrices related to the vector representation of non-exceptional Lie algebras, ii) construction of representations of $\hat{U}(\mathfrak{g})$ for $\mathfrak{g} = \mathfrak{gl}(n+1)$, and iii) existence of R for a class of representations of $\mathfrak{gl}(n+1)$.

§1. Introduction

Let V be a finite dimensional complex vector space. By the quantum Yang-Baxter equation (QYBE) we mean the following functional equation for a matrix valued function $R(u) \in \mathrm{End}(V \otimes V)$ of $u \in \mathbb{C}$:

$$R_{12}(u-v)R_{13}(u)R_{23}(v) = R_{23}(v)R_{13}(u)R_{12}(u-v). \qquad (1.1)$$

Here $R_{ij}(u) \in \mathrm{End}(V_1 \otimes V_2 \otimes V_3)$ $(V_1=V_2=V_3 \equiv V)$ signifies the matrix $R(u)$ on the space $V_i \otimes V_j$, acting as identity on the third space; e.g. $R_{12}(u) = R(u) \otimes I$. As is well known [1,2], the QYBE plays a central role in integrable quantum field theory and statistical mechanics.

In the paper [3], Kulish-Reshetikhin-Sklyanin initiated an algebraic approach to the QYBE via representation theory. Suppose a solution $R(u) = R(u,\hbar)$ of (1.1) contains an extra parameter $\hbar$, in such a way that

$$R(u,h) = \kappa(u,\hbar)(1+\hbar r(u)+\cdots) \quad \text{as } \hbar \to 0 \qquad (1.2)$$

holds with some scalar $\kappa(u,\hbar)$. The coefficient $r(u)$, called the classical limit of $R(u,\hbar)$, then satisfies the classical Yang-Baxter equation (CYBE) [1]

$$[r_{12}(u-v),r_{13}(u)]+[r_{12}(u-v),r_{23}(v)]+[r_{13}(u),r_{23}(v)] = 0. \qquad (1.3)$$

The characteristic feature of (1.3) is that, being written in terms of commutators, it makes sense for a $\mathfrak{g}\otimes\mathfrak{g}$-valued function $r(u)$ where $\mathfrak{g}$ is an abstract Lie algebra. (This becomes manifest if we write down a typical term of (1.3) in a basis $\{X_a\}$ of $\mathfrak{g}$,

$$[r_{13}(u),r_{23}(v)] = \sum r^{ab}(u)r^{cd}(v)X_a\otimes X_c\otimes[X_b,X_d],$$

where we have set $r(u) = \sum r^{ab}(u)X_a\otimes X_b$.) On the other hand, to each such $\mathfrak{g}\otimes\mathfrak{g}$-valued solution $r(u)$ there corresponds a family of solution matrices $r_{ij}(u) = (\rho_i\otimes\rho_j)(r(u))\in \mathrm{End}(V_i\otimes V_j)$ by specifying irreducible representations (ρ_i,V_i) of $\mathfrak{g}$. Kulish-Reshetikhin-Sklyanin proposed to "quantize" these solutions, namely to find a family of matrices $R_{ij}(u,\hbar)\in \mathrm{End}(V_i\otimes V_j)$ satisfying the QYBE and having $r_{ij}(u)$ as their classical limit. (Here the meaning of (1.1) is slightly generalized in that the V_i's are allowed to be distinct.) In the case $\mathfrak{g}$ is a simple Lie algebra, classical r matrices (= solutions of the CYBE) have been classified by Belavin-Drinfel'd (see [4] as to the precise statement). In the quantum case (1.1), such classification is still unknown. For this reason the "quantization" problem above seems to be of particular interest.

It has been recognized, through the pioneering works [5,6] for $\mathfrak{g} = \mathfrak{sl}(2)$, that in order to carry out this program one is forced to introduce a novel algebraic object, an associative algebra which "quantize" the Lie algebra structure in some sense. Recently Drinfel'd [7] gave a general formulation of such an algebra for a class of rational solutions

$$r(u) = \frac{t}{u}, \qquad t = \sum_a I_a\otimes I_a, \qquad (1.4)$$

where $\{I_a\}$ denotes an orthonormal basis of a simple Lie algebra $\mathfrak{g}$. He introduced an associative algebra $Y(\mathfrak{g})$ as a quantum counterpart of the universal enveloping algebra $U(\mathfrak{g})$, constructed the R matrix in $Y(\mathfrak{g})\otimes Y(\mathfrak{g})$ and studied the representations of $Y(\mathfrak{g})$.

The present article concerns with a similar problem for a class of trigonometric r matrices of the type (cf. §2, (2.11) with $k=1$,

$x=e^{2u}$)

$$r(u) = r_0 - t \coth u, \quad r_0 = \sum \operatorname{sgn}\alpha\, X_\alpha \otimes X_{-\alpha},$$

where X_α are root vectors normalized as $(X_\alpha, X_{-\alpha})=1$. It is related to the generalized Toda system (GTS) and includes (1.4) as a limiting case $-\lim_{\varepsilon\to 0} \varepsilon\, r(\varepsilon u)$. The relevant algebra in this case is a q-analogue $\hat{U}(\mathfrak{g})$ of $U(\mathfrak{g})$, introduced in [6] for $\mathfrak{g}=\mathfrak{sl}(2)$ and in [7,8] for $\mathfrak{g}$ general. We wish to give here a coherent account of the results scattered in [8-10] about the representations of $\hat{U}(\mathfrak{g})$ and the existence of the R matrix.

The plan of this paper is as follows. §2 is of preliminary nature. We describe the GTS and the corresponding classical r matrix. In §3, we give the quantum R matrix explicitly, assuming that the Lie algebra in question is of non-exceptional type and that (ρ_1,V_1) is its vector representation. The results given here were obtained independently in [9,11] through different methods. Part of the formulas are given in the Appendix. In §4 we show how the QYBE related to the GTS is reduced to a linear system for $R(u)$ by using the algebra $\hat{U}(\mathfrak{g})$. §5 is devoted to the construction of finite dimensional representations of $\hat{U}(\mathfrak{gl}(n+1))$. Here we give two methods: one is to consider the irreducible decomposition of the tensor product of $V=\mathbb{C}^{n+1}$ with the aid of the Hecke algebra, and another is to give explicit matrix form of the generators of $\hat{U}(\mathfrak{gl}(n+1))$ relative to the Gelfand-Zetlin basis [12]. Applying these results we show in §6 the existence of $R(u)\in \mathrm{End}(V_1\otimes V_2)$ for $\mathfrak{g}=\mathfrak{gl}(n+1)$ in the case when one of (ρ_1,V_1) is the (anti-)symmetric tensor representation and another is arbitrary. In the special case $\mathfrak{g}=\mathfrak{gl}(2)$ and $V_1=V_2$, and explicit formula is available in terms of an infinite product involving the q-analogue of the Casimir operator. Finally in §7 we raise some further problems.

Throughout the text we shall use the multiplicative parameter $x=e^u$ and write $R(x)$, $r(x)$ for $R(u)$, $r(u)$, respectively. The following notations are of frequent use:

$\mathfrak{g}$: complex simple Lie algebra ($\dim \mathfrak{g} < \infty$)
$U(\mathfrak{g})$: the universal enveloping algebra
$\hat{U}(\mathfrak{g})$: q-analogue of $U(\mathfrak{g})$ (§4)
E_{ij}: standard basis of $\mathfrak{gl}(n+1)$
E^0_{ij}: matrix unit $(\delta_{ai}\delta_{bj})$
P: transposition ($P\in \mathrm{End}(V\otimes V)$, $Pu\otimes v=v\otimes u$)
$\check{R}(x) = PR(x)$ (if $V_1=V_2$)

$$q = e^{\hbar}$$

$$[n] = \frac{q^n - q^{-n}}{q - q^{-1}} \quad .$$

§2. Classical r matrix for the GTS

In this section we summarize known results about the classical r matrix related to the GTS, thereby fixing notations for later reference.

Although this article concerns with the GTS associated with an affine (Kac-Moody) Lie algebra, it is instructive to consider first the case of a finite dimensional simple Lie algebra $\mathfrak{g}$. Let $\mathfrak{h}$ be a Cartan subalgebra of $\mathfrak{g}$, and let Π denote the set of simple roots. The GTS is a dynamical system for a $\mathfrak{h}$-valued function $\phi=\phi(t)$

$$\frac{d^2}{dt^2}\phi = -\sum_{\alpha\in\Pi} 2\,\alpha\, e^{\alpha(\phi)}. \tag{2.1}$$

In (2.1) we identified $\mathfrak{h}^*$ with $\mathfrak{h}$ via an invariant bilinear form $(\ ,\ )$ on $\mathfrak{g}$. If $\mathfrak{g} = \mathfrak{sl}(n+1)$ and $\phi = \mathrm{diag}(\phi_0,\dots,\phi_n)$ with $\sum_{i=0}^{n}\phi_i = 0$, then (2.1) is the familiar non-periodic Toda equation

$$\frac{d^2\phi_i}{dt^2} = 2e^{2(\phi_{i-1}-\phi_i)} - 2e^{2(\phi_i-\phi_{i+1})}, \quad i=0,1,\dots,n \tag{2.2}$$

with the convention $\phi_{-1} = -\infty$, $\phi_{n+1} = +\infty$. Equation (2.1) has the well-known Lax representation

$$\frac{d}{dt} L = [A, L],$$

with

$$L = \pi + \sum_{\alpha\in\Pi} e^{\alpha(\phi)}(e_\alpha + f_\alpha) \tag{2.3}$$

$$A = \sum_{\alpha\in\Pi} e^{\alpha(\phi)}(-e_\alpha + f_\alpha).$$

Here $\pi = \frac{d\phi}{dt}$, and $\{e_\alpha, f_\alpha, h_\alpha = [e_\alpha, f_\alpha]\}$ signify the Chevalley generators relative to Π. Take a basis $\{h_i\}$ of $\mathfrak{h}$, its dual basis $\{h_i^*\}$ and write $\phi = \sum\phi_i h_i$, $\pi = \sum\pi_i h_i^*$. We introduce the Poisson bracket $\{\ ,\ \}$ by letting

$$\{\pi_i, \phi_j\} = \delta_{ij}. \tag{2.4}$$

For a $\mathfrak{g}$-valued function $L = \sum f_a X_a$ ($\{X_a\}$: a basis of $\mathfrak{g}$) the customary notation $\{L \underset{,}{\otimes} L\}$ is meant to be $\sum\{f_a, f_b\}X_a \otimes X_b \in \mathfrak{g} \otimes \mathfrak{g}$. In the standard inverse spectral method (see e.g. [2]) the starting point is the existence of a tensor $r \in \mathfrak{g} \otimes \mathfrak{g}$ satisfying the identity

$$\{L \underset{,}{\otimes} L\} = [r, L\otimes 1 + 1\otimes L]. \tag{2.5}$$

Upon substitution of (2.3), (2.5) reduces to the following:

$$[r, h\otimes 1 + 1\otimes h] = 0 \qquad (h \in \mathfrak{h}) \tag{2.6}$$

$$[r, e_\alpha \otimes 1 + 1\otimes e_\alpha] = -e_\alpha \otimes h_\alpha + h_\alpha \otimes e_\alpha \qquad (\alpha \in \Pi)$$

$$[r, f_\alpha \otimes 1 + 1\otimes f_\alpha] = -f_\alpha \otimes h_\alpha + h_\alpha \otimes f_\alpha \qquad (\alpha \in \Pi).$$

To describe the solutions of (2.6), we need some more notations. For an orthonormal basis $\{I_a\}$ of $\mathfrak{g}$, set

$$t = \sum I_a \otimes I_a. \tag{2.7}$$

Denote by $\mathcal{R}$ the set of roots, and let X_α ($\alpha \in \mathcal{R}$) be root vectors normalized as $(X_\alpha, X_{-\alpha}) = 1$. Put

$$r_0 = \sum_{\alpha \in \mathcal{R}} \operatorname{sgn}\alpha \, X_\alpha \otimes X_{-\alpha} \tag{2.8}$$

where $\operatorname{sgn}\alpha = \pm 1$ according as the root α is positive or negative.

Proposition 2.1. The general solution of (2.6) is $r_\kappa = r_0 + \kappa t$ ($\kappa \in \mathbb{C}$). It satisfies the modified CYBE [13]

$$[(r_\kappa)_{12}, (r_\kappa)_{13}] + [(r_\kappa)_{12}, (r_\kappa)_{23}] + [(r_\kappa)_{13}, (r_\kappa)_{23}]$$

$$= (\kappa^2 - 1) \sum c_{abc} \, I_a \otimes I_b \otimes I_c$$

where $c_{abc} = ([I_a, I_b], I_c)$.

In particular, $r_{\pm 1}$ satisfies the CYBE (1.3) with no u-dependence. Note that r_κ is unitary only if $\kappa = 0$: $\mathcal{P}(r_0) = -r_0$ where $\mathcal{P} \in \operatorname{End}(\mathfrak{g} \otimes \mathfrak{g})$, $\mathcal{P}(X\otimes Y) = Y\otimes X$.

These considerations carry over to the case of an affine Lie algebra $\hat{\mathfrak{g}}$ in place of $\mathfrak{g}$. The formulas (2.1) through (2.6) are valid with $\mathfrak{h}, \Pi, \cdots$ replaced by their counterparts $\hat{\mathfrak{h}}, \hat{\Pi}, \cdots$. For $= A_n^{(1)}$, (2.1) becomes the Toda equation (2.2) with the periodic boundary condition $\phi_{i+n+1} = \phi_i$. On the other hand, the formulas (2.7), (2.8) need justification, since the set of roots $\hat{R}$ is now infinite. To this end we employ the realization of $\hat{\mathfrak{g}}$ in terms of a loop algebra.

Recall that an affine Lie algebra $\hat{\mathfrak{g}}$ is in one-to-one correspondence with a pair $(\mathfrak{g},\sigma)$, where $\mathfrak{g}$ is a finite dimensional simple Lie algebra and σ is its diagram automorphism [14]. Let $\mathfrak{g} = \bigoplus_{j=0}^{k-1} \mathfrak{g}_j$, $\mathfrak{g}_j = \{X\in\mathfrak{g} \mid \sigma(X)=\omega^j X\}$, denote the gradation associated with σ, where $k = \mathrm{ord}\,\sigma\ (=1,2,3)$ and ω is a primitive k-th root of unity. Then the loop algebra $\mathfrak{g}^{(k)}[\lambda,\lambda^{-1}] = \bigoplus_{j\in\mathbb{Z}} \lambda^j \mathfrak{g}_{j \bmod k}$ gives the homogeneous realization of $\tilde{\mathfrak{g}} = \hat{\mathfrak{g}} \bmod \mathbb{C}c\oplus\mathbb{C}d$: $\tilde{\mathfrak{g}} \cong \mathfrak{g}^{(k)}[\lambda,\lambda^{-1}]$. (Here c is a central element and $d = \lambda\frac{d}{d\lambda}$ is a distinguished derivation; see [14].) Accordingly we identify $\tilde{\mathfrak{g}}\otimes\tilde{\mathfrak{g}}$ with $\mathfrak{g}^{(k)}[\lambda,\lambda^{-1}]\otimes\mathfrak{g}^{(k)}[\mu,\mu^{-1}]$. Let θ denote the highest weight in the $\mathfrak{g}_0$-module $\mathfrak{g}_{-1}$. Choose $e_0'\in\mathfrak{g}_{1,-\theta}$, $f_0'\in\mathfrak{g}_{-1,\theta}$ and normalize them as $(e_0',f_0') = 2/(\theta,\theta)$. Then the Chevalley generators for $\hat{\mathfrak{g}}$ are given by

$$e_0 = \lambda e_0', \quad f_0 = \lambda^{-1} f_0', \quad h_0 = [e_0, f_0] \tag{2.9}$$

together with the ones $\{e_i, f_i, h_i\}_{1\le i\le n}$ for $\mathfrak{g}_0$. The L-operator (2.3) now depends on λ through (2.9). The equations (2.6) for r read as follows:

$$[r, h\otimes 1 + 1\otimes h] = 0 \qquad (h\in\mathfrak{h}_0) \tag{2.10}$$

$$[r, \lambda e_0'\otimes 1 + 1\otimes\mu e_0'] = -\lambda e_0'\otimes h_0 + h_0\otimes\mu e_0'$$

$$[r, e_i\otimes 1 + 1\otimes e_i] = -e_i\otimes h_i + h_i\otimes e_i \qquad (1\le i\le n)$$

$$[r, \lambda^{-1} f_0'\otimes 1 + 1\otimes\mu^{-1} f_0'] = -\lambda^{-1} f_0'\otimes h_0 + h_0\otimes\mu^{-1} f_0'$$

$$[r, f_i\otimes 1 + 1\otimes f_i] = -f_i\otimes h_i + h_i\otimes f_i \qquad (1\le i\le n).$$

Next define a $\mathfrak{g}\otimes\mathfrak{g}$-valued rational function

$$r(x) = r_0 + \frac{1+x^k}{1-x^k} t_0 + \sum_{j=1}^{k-1} \frac{2x^j}{1-x^k} t_j \ . \tag{2.11}$$

Here r_0 signifies the tensor (2.8) with respect to the Lie algebra $\mathfrak{g}_0$, while $t = \sum_{j=0}^{k-1} t_j$ with $(\sigma\otimes 1)t_j = \omega^j t_j$ denotes the one (2.7) with respect to $\mathfrak{g}$. In the affine case, the tensors (2.7), (2.8) are formal Laurent series in λ, μ. Performing the sum we find that they are given respectively by $\delta(\lambda/\mu)t$ and $r(\lambda/\mu) - \delta(\lambda/\mu)t$, where $\delta(\lambda/\mu) = \sum_{j\in\mathbf{Z}} (\lambda/\mu)^j$.

Proposition 2.2. The r matrix (2.11) satisfies the CYBE by setting $x = e^u$. It is the unique solution of (2.10) as a $\mathfrak{g}\otimes\mathfrak{g}$-valued rational function in $x = \lambda/\mu$.

Henceforth we regard the rational function (2.11) as the classical r matrix for the GTS in the affine case.

Remark. Often in the literature [4,11] the r matrix (2.8) is given in the principal realization. For this point see the remark at the end of §3.

§3. R matrix in the vector representation

Let us describe the explicit form of the quantum R matrix corresponding to the classical solution (2.11), when $\hat{\mathfrak{g}}$ is (i.e. both $\mathfrak{g}$ and $\mathfrak{g}_0$ are) non-exceptional and $\rho: \mathfrak{g} \to \mathrm{End}(V)$ is the vector representation. The degree $N = \dim V$ of the representation is listed in Table 1.

$\hat{\mathfrak{g}}$	$A_n^{(1)}$	$B_n^{(1)}$	$C_n^{(1)}$	$D_n^{(1)}$	$A_{2n}^{(2)}$	$A_{2n-1}^{(2)}$	$D_{n+1}^{(2)}$
N	$n+1$	$2n+1$	$2n$	$2n$	$2n+1$	$2n$	$2n+2$
ξ	/	q^{2n-1}	q^{2n+2}	q^{2n-2}	$-q^{2n+1}$	$-q^{2n}$	q^n

Table 1.

In the sequel the indices $\alpha, \beta, \ldots$ are understood to run over $1, 2, \ldots, N$. Put $R(x) = \sum R_{\alpha\beta\gamma\delta}(x) E^0_{\alpha\gamma} \otimes E^0_{\beta\delta}$. The $(\alpha\beta, \gamma\delta)$-element $R_{\alpha\beta\gamma\delta}(x)$, represented graphically as

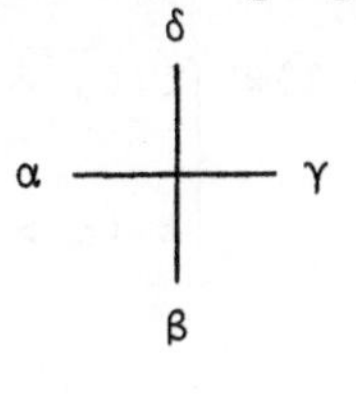

$= R_{\alpha\beta\gamma\delta}(x)$,

gives the vertex weight of the corresponding model in statistical mechanics (see e.g. [15]). Though the results for $\hat{\mathfrak{g}} = A_n^{(1)}$ [16] and $A_2^{(2)}$ [17] have been known for some time, we include it below for comparison.

(I) Case $\hat{\mathfrak{g}} = A_n^{(1)}$

In this case the non-zero weights occur for the following configurations:

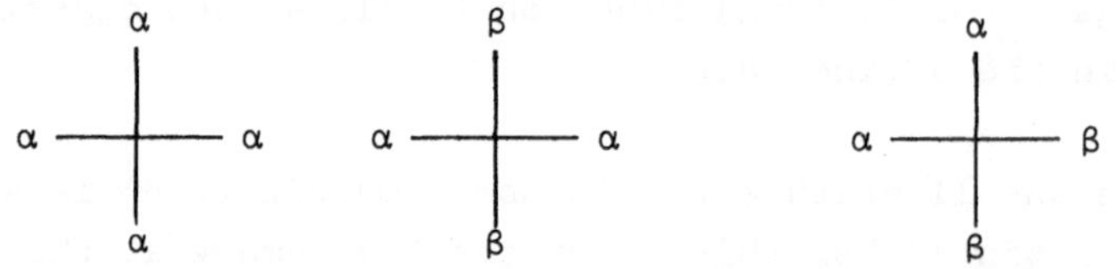

$(\alpha\neq\beta)$ $(\alpha\neq\beta)$

$$q^2x - 1 \qquad q(x-1) \qquad (q^2-1)\times\begin{cases}1 & (\alpha<\beta)\\ x & (\alpha>\beta).\end{cases}$$

This means that the matrix $R(x)$ is a direct sum of 1×1 and 2×2 pieces

$$R(x) = \Big(\bigoplus_{\alpha=1}^{n+1} \alpha\alpha)(q^2x \overset{\alpha\alpha}{-} 1)\Big) \oplus \Big(\bigoplus_{\alpha<\beta} \begin{matrix}\alpha\beta)\\ \beta\alpha)\end{matrix}\begin{bmatrix} q(x \overset{\alpha\beta}{-} 1) & q^2 \overset{\beta\alpha}{-} 1 \\ (q^2-1)x & q(x-1)\end{bmatrix}\Big).$$

(II) Case $\hat{\mathfrak{g}} = B_n^{(1)}, C_n^{(1)}, D_n^{(1)}, A_{2n}^{(2)}, A_{2n-1}^{(2)}$

Put $\alpha' = N + 1 - \alpha$, and define ξ as in Table 1. The structure of the weights are quite the same as above:

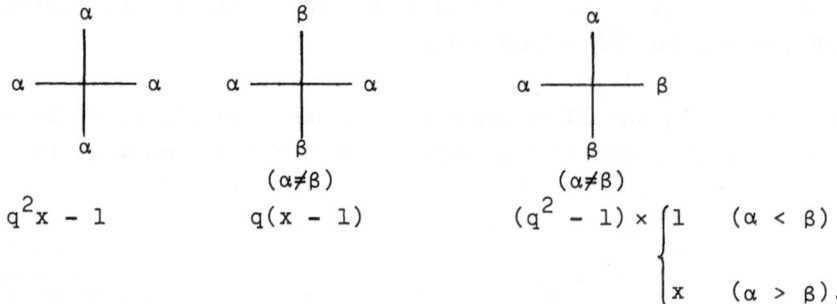

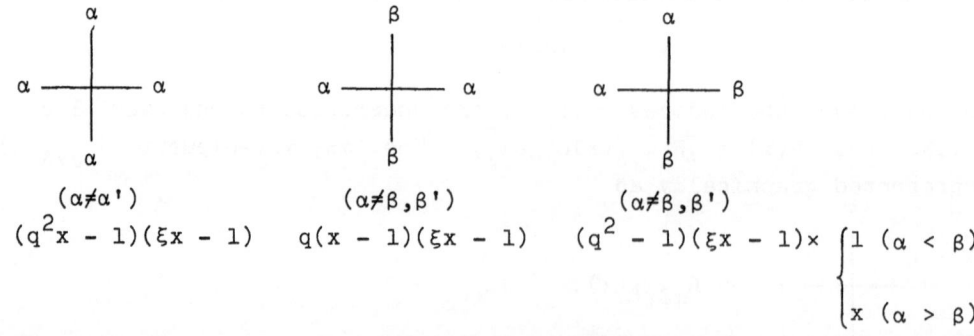

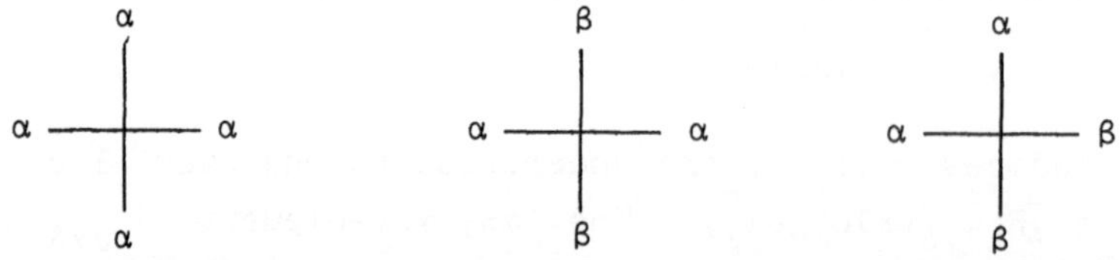

$(\alpha\neq\alpha')$ $(\alpha\neq\beta,\beta')$ $(\alpha\neq\beta,\beta')$

$$(q^2x-1)(\xi x-1) \qquad q(x-1)(\xi x-1) \qquad (q^2-1)(\xi x-1)\times\begin{cases}1\ (\alpha<\beta)\\ x\ (\alpha>\beta)\end{cases}$$

except that there appears a new type of weights

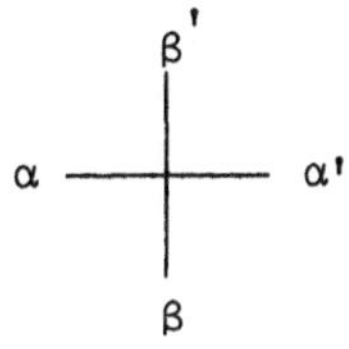

$= a_{\alpha\beta}(x)$ (3.1)

which gives rise to an $N\times N$ block $(a_{\alpha\beta}(x))$ of $R(x)$. Explicit form of $(a_{\alpha\beta}(x))$ is given in the Appendix.

(III) Case $\mathfrak{g} = D^{(2)}_{n+1}$

This is the most complicated case. Retaining the notation as above, let α,β represent the indices $\neq n+1, n+2$ and let $\gamma,\delta = n+1, n+2$. The non-zero weights are

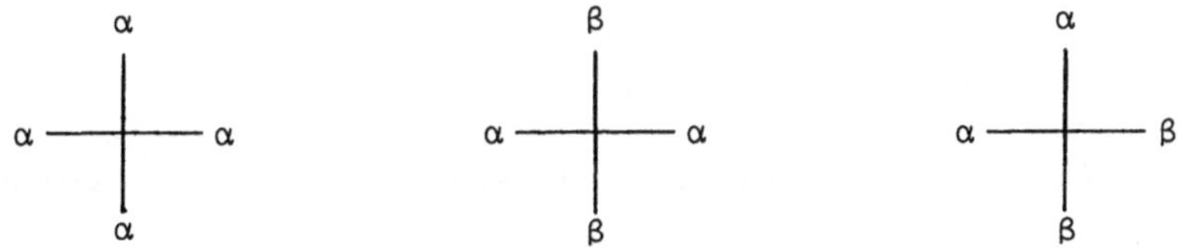

$(q^2x^2-1)(\xi^2x^2-1)$ $\quad$ $(\alpha\neq\beta,\beta')$ $q(x^2-1)(\xi^2x^2-1)$ $\quad$ $(\alpha\neq\beta,\beta')$ $(q^2-1)(\xi^2x^2-1)\times\begin{cases}1 & (\alpha<\beta)\\ x^2 & (\alpha>\beta)\end{cases}$

corresponding to 1×1 and 2×2 blocks,

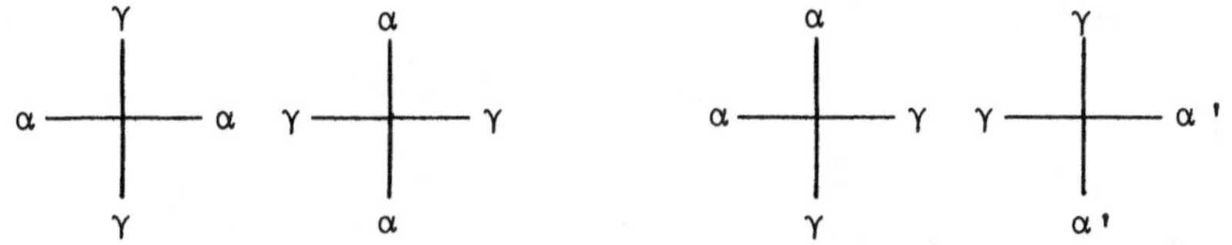

$q(x^2-1)(\xi^2x^2-1)$ $\quad$ $\frac{1}{2}(q^2-1)(\xi^2x^2-1)(x+1)\times\begin{cases}1 & (\alpha<n+1)\\ x & (\alpha>n+2)\end{cases}$

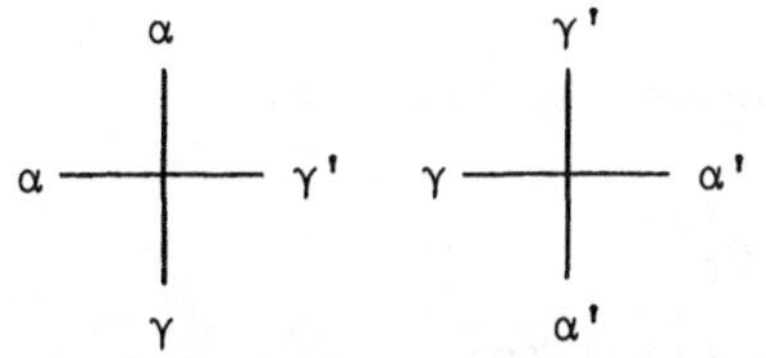

$\frac{1}{2}(q^2-1)(\xi^2x^2-1)(x-1)\times\begin{cases}-1 & (\alpha < n+1)\\ x & (\alpha > n+2)\end{cases}$

which constitute 4×4 blocks

$$\begin{bmatrix} q(x^2-1) & 0 & -\frac{1}{2}(q^2-1)(x-1) & \frac{1}{2}(q^2-1)(x+1) \\ 0 & q(x^2-1) & \frac{1}{2}(q^2-1)(x+1) & -\frac{1}{2}(q^2-1)(x-1) \\ \frac{1}{2}(q^2-1)x(x-1) & \frac{1}{2}(q^2-1)x(x+1) & q(x^2-1) & 0 \\ \frac{1}{2}(q^2-1)x(x+1) & \frac{1}{2}(q^2-1)x(x-1) & 0 & q(x^2-1) \end{bmatrix} \times(\xi^2x^2-1) ,$$

and the types

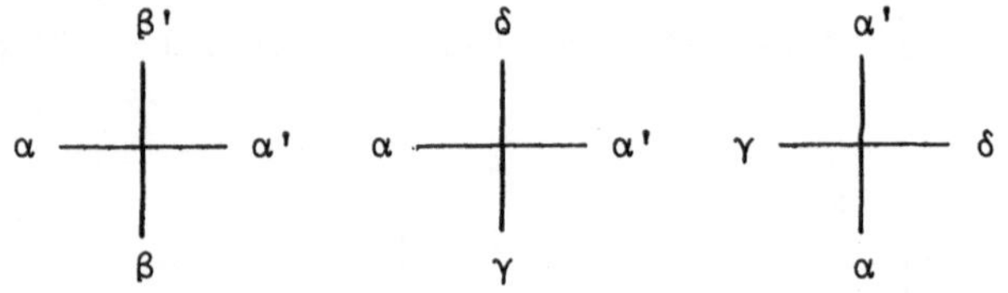

(3.2)

giving rise to an $(N+2)\times(N+2)$ block. The weights (3.2) are also given in the Appendix.

These solutions enjoy the following properties.

$$(e^h\otimes e^h)R(x)(e^h\otimes e^h)^{-1} = R(x) \qquad (h\in \mathfrak{h}_0)$$

$$(S\otimes S)R(x) = {}^t((S\otimes S)R(x)), \qquad S = \begin{pmatrix} & & 1 \\ & ⋰ & \\ 1 & & \end{pmatrix}$$

$$R(x^{-1},q^{-1}) = \gamma^{-1}\,{}^tR(x,q) \tag{3.3}$$

$${}^t\check{R}(x) = \check{R}(x)$$

$$\check{R}(1) = \rho(1)^{1/2}I$$

$$\check{R}(x)\check{R}(x^{-1}) = \rho(x)I .$$

Here $\mathfrak{h}_0$ denotes the set of diagonal matrices of the form

$$\mathrm{diag}(\epsilon_0,\epsilon_1,\dots,\epsilon_n),\ \sum_{i=0}^{n}\epsilon_i=0 \qquad (\hat{\mathfrak{g}} = A_n^{(1)})$$

$$\mathrm{diag}(\epsilon_1,\dots,\epsilon_n,0,-\epsilon_n,\dots,-\epsilon_1) \qquad (\hat{\mathfrak{g}} = B_n^{(1)},\ A_{2n}^{(2)})$$

$$\mathrm{diag}(\epsilon_1,\dots,\epsilon_n,-\epsilon_n,\dots,-\epsilon_1) \qquad (\hat{\mathfrak{g}} = C_n^{(1)},\ D_n^{(1)},\ A_{2n-1}^{(2)})$$

$$\mathrm{diag}(\epsilon_1,\dots,\epsilon_n,0,0,-\epsilon_n,\dots,-\epsilon_1) \ (\hat{\mathfrak{g}} = D_{n+1}^{(2)}),$$

and the factors $\gamma, \rho(x)$ are given by

$$\gamma = -q^2x,\ q^2\xi x^2,\ q^2\xi^2x^4 \quad \text{for (I), (II), (III)}$$

and $\rho(x) = R_{\alpha\alpha\alpha\alpha}(x)R_{\alpha\alpha\alpha\alpha}(x^{-1})$.

Under the first condition of (3.3), it can be shown that

$$R_h(x) = (x^h \otimes 1)R(x)(x^h \otimes 1)^{-1}$$

is also a solution of the QYBE for any $h \in \mathfrak{h}_0$. This freedom corresponds to different choices of realization of the affine Lie algebra $\hat{\mathfrak{g}}$ in terms of Laurent polynomials. Suitable choice of h leads to the principal realization employed in [11]. The present choice, homogeneous realization, has the advantages

i) deg $R(x)$ (=1,2,4 for (I), (II), (III)) becomes independent of the rank of $\hat{\mathfrak{g}}$,
ii) $[\check{R}(x), \check{R}(y)] = 0$ for any x,y (3.4)
except in the case (III).

Actually ii) follows from the last two properties of (3.3) and that deg $\check{R}(x) \leqq 2$. The drawback is, the $\mathbb{Z}_{n+1}$-symmetry for (I) is not manifest in the homogeneous realization.

§4. The algebra $\hat{U}(\mathfrak{g})$

We now proceed to the method of constructing the quantum R matrix for the GTS. We are going to introduce an algebra $\hat{U}(\mathfrak{g})$ and write down a system of linear equations for R similar to §2, (2.6). As a motivation, we shall begin with some heuristic considerations based upon the approximate L-operator [6].

Returning to the setting of §2, let now π_i, ϕ_i in (2.4) stand for operators satisfying

$$[\pi_i, \phi_j] = \frac{\hbar}{2}\delta_{ij}. \tag{4.1}$$

Next let $\{\hat{e}_\alpha, \hat{f}_\alpha\}$ be elements of some algebra that satisfy the same commutation relations with $h \in \mathfrak{h}$ as do the Chevalley generators:

$$[h, \hat{e}_\alpha] = \alpha(h)\hat{e}_\alpha, \ [h,\hat{f}_\alpha] = -\alpha(h)\hat{f}_\alpha. \tag{4.2}$$

At this stage the relations among $\hat{e}$'s and $\hat{f}$'s are left unspecified. As a substitute for the L-operator (2.3) in the classical case, we introduce

$$\begin{aligned} L_\varepsilon &= e^\pi(1 + \varepsilon\sum e^{\alpha(\phi)}(\hat{e}_\alpha+\hat{f}_\alpha))e^\pi \\ &= (1 + \varepsilon\sum e^{\alpha(\phi)}q^{\alpha/2}(e^{\alpha(\pi)}\hat{e}_\alpha+e^{-\alpha(\pi)}\hat{f}_\alpha))e^{2\pi}. \end{aligned}$$

Here we have set $q = e^\hbar$ and ε is a parameter. In the second line the operators π,ϕ are normal-ordered by using (4.1), (4.2). We require for the quantum R matrix to satisfy the equation

$$R(L_\varepsilon\otimes 1)(1\otimes L_\varepsilon) \equiv (1\otimes L_\varepsilon)(L_\varepsilon\otimes 1)R \quad \mod \varepsilon^2.$$

Reducing the expressions $(L_\varepsilon\otimes 1)(1\otimes L_\varepsilon)$, $(1\otimes L_\varepsilon)(L_\varepsilon\otimes 1)$ into normal ordered forms and comparing the coefficients of ε^0, $\varepsilon^1 e^{\alpha(\phi)}e^{\pm\alpha(\pi)}$, we are led to the linear equations

$$\begin{aligned} &[R, h\otimes 1+1\otimes h] = 0 \qquad (h\in\mathfrak{h}) \\ &R(\hat{e}_\alpha\otimes q^{-\alpha/2}+q^{\alpha/2}\otimes\hat{e}_\alpha) = (\hat{e}_\alpha\otimes q^{\alpha/2}+q^{-\alpha/2}\otimes\hat{e}_\alpha)R \\ &R(\hat{f}_\alpha\otimes q^{-\alpha/2}+q^{\alpha/2}\otimes\hat{f}_\alpha) = (\hat{f}_\alpha\otimes q^{\alpha/2}+q^{-\alpha/2}\otimes\hat{f}_\alpha)R. \end{aligned} \tag{4.3}$$

Note that, in the limit $\hbar\to 0$, (4.3) reduces to the linear equations for r (2.6) by setting

$$R = 1 + \hbar r + \cdots, \ \hat{e}_\alpha = e_\alpha + \hbar e_\alpha^{(1)} + \cdots \ \text{and} \ \hat{f}_\alpha = f_\alpha + \hbar f_\alpha^{(1)} + \cdots.$$

The relations among $\hat{e}$'s and $\hat{f}$'s should be so chosen that these equations (notably the last two) become consistent. We are thus led to the following definition.

Let $A=(a_{ij})_{1\leq i,j\leq N}=DB$ be a symmetrizable generalized Cartan matrix [14], where $D = \mathrm{diag}(d_1,\ldots,d_N)$, $d_i\neq 0$, and $B = (b_{ij}) = {}^tB$. As in [14], denote by $(\ ,\ )$ the invariant bilinear form on $\mathfrak{h}^*$ such that $(\alpha_i,\alpha_j) = b_{ij}$. Let further q be a non-zero parameter. We define an associative algebra $U(A)=U(\mathfrak{g}(A))$ to be the one generated by the symbols $\{\hat{e}_i, \hat{f}_i, q^{\pm\alpha_i/2}\}_{1\leq i\leq N}$ under the following relations:

$$q^{\alpha_i/2} q^{\alpha_j/2} = q^{\alpha_j/2} q^{\alpha_i/2}, \quad q^{\alpha_i/2} q^{-\alpha_i/2} = q^{-\alpha_i/2} q^{\alpha_i/2} = 1 \qquad (4.4)$$

$$q^{\alpha_i/2} \hat{e}_j q^{-\alpha_i/2} = q^{(\alpha_i,\alpha_j)/2} \hat{e}_j, \quad q^{\alpha_i/2} \hat{f}_j q^{-\alpha_i/2} = q^{-(\alpha_i,\alpha_j)/2} \hat{f}_j$$

$$[\hat{e}_i, \hat{f}_j] = \delta_{ij}(q^{\alpha_i} - q^{-\alpha_i})/(q_i - q_i^{-1})$$

$$\sum_{\nu=0}^{1-a_{ij}} (-)^\nu \begin{bmatrix} 1-a_{ij} \\ \nu \end{bmatrix}_{q_i} \hat{e}_i^{1-a_{ij}-\nu} \hat{e}_j \hat{e}_i^\nu = 0 \qquad (i\neq j)$$

$$\sum_{\nu=0}^{1-a_{ij}} (-)^\nu \begin{bmatrix} 1-a_{ij} \\ \nu \end{bmatrix}_{q_i} \hat{f}_i^{1-a_{ij}-\nu} \hat{f}_j \hat{f}_i^\nu = 0 \qquad (i\neq j).$$

Here we have set $q_i = q^{(\alpha_i,\alpha_i)/2}$. For q fixed, the symbol $\begin{bmatrix} m \\ n \end{bmatrix}_q =$ $\begin{bmatrix} m \\ n \end{bmatrix}$ is defined by

$$\begin{bmatrix} m \\ n \end{bmatrix} = \frac{[m]!}{[n]![m-n]!}$$

$$[\nu]! = [\nu][\nu-1]\cdots[1], \quad [\nu] = \frac{q^\nu - q^{-\nu}}{q - q^{-1}} .$$

In the limit $q\to 1$, these relations tend to the defining relations among the Chevalley generators of the Kac-Moody Lie algebra $\mathfrak{g}(A)$. Thus this algebra $\hat{U}(\mathfrak{g}(A))$ may be regarded as a deformation (or a q-analogue) of the universal enveloping algebra $U(\mathfrak{g}(A))$.

We have deliberately introduced the square root of q^{α_i} for the following reason:

Proposition 4.1. There is an algebra homomorphism $\Delta: \hat{U} \to \hat{U}\otimes\hat{U}$ ($\hat{U}=\hat{U}(\mathfrak{g}(A))$) such that

$$\Delta(q^{\alpha_i/2}) = q^{\alpha_i/2} \otimes q^{\alpha_i/2} \qquad (4.5)$$

$$\Delta(\hat{e}_i) = \hat{e}_i \otimes q^{-\alpha_i/2} + q^{\alpha_i/2} \otimes \hat{e}_i$$

$$\Delta(\hat{f}_i) = \hat{f}_i \otimes q^{-\alpha_i/2} + q^{\alpha_i/2} \otimes \hat{f}_i .$$

Interchanging the tensor components we get another homomorphism $\bar{\Delta} = \mathcal{P}\Delta$,

$\mathcal{P}(a\otimes b) = b\otimes a \ (a,b \in \hat{U})$.

Note that the expressions appearing in (4.4) are precisely the images of Δ and $\bar{\Delta}$.

Going back to the case of an affine Lie algebra $\hat{\mathfrak{g}} = (\mathfrak{g},\sigma)$, let (ρ,V) be a finite dimensional irreducible representation of $\mathfrak{g}$. By virtue of the realization in §2, this gives rise to a representation of the loop algebra $\tilde{\mathfrak{g}} = \mathfrak{g}^{(k)}[\lambda,\lambda^{-1}]$ on $V\otimes\mathbb{C}[\lambda,\lambda^{-1}]$. We say that (ρ,V) is q-liftable to $\hat{U} = \hat{U}(\mathfrak{g})$ (resp. $\hat{U}(\hat{\mathfrak{g}})$) if there exists a representation $\hat{\rho} : \hat{U} \to \mathrm{End}(V)$ (resp. $\mathrm{End}(V)\otimes\mathbb{C}[\lambda,\lambda^{-1}]$) such that $\hat{\rho} \to \rho$ as $q \to 1$ (i.e. $\hat{\rho}(\hat{e}_i) \to \rho(e_i)$, etc.)

Example. Let $\hat{\mathfrak{g}} = A_n^{(1)}$, i.e. $\mathfrak{g} = \mathfrak{sl}(n+1)$ and σ = identity. In the vector representation $V = \mathbb{C}^{n+1}$, the eigenvalues of α_i $(= h_i)$ are 0 and ±1, so that $\frac{q^{\alpha_i}-q^{-\alpha_i}}{q-q^{-1}} = \alpha_i = h_i$. In fact, $\hat{\rho}(\hat{e}_0) = \lambda E^0_{n0}$, $\hat{\rho}(\hat{f}_0) = \lambda^{-1}E^0_{0n}$, $\hat{\rho}(\hat{e}_i) = E^0_{i-1\ i}$ and $\hat{\rho}(\hat{f}_i) = E^0_{i\ i-1}$ $(E^0_{ij} = (\delta_{ia}\delta_{jb}))$ gives a representation $\hat{\rho}: \hat{U}(\hat{\mathfrak{g}}) \to \mathrm{End}(\mathbb{C}^{n+1})$. In other words the fundamental representation of $\hat{U}(\mathfrak{g})$ is identical with that of $\mathfrak{g}$ (hence is trivially q-liftable to $\hat{U}(\hat{\mathfrak{g}})$.) The same is true of all the non-exceptional affine Lie algebras in their vector representation treated in §3.

Theorem 4.2 [9]. Let (ρ_i,V_i) $(i=1,2,3)$ be irreducible representations of $\mathfrak{g}$ which are q-liftable to $\hat{U}(\hat{\mathfrak{g}})$. Then the solution space of the equations (4.3), viewed as the ones in $\mathrm{End}(V_i\otimes V_j)$, is at most one dimensional. Moreover, if there are non-trivial solutions R_{ij}, then they satisfy the QYBE (1.1) and have $r_{ij} = (\rho_i\otimes\rho_j)(r(x))$ ($r(x)$ given in (2.11)) as their classical limit in the sense of (1.2).

Hence the construction of the quantum R matrix is reduced to the following:

Question 1. When is (ρ,V) q-liftable?

Question 2. Supposing (ρ_i,V_i) to be q-liftable, is there a non-trivial solution R in $\mathrm{End}(V_1\otimes V_2)$?

The results of §3 have been obtained by solving explicitly the linear equations (4.3) (cf. Example above). In the next section we show that, in the case $\hat{\mathfrak{g}} = A_n^{(1)}$, all the irreducible representations of $\mathfrak{g} = \mathfrak{sl}(n+1)$ are q-liftable to $\hat{U}(\mathfrak{g})$ and to $\hat{U}(\hat{\mathfrak{g}})$. Incidentally, we find it more convenient in this case to deal with $\mathfrak{g} = \mathfrak{gl}(n+1)$ by setting $\alpha_i = \epsilon_{i-1}-\epsilon_i$ $(1 \le i \le n)$ and adjoining to $U(\mathfrak{sl}(n+1))$ the elements $q^{\pm\epsilon_i/2}$ such that $q^{\epsilon_0+\cdots+\epsilon_n}$ commutes with everything. See §5 (5.1) below.

Remark. Recently Drinfel'd [7] studied the quantization of rational r matrix (1.4) by introducing an algebra $Y(\mathfrak{g})$. (The present algebra $\hat{U}(\mathfrak{g})$ is its trigonometric counterpart.) He showed that

1) $R(u,\hbar)$ exists in $Y(\mathfrak{g})\otimes Y(\mathfrak{g})$ for all $\mathfrak{g}$.
2) If $\mathfrak{g} = \mathfrak{sl}(n+1)$, then any irreducible representation of $\mathfrak{g}$ is liftable to $Y(\mathfrak{g})$.
   If $\mathfrak{g} \ne \mathfrak{sl}(n+1)$, then some representations of $\mathfrak{g}$ (e.g. the adjoint representation) are <u>not</u> liftable to $Y(\mathfrak{g})$.

See [7] for a more precise statement.

§5. Representations of $\hat{U}(\mathfrak{gl}(n+1))$ and the Hecke algebra

In this and the next section, $\mathfrak{g}$ will always stand for $\mathfrak{gl}(n+1)$. The goal of this section is the following.

Theorem 5.1 [10]. Any irreducible finite dimensional representation of $\mathfrak{g}$ is q-liftable to $\hat{U}(\mathfrak{g})$.

To begin with, let us write down the commutation rules for the generators $q^{\pm\epsilon_i/2}$ $(0 \le i \le n)$, $\hat{e}_i$ and $\hat{f}_i$ $(1 \le i \le n)$ of $\hat{U} = \hat{U}(\mathfrak{g})$.

$$q^{\epsilon_i/2}\hat{e}_j q^{-\epsilon_i/2} = \begin{cases} q\hat{e}_j & (i=j-1) \\ q^{-1}\hat{e}_j & (i=j) \\ \hat{e}_j & (i \ne j-1,j) \end{cases}, \quad q^{\epsilon_i/2}\hat{f}_j q^{-\epsilon_i/2} = \begin{cases} q^{-1}\hat{f}_j & (i=j-1) \\ q\hat{f}_j & (i=j) \\ \hat{f}_j & (i \ne j-1,j) \end{cases}$$

$$[\hat{e}_i,\hat{f}_j] = \delta_{ij}\frac{q^{\alpha_i}-q^{-\alpha_i}}{q-q^{-1}} \quad (\alpha_i = \epsilon_{i-1}-\epsilon_i) \tag{5.1}$$

$$\hat{e}_i\hat{e}_j = \hat{e}_j\hat{e}_i, \; \hat{f}_i\hat{f}_j = \hat{f}_j\hat{f}_i \quad (|i-l| \ge 2)$$

$$\hat{e}_i^2\hat{e}_{i\pm1}-(q+q^{-1})\hat{e}_i\hat{e}_{i\pm1}\hat{e}_i+\hat{e}_{i\pm1}\hat{e}_i^2 = 0 \quad (1 \le i,\; i\pm1 \le n)$$

$$\hat{f}_i^2\hat{f}_{i\pm1}-(q+q^{-1})\hat{f}_i\hat{f}_{i\pm1}\hat{f}_i+\hat{f}_{i\pm1}\hat{f}_i^2 = 0 \quad (1\leq i,\ i\pm1\leq n).$$

Here $q^{\epsilon_i/2}$'s are understood to be invertible and commutative among themselves. Let $\hat{E}_{ij}\in\hat{U}$ $(i\neq j)$ be defined recursively by $E_{i-1i} = \hat{e}_i$, $\hat{E}_{ii-1} = \hat{f}_i$ and

$$\hat{E}_{ij} = \hat{E}_{ik}\hat{E}_{kj}-q^{\pm1}\hat{E}_{kj}\hat{E}_{ik} \qquad (i\lessgtr k\lessgtr j). \tag{5.2}$$

(The RHS of (5.2) is shown to be independent of k). Put

$$\hat{e}_0' = q^{\epsilon_0+\epsilon_n-\epsilon}\hat{E}_{n0},\ \hat{f}_0' = q^{-\epsilon_0-\epsilon_n+\epsilon}E_{0n},\ \epsilon = \epsilon_0+\epsilon_1+\cdots+\epsilon_n. \tag{5.3}$$

We have then the following "homogeneous realization" of $\hat{U}(A_n^{(1)})$.

Proposition 5.2. There is an algebra homomorphism φ: $\hat{U}(A_n^{(1)}) \to \hat{U}\otimes\mathbb{C}[\lambda,\lambda^{-1}]$ such that $\varphi(\hat{e}_0) = \lambda\hat{e}_0'$, $\varphi(\hat{f}_0) = \lambda^{-1}\hat{f}_0'$ and $\varphi(\hat{e}_i) = \hat{e}_i$, $\varphi(\hat{f}_i) = \hat{f}_i$ $(1\leq i\leq n)$.

Thanks to this proposition, the question of q-liftability to $\hat{U}(A_n^{(1)})$ is reduced to that to $\hat{U}(\mathfrak{gl}(n+1))$.

Warning: The map φ is <u>not</u> compatible with Δ of (4.5) in the sense that $\Delta_{A_n}\varphi\neq(\varphi\otimes\varphi)\Delta_{A_n^{(1)}}$.

Written out in full, the linear equations for R (4.3) now read as follows $(x = \lambda/\mu)$:

$$R(x)\Delta(\hat{X}) = \bar{\Delta}(\hat{X})R(x),\ \hat{X}\in\hat{U}$$

$$R(x)(\lambda\hat{e}_0'\otimes q^{-\alpha_0/2}+q^{\alpha_0/2}\otimes\mu\hat{e}_0') = (\lambda\hat{e}_0'\otimes q^{\alpha_0/2}+q^{-\alpha_0/2}\otimes\mu\hat{e}_0')R(x) \tag{5.4}$$

$$(\hat{e}_0' \leftrightarrow \hat{f}_0',\ \lambda \leftrightarrow \lambda^{-1},\ \mu \leftrightarrow \mu^{-1}).$$

Note that, if we fix a representation space $V = V_1 = V_2$, then the first equation of (5.4) means the invariance condition

$$[\check{R}(x),\ \Delta(\hat{X})] = 0,\ \hat{X}\in\hat{U}. \tag{5.4$'$}$$

In the sequel we shall concentrate on the representations of $\hat{U}(\mathfrak{g})$. One way to construct irreducible representations of $\mathfrak{g}$ (or $U(\mathfrak{g})$) is to decompose the tensor representation $V^{(m)} = V_1 \otimes \ldots \otimes V_m$ of $V_i = \mathbb{C}^{n+1}$ via the natural action of the symmetric group $\mathfrak{S}_m$ (or its group algebra $\mathbb{C}[\mathfrak{S}_m]$). Here we shall follow the same idea, replacing $U(\mathfrak{g})$ by $\hat{U}(\mathfrak{g})$ and $\mathbb{C}[\mathfrak{S}_m]$ by its q-analogue $\mathcal{H}_m$, the Hecke algebra of type A_{m-1} [18].

By definition, $\mathcal{H}_m$ is an algebra generated by a_i $(i=1,\ldots,m-1)$ under the defining relations

$$\begin{cases} (a_i - q^*)(a_i + 1) = 0 \\ a_i a_{i+1} a_i = a_{i+1} a_i a_{i+1} \\ a_i a_j = a_j a_i \qquad (|i-j| \geq 2). \end{cases} \tag{5.5}$$

Here $q^* \neq 0$ is a parameter. $\mathcal{H}_m$ is spanned by basis elements a_w indexed by $w \in \mathfrak{S}_m$ with the properties: $a_{(i,i+1)} = a_i$, and $a_{ww'} = a_w a_{w'}$ if $\ell(ww') = \ell(w) + \ell(w')$. Here $\ell(w) = \min\{p \mid w = s_1 s_2 \ldots s_p,\ s_i \in S\}$ signifies the length of w with respect to the set of simple transpositions $S = \{(12),(23),\ldots,(m-1\ m)\}$. If $q^* = 1$, then $a_i^2 = 1$ and $\mathcal{H}_m$ is nothing but the group algebra $\mathbb{C}[\mathfrak{S}_m]$. It is known (see [19] and references therein) that, if q^* is not a root of unity, then one still has $\mathcal{H}_m \cong \mathbb{C}[\mathfrak{S}_m]$ as abstract algebras (though explicit isomorphism seems unavailable).

To see how $\mathcal{H}_m$ comes into picture, let us return to the R matrix in the vector representation in §3. After changing x to x^2 and multiplying $(qx)^{-1}$, the $\check{R}$ matrix is given by $\check{R}(x) = xT - x^{-1}T^{-1}$ with

$$T^{\pm 1} = q^{\pm 1} I - \sum_{i \neq j} (q^{\epsilon(i-j)} E_{ij}^0 \otimes E_{jj}^0 - E_{ji}^0 \otimes E_{ij}^0).^{\dagger)}$$

Consider the tensor representation $\Delta^{(m)}: \hat{U} \to \mathrm{End}(V^{(m)})$ obtained by iterating Δ (4.5), and let T_i be the T-matrix corresponding to $\check{R}_{ii+1}(x)$.

Proposition 5.3. There exists a representation $\sigma^{(m)}: \mathcal{H}_m \to \mathrm{End}(V^{(m)})$ such that $\sigma^{(m)}(a_i) = qT_i$, with $q^* = q^2$.

In fact, the first two relations of (5.5) are equivalent to the QYBE for $\check{R}_{ii+1}(x)$. On the other hand, (5.4)' implies

$$[T_i, \Delta^{(m)}(\hat{U})] = 0,$$

---

†) Our T here is T^{-1} in [10].

so the action of $\mathcal{H}_m$ commutes with that of $\hat{U}$. It can be shown further that the two subalgebras $A = \sigma^{(m)}(\mathcal{H}_m)$ and $A' = \Delta^{(m)}(\hat{U})$ of $\mathrm{End}(V^{(m)})$ are commutant to each other: $A' = \mathrm{End}_A(V^{(m)})$, $A = \mathrm{End}_{A'}(V^{(m)})$. From the general theory of semi-simple algebras [20], the irreducible decomposition of $V^{(m)}$ with respect to $\hat{U}$ now reduces to that with respect to $\mathcal{H}_m$. In practice, the latter is achieved by using the q-analogue of Young symmetrizers [21]. Here we are content to give the formulas for the simplest (anti-)symmetrizers $s_m^{\pm} = (s_m^{\pm})^2$:

$$s_m^{\pm} = \frac{1}{[m]!} \sum (\pm)^{\ell(w)} q^{\pm(\ell(w)-m(m-1)/2)} T_w \tag{5.6}$$

where $T_w = q^{-\ell(w)} \sigma^{(m)}(a_w)$. In this way we arrive at the conclusion of Theorem 5.1.

There is also an alternative and more direct way to show Theorem 5.1, that is, to give explicit matrix representations of $\hat{e}_i$ and $\hat{f}_i$ relative to a suitable basis.

Let $V(\underline{\lambda})$ be an irreducible $\mathfrak{gl}$-module with highest weight $\underline{\lambda} = (\lambda_0, \lambda_1, \ldots, \lambda_n)$ $(\lambda_0 \geq \lambda_1 \geq \cdots \geq \lambda_n,\ \lambda_i \in \mathbb{Z})$. It is known [12] that $V(\underline{\lambda})$ has a natural orthonormal basis

$$|\lambda_{\alpha\beta}\rangle = \left| \begin{matrix} \lambda_{0n} & \lambda_{1n} & \cdots & \lambda_{nn} \\ \lambda_{0n-1} & \lambda_{1n-1} & \cdots & \lambda_{n-1n-1} \\ & \ddots & & \\ & \lambda_{01} & \lambda_{11} & \\ & & \lambda_{00} & \end{matrix} \right\rangle , \tag{5.7}$$

which are indexed by integers $\{\lambda_{\alpha\beta}\}$ satisfying the conditions i) $\lambda_{\alpha n} = \lambda_\alpha$ $(0 \leq \alpha \leq n)$ and ii) $\lambda_{\alpha\beta} \geq \lambda_{\alpha\beta-1} \geq \lambda_{\alpha+1\beta}$ for $0 \leq \alpha < \beta \leq n$. Consider the tower of subalgebras $\mathfrak{gl}(n+1) \supset \mathfrak{gl}(n) \supset \ldots \supset \mathfrak{gl}(1)$ (natural embedding). Then the vector (5.7) is characterized (up to constant multiple) by the property that, for any β, it belongs to the irreducible component of highest weight $(\lambda_{0\beta}, \lambda_{1\beta}, \ldots, \lambda_{\beta\beta})$ with respect to the subalgebra $\mathfrak{gl}(\beta+1)$. In particular,

$$E_{ii}|\lambda_{\alpha\beta}\rangle = \left(\sum_{\gamma=0}^{i} \lambda_{\gamma i} - \sum_{\gamma=0}^{i-1} \lambda_{\gamma i-1}\right)|\lambda_{\alpha\beta}\rangle.$$

Define the operators $\hat{E}'_{i-1i} \in \mathrm{End}(V(\underline{\lambda}))$ by setting

$$\hat{E}'_{i-1i}|\lambda_{\alpha\beta}\rangle = \sum_{\nu=0}^{i-1} \hat{c}_{\nu i-1}(\ell_{\alpha\beta})|\lambda_{\alpha\beta}+\delta_{\nu\alpha}\delta_{i-1\beta}\rangle \tag{5.8}$$

$$\hat{c}_{\nu i-1}(\ell_{\alpha\beta}) = \left\{ - \frac{\prod_{s=0}^{i-2}[\ell_{si-2}-\ell_{\nu i-1}-1]\prod_{t=0}^{i}[\ell_{ti}-\ell_{\nu i-1}]}{\prod_{\substack{\mu(\neq\nu)\\ 0\leqq\mu\leqq i-1}}[\ell_{\mu i-1}-\ell_{\nu i-1}-1][\ell_{\mu i-1}-\ell_{\nu i-1}]} \right\}^{1/2}$$

where $\ell_{\alpha\beta} = \lambda_{\alpha\beta}-\alpha$.

Proposition 5.4. There exists a representation $\hat{\rho}: \hat{U}(\mathfrak{g}) \to \mathrm{End}(V(\underline{\lambda}))$ given by $\hat{\rho}(q^{\epsilon_i}) = q^{E_{ii}}$, $\hat{\rho}(\hat{e}_i) = \hat{E}'_{i-1i}$ and $\hat{\rho}(\hat{f}_i) = {}^t\hat{E}'_{i-1i}$.

In the limit $q \to 1$, (5.8) reduces to the formula for the Lie algebra generators E_{i-1i} given in [12].

Example. When $V(\underline{\lambda})$ is the symmetric tensor representation ($\underline{\lambda} = (\lambda_0,0,\dots,0)$), we can eliminate the basis vectors from (5.8) to find

$$\hat{\rho}(\hat{e}_i) = F(E_{ii}+1)F(E_{i-1i-1})e_i$$

$$\hat{\rho}(\hat{f}_i) = f_iF(E_{ii}+1)F(E_{i-1i-1}) \qquad (1\leq i\leq n)$$

where $F(z) = \sqrt{(q^z-q^{-z})/z(q-q^{-1})}$. More generally, we have by induction

$$\hat{\rho}(\hat{E}_{ij}) = q^{c_{ij}}F(E_{ii})F(E_{jj}+1)E_{ij} \qquad (i\neq j) \tag{5.9}$$

with $c_{ij} = \mp \sum_{i\lessgtr k\lessgtr j} E_{kk}$ for $i\lessgtr j$.

§6. R matrix related to higher representation of $\mathfrak{gl}(n+1)$

Having settled the question about representations of $\hat{U} = \hat{U}(\mathfrak{g})$ ($\mathfrak{g} = \mathfrak{gl}(n+1)$), let us consider the existence of the corresponding R matrix.

Proposition 6.1. Define

$$\hat{E}_{ij}(x) = \begin{cases} (xq^{(\epsilon_i+\epsilon_j-1)/2})^{\mp 1}\hat{E}_{ij} & (i \lessgtr j) \\ \dfrac{xq^{\epsilon_i}-x^{-1}q^{-\epsilon_i}}{q-q^{-1}} & (i=j) \end{cases}$$

where $\hat{E}_{ij}$ is given in (5.2). Then

$$R(x) = \sum \hat{E}_{ij}(x) \otimes E^0_{ji} \tag{6.1}$$

satisfies the linear equations (5.4) (with $\lambda/\mu = x^2$) in $\hat{U} \otimes \mathrm{End}(\mathbb{C}^{n+1})$.

In the case of the symmetric tensor representation, the result above together with (5.9) reproduces Babelon's result [22].

With the help of the Hecke algebra, it is possible to construct R matrices out of (6.1) by the "fusion" process introduced in [3] in the rational case. Let $R_{0i}(x) \in \hat{U} \otimes \mathrm{End}(V_i)$ $(V_i = \mathbb{C}^{n+1})$ denote a copy of (6.1). We set

$$\begin{aligned} R^{\pm}_{0,\{1\ldots m\}}(x) &= R_{0m}(x)R_{0m-1}(xq^{\pm 1})\ldots R_{01}(xq^{\pm(m-1)})s^{\pm}_m \\ &= s^{\pm}_m R_{0m}(xq^{\pm(m-1)})R_{0m-1}(xq^{\pm(m-2)})\ldots R_{01}(x) \end{aligned} \tag{6.2}$$

where $s^{\pm}_m$ is given in (5.6).

Proposition 6.2. $R^{\pm}_{0,\{1\ldots m\}}(x)$ solves (5.4) in $\hat{U} \otimes \mathrm{End}(s^{\pm}_m V^{(m)})$.

Obviously, similar construction is possible in $\mathrm{End}(s^{\pm}_m V^{(m)}) \otimes \hat{U}$. Combining this with the results of §5, we conclude that the existence of $R(x) \in \mathrm{End}(V_1 \otimes V_2)$ is established in the case one of V_i is the symmetric or anti-symmetric tensor space, and the other is arbitrary.

As a byproduct of Proposition 6.2, we obtain an expression for central elements of $\hat{U}$ (cf. the quantum determinant [2]). Define

$$z(x) \underset{\mathrm{def}}{=} \sum_{w \in \mathfrak{S}_{n+1}} (-q)^{\ell(w)} \hat{E}_{nw(n)}(xq^{-n}) \ldots \hat{E}_{1w(1)}(xq^{-1}) \hat{E}_{0w(0)}(x). \tag{6.3}$$

The RHS is reminiscent of the determinant and so is written symbolically as

$$\text{q-}\underset{\rightarrow}{\det} \begin{bmatrix} \hat{E}_{nn}(xq^{-n}) & \ldots & \hat{E}_{0n}(x) \\ \vdots & & \vdots \\ \hat{E}_{n0}(xq^{-n}) & \ldots & \hat{E}_{00}(x) \end{bmatrix}.$$

Proposition 6.3. $z(x)$ belongs to the center of $\hat{U}(\mathfrak{g})$ for any x. On the highest weight module $V(\underline{\lambda})$, $z(x) = \prod_{i=0}^{n} \left(\frac{xq^{\lambda_i - i} - x^{-1}q^{-\lambda_i + i}}{q-q^{-1}} \right)$.

For instance, if $n = 1$, then

$$z(x) = \frac{1}{(q-q^{-1})^2}\left(x^2 q^{\epsilon_0+\epsilon_1-1} + x^{-2} q^{-\epsilon_0-\epsilon_1+1} - \hat{C}\right)$$

$$\hat{C} = q^{\alpha+1}+q^{-\alpha-1}+(q-q^{-1})^2 \hat{f}\hat{e}. \tag{6.4}$$

Formula (6.4) is a q-analogue of the Casimir element for $s\ell(2)$. If we set in (6.3) $x = q^u$ and let $q \to 1$, then $z(x)$ tends to

$$\underset{\rightarrow}{\det}\begin{bmatrix} u-n+E_{nn} & \cdots & E_{0n} \\ \vdots & & \vdots \\ E_{n0} & \cdots & u+E_{00} \end{bmatrix}.$$

It is known [23] that the coefficients of u^j provides the generators of the center of the universal enveloping algebra $U(\mathfrak{g})$. We suspect the same is true of $z(x)$ and $\hat{U}(\mathfrak{g})$.

In the special case $\mathfrak{g} = s\ell(2)$, the symmetric tensors exhaust its irreducible representations. Let us consider the case $V_1 = V_2 = \boxed{\cdots}$ (λ-fold symmetric tensor) for which $\check{R}(x) = PR(x)$ makes sense. The usual Clebsch-Gordan law applies to the decomposition of $V_1 \otimes V_2$ with respect to $\Delta(\hat{U})$:

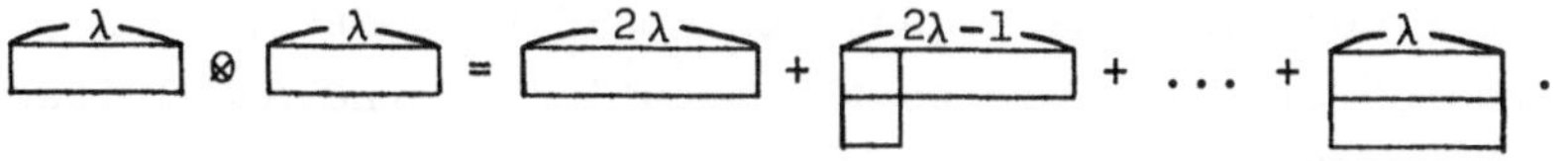

Since $\check{R}(x)$ commutes with $\Delta(\hat{U})$, one must have

$$\check{R}(x) = \sum_{\nu=0}^{\lambda} \rho_\nu(x) P_\nu \tag{6.5}$$

where P_ν denotes the projector onto the component 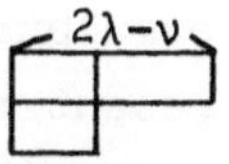

.

Proposition 6.4. The eigenvalues $\rho_\nu(x)$ are given by

$$\rho_\nu(x) = \rho_\lambda(x) x^{\nu-\lambda} \prod_{j=1}^{\lambda-\nu} \frac{1-xq^{2j}}{1-x^{-1}q^{2j}} = \rho_0(x) x^{-\nu} \prod_{j=\lambda-\nu+1}^{\lambda} \frac{1-xq^{-2j}}{1-x^{-1}q^{-2j}}.$$

It follows that, by adjusting a constant multiple, the matrix (6.5) has the universal form (cf. [3])

$$\check{R}(x)^2 = \prod_{j=1,3,5,\dots} \frac{1-x^{-1}q^j\Delta(\hat{C})+x^{-2}q^{2j}}{1-xq^j\Delta(\hat{C})+x^2q^{2j}}$$

where $\hat{C}$ is given by (6.4).

§7. Discussions

A number of questions are left unanswered. Those directly related to the present article are

1) Representation theory of $\hat{U}(\mathfrak{g})$ for $\mathfrak{g}$ general,

2) Explicit description of $R(x)$,

3) Generalization to the elliptic case for $\mathfrak{g} = \mathfrak{sl}(n+1)$.

It seems likely that, after appropriate completion, the algebra $\hat{U}(\mathfrak{g})$ becomes isomorphic to $U(\mathfrak{g})$ if q is not a root of unity, much as the Hecke algebra is to the group algebra of the Weyl group. Representation theory in the case q = a root of unity seems also interesting (cf. [24]). Point 2) includes the problem of computing eigenvalues of $R(x)$. (In the rational case with $\mathfrak{g} = \mathfrak{sl}(3)$, see [25].) It seems that the commuting property of $\check{R}(x)$ (3.4) cannot be expected in general. In the elliptic case, the structure of the "quantum algebra" becomes much involved [5,26], and one does not know how to define the co-multiplication Δ. Perhaps a more important question would be whether or not the "quantum algebra" plays a role to the solution of the model itself defined by the R matrix.

Finally we would like to emphasize that there are a number of solutions to the QYBE which do not fall within the category of "quasi-classical" R matrix (i.e. one that satisfies (1.2)). There is also another class, the IRF model [27] for which even the classical limit can not be formulated straightforwardly. The reader is referred to [28-31] for recent progress in the latter framework.

References

[1] P. P. Kulish and E. K. Sklyanin, J. Soviet Math. 19 (1982) 1596.

[2] ________, in : Integrable quantum field theories, Lecture Notes in Physics 151, Springer (1982) 61.

[3] P. P. Kulish, N. Yu. Reshetikhin and E. K. Sklyanin, Lett. Math. Phys. 5 (1981) 393.

[4] A. A. Belavin and V. G. Drinfel'd, Funct. Anal. and Appl. 17 (1983) 220.

[5] E. K. Sklyanin, Funct. Anal. and Appl. 16 (1982) 263, 17 (1983) 273.

[6] P. P. Kulish and N. Yu. Reshetikhin, J. Soviet Math. 23 (1983) 2435.

[7] V. G. Drinfel'd, Dokl. Akad. Nauk SSSR (1985).

[8] M. Jimbo, Lett. Math. Phys. 10 (1985) 63.

[9] ________, Comm. Math. Phys., to appear.

[10] ________, A q-analogue of U((N+1)), Hecke algebra and the Yang-Baxter equation, RIMS preprint 517, Kyoto Univ. (1985).

[11] V. V. Bazhanov, Trigonometric solutions of the triangle equation and classical Lie algebras (in Russian), preprint ИФЗЭ 85-18, Serpukov, 1985.

[12] I. M. Gelfand and M. L. Zetlin, Dokl. Akad. Nauk SSSR 71 (1950) 825.

[13] M. A. Semenov-Tyan-Shanskii, Funct. Anal. and Appl. 17 (1984) 259.

[14] V. G. Kac, Infinite dimensional Lie algebras, Birkhauser, Boston, Mass. 1983.

[15] J. H. H. Perk and C. L. Schultz, in : Non-linear integrable systems, World Scientific, Singapore 1983.

[16] O. Babelon, H. J. de Vega and C. M. Viallet, Nucl. Phys. B190 (1981) 542.
I. V. Cherednik, Theor. Math. Phys. 43 (1980) 356.
D. V. Chudnovsky and G. V. Chudnovsky, Phys. Lett. 79A (1980) 36.
C. L. Schultz and J. H. H. Perk, Phys. Lett. 84 A (1981) 407.

[17] A. G. Izergin and V. E. Korepin, Comm. Math. Phys. 79 (1981) 303.

[18] N. Bourbaki, Groupes et algebres de Lie, Chap. 4, Exerc. 22-24, Hermann, Paris, 1968.

[19] C. W. Curtis, N. Iwahori and R. Kilmoyer, I.H.E.S. Publ. Math. 40 (1972) 81.

[20] N. Iwahori, Representation theory of the symmetric group and the general linear group (in Japanese), Iwanami, Tokyo 1978.

[21] A. Gyoja, A q-analogue of Young symmetrizer, preprint, Osaka Univ. (1985).

[22] O. Babelon, Nucl. Phys. B230 [FS 10] (1984) 241.

[23] H. Weyl, The classical groups, Princeton University Press, Princeton 1953.

[24] V. F. R. Jones, Braid groups, Hecke algebras and type II_1 factors, to appear in Japan-U.S. Conf. Proc. 1983.

[25] P. P. Kulish and S. I. Alishauskas, Zapiski Nauch. Semin. LOMI 145 (1985) 3.

[26] I. V. Cherednik, Funct. Anal. and Appl. 19 (1985) 89.

[27] R. J. Baxter, Exactly solved models in statistical mechanics, Academic, London 1982.

[28] V. V. Bazhanov and Yu. G. Stroganov, Nucl. Phys. B205 [FS 5] (1982) 505.

[29] G. E. Andrews, R. J. Baxter and P. J. Forrester, J. Stat. Phys. 35 (1984) 193.
M. Jimbo and T. Miwa, Physica 15 D (1985) 335.

[30] M. Jimbo and T. Miwa, Nucl. Phys. B 257 [FS 14] (1985) 1.

[31] T. Miwa, Multi-state solutions to the star-triangle relation with Abelian symmetries, preprint (1985).

Appendix.

We give below the explicit formulas for the matrix $(A_{\alpha\beta}(x))$ (3.1) and the weights (3.2) for $D_{n+1}^{(2)}$ in §3. For definiteness we shall write them down in the case $n = 3$, but the reader can easily see the rules in the general case (cf. [9]). Except in the case $\hat{\mathfrak{g}} = D_{n+1}^{(2)}$ we set

$$a = (\xi x - q^2)(x-1)$$

$$b = (q^2-1)\xi(x-1)$$

$$c = (q^2-1)(\xi x-1).$$

The parameter ξ is given in §3, Table 1.

$\hat{\mathfrak{g}} = B_3^{(1)},\ A_6^{(2)}$ $\quad (\xi = q^5, -q^7)$

$$\left[\begin{array}{ccc|c|ccc}
a & -qb & -q^2b & -q^{5/2}b & -q^3b & -q^4b & -q^5b+c \\
-\xi q^{-1}xb & a & -qb & -q^{3/2}b & -q^2b & -q^3b+c & -q^4b \\
-\xi q^{-2}xb & -\xi q^{-1}xb & a & -q^{1/2}b & -qb+c & -q^2b & -q^3b \\
\hline
-\xi q^{-5/2}xb & -\xi q^{-3/2}xb & -\xi q^{-1/2}xb & s & -q^{1/2}b & -q^{3/2}b & -q^{5/2}b \\
\hline
-\xi q^{-3}xb & -\xi q^{-2}xb & -x(\xi q^{-1}b-c) & -\xi q^{-1/2}xb & a & -qb & -q^2b \\
-\xi q^{-4}xb & -x(\xi q^{-3}b-c) & -\xi q^{-2}xb & -\xi q^{-3/2}xb & -\xi q^{-1}xb & a & -qb \\
-x(\xi q^{-5}b-c) & -\xi q^{-4}xb & -\xi q^{-3}xb & -\xi q^{-5/2}xb & -\xi q^{-2}xb & -\xi q^{-1}xb & a
\end{array}\right]$$

Here $s = q(\xi x-1)(x-1)+(\xi-1)(q^2-1)x$.

This matrix has the following structure:

(i) Diagonal elements are all the same except in the middle.

(ii) Elements in the upper (resp. lower) triangle are $-q^i b$ (resp. $-\xi q^{-i}xb$). The power i is incremented by one to the right along the row (resp. down the column) in general, and by 1/2 at the middle.

(iii) Anti-diagonal elements contain an extra term c or xc.

Similar rules apply to other cases as well.

$\hat{\mathfrak{g}} = D_3^{(1)},\ A_5^{(2)}$ $\quad (\xi = q^4, -q^6)$

$$\left[\begin{array}{ccc|ccc}
a & -qb & -q^2b & -q^2b & -q^3b & -q^4b+c \\
-\xi q^{-1}xb & a & -qb & -qb & -q^2b+c & -q^3b \\
-\xi q^{-2}xb & -\xi q^{-1}xb & a & -b+c & -qb & -q^2b \\
\hline
-\xi q^{-2}xb & -\xi q^{-1}xb & -x(\xi b-c) & a & -qb & -q^2b \\
-\xi q^{-3}xb & -x(\xi q^{-2}b-c) & -\xi q^{-1}xb & -\xi q^{-1}xb & a & -qb \\
-x(\xi q^{-4}b-c) & -\xi q^{-3}xb & -\xi q^{-2}xb & -\xi q^{-2}xb & -\xi q^{-1}xb & a
\end{array}\right]$$

$\hat{\mathfrak{g}} = C_3^{(1)} \quad (\xi = q^8)$

$$\left[\begin{array}{ccc|ccc}
a & -qb & -q^2b & q^4b & q^5b & q^6b+c \\
-\xi q^{-1}xb & a & -qb & q^3b & q^4b+c & q^5b \\
-\xi q^{-2}xb & -\xi q^{-1}xb & a & q^2b+c & q^3b & q^4b \\
\hline
\xi q^{-4}xb & \xi q^{-3}xb & x(\xi q^{-2}b+c) & a & -qb & -q^2b \\
\xi q^{-5}xb & x(\xi q^{-4}b+c) & \xi q^{-3}xb & -\xi q^{-1}xb & a & -qb \\
x(\xi q^{-6}b+c) & \xi q^{-5}xb & \xi q^{-4}xb & -\xi q^{-2}xb & -\xi q^{-1}xb & a
\end{array}\right]$$

$\hat{\mathfrak{g}} = D_4^{(2)} \quad (\xi = q^3)$

We write down the matrix $(R_{\alpha\beta\gamma\delta}(x))$ where $(\alpha\beta)$, $(\gamma\delta)$ run over (18),(27),(36),(44),(45),(54),(55),(63),(72),(81) in this order.

$$\left[\begin{array}{ccc|cc|cc|ccc}
a & -qb & -q^2b & q^{5/2}b_- & -q^{5/2}b_+ & -q^{5/2}b_+ & q^{5/2}b_- & -q^3b & -q^4b & -q^5b+c \\
-q^5x^2b & a & -qb & q^{3/2}b_- & -q^{3/2}b_+ & -q^{3/2}b_+ & q^{3/2}b_- & -q^2b & -q^3b+c & -q^4b \\
-q^4x^2b & -q^5x^2b & a & q^{1/2}b_- & -q^{1/2}b_+ & -q^{1/2}b_+ & q^{1/2}b_- & -qb+c & -q^2b & -q^3b \\
\hline
-q^{1/2}xb_- & -q^{3/2}xb_- & -q^{5/2}xb_- & c_- & 0 & 0 & d_- & q^{1/2}b_- & q^{3/2}b_- & q^{5/2}b_- \\
-q^{1/2}xb_+ & -q^{3/2}xb_+ & -q^{5/2}xb_+ & 0 & c_+ & d_+ & 0 & -q^{1/2}b_+ & -q^{3/2}b_+ & -q^{5/2}b_+ \\
\hline
-q^{1/2}xb_+ & -q^{3/2}xb_+ & -q^{5/2}xb_+ & 0 & d_+ & c_+ & 0 & -q^{1/2}b_+ & -q^{3/2}b_+ & -q^{5/2}b_+ \\
-q^{1/2}xb_- & -q^{3/2}xb_- & -q^{5/2}xb_- & d_- & 0 & 0 & c_- & q^{1/2}b_- & q^{3/2}b_- & q^{5/2}b_- \\
\hline
-q^3x^2b & -q^4x^2b & -x^2(q^5b-c) & -q^{5/2}xb_- & -q^{5/2}xb_+ & -q^{5/2}xb_+ & -q^{5/2}xb_- & a & -qb & -q^2b \\
-q^2x^2b & -x^2(q^3b-c) & -q^4x^2b & -q^{3/2}xb_- & -q^{3/2}xb_+ & -q^{3/2}xb_+ & -q^{3/2}xb_- & -q^5x^2b & a & -qb \\
-x^2(qb-c) & -q^2x^2b & -q^3x^2b & -q^{1/2}xb_- & -q^{1/2}xb_+ & -q^{1/2}xb_+ & -q^{1/2}xb_- & -q^4x^2b & -q^5x^2b & a
\end{array}\right]$$

Here we have set

$$a = (\xi^2x^2-q^2)(x^2-1)$$

$$b = (q^2-1)(x^2-1)$$

$$c = (q^2-1)(\xi^2x^2-1)$$

$$b_{\pm} = \tfrac{1}{2}(q^2-1)(x^2-1)(\xi x\pm 1)$$

$$c_{\pm} = \mp\tfrac{1}{2}(q^2-1)(\xi+1)x(x\mp 1)(\xi x\pm 1)+q(x^2-1)(\xi^2x^2-1)$$

$$d_{\pm} = \pm\tfrac{1}{2}(q^2-1)(\xi-1)x(x\pm 1)(\xi x\pm 1).$$

SOLUTION OF THE MULTICHANNEL KONDO-PROBLEM

Natan Andrei
Serin Physics Laboratory
Rutgers University
Piscataway, NJ 08854

In this talk I will present some work I have done with C. Destri to solve a model proposed originally by P. Nozières and A. Blandin to describe the physics of magnetic impurities in a metal when their orbital structure is taken into account. I will describe the "physical thinking" going into the solution, rather than the mathematical details, which are available in the literature.

The "canonical model", the so called S-d model

$$H = \sum q\, C^{*}_{qa} C_{qa} + J \sum_{qq'} C^{*}_{qa} \sigma^{i}_{ab} C_{q'b} S^{i} \qquad (1)$$

is obtained through the following consideration: Given an impurity S located at r=0, it is convenient to express the electron field $C_{k,a}$ in a basis appropriate to the spherical symmetry around the impurity — $C_{k\ell m;a}$. Here k is the magnitude of the momentum, ℓ and m the angular quantum numbers and a is the spin component. One then linearizes the spectrum around the Fermi momentum, $k = k_F+q$. If one assumes the impurity to be in a s-wave state, only the $\ell=0$ $m=0$ modes of the electron field would couple and writing $C_{qa} = C_{k_F+q,\ell=0,m=0;a}$ one obtains the S-d or Kondo hamiltonian after dropping the contribution of the Fermi sphere. In real metals, however, the impurities are electrons in atomic orbitals around some nucleus. If they are in an orbital characterized by angular momentum ℓ_a, we can couple it only to the corresponding mode in the electron field around it — $C_{q\ell_a m;a}$, $\ell_a<m<\ell_a$. In the case an orbital singlet splits off one is led to consider the following hamiltonian (The quantum number ℓ_a is fixed throughout the problem and is not made explicit)

$$H = \sum q\, C^{*}_{qm;a} C_{qm;a} + J \sum_{q,q'm} C^{*}_{qm;a} \sigma^{i}_{ab} C_{q'm;b} S^{i}$$

which differs from the canonical model only in that new quantum number $m = -\ell_a,..,\ell_a$ have been added. I'll refer to these quantum numbers as flavor degrees of freedom since they play a role very similar to that played by flavor in QCD. The spin degrees of freedom are analogous to color, of course.

What is the effect of flavor? To put it in perspective let me briefly review the results of the canonical model. That model exhibits the Kondo effect which is a smooth crossover between weak and strong coupling regime without any singularities.

This can be illustrated by the behavior of the impurity susceptibility as a function of the temperature

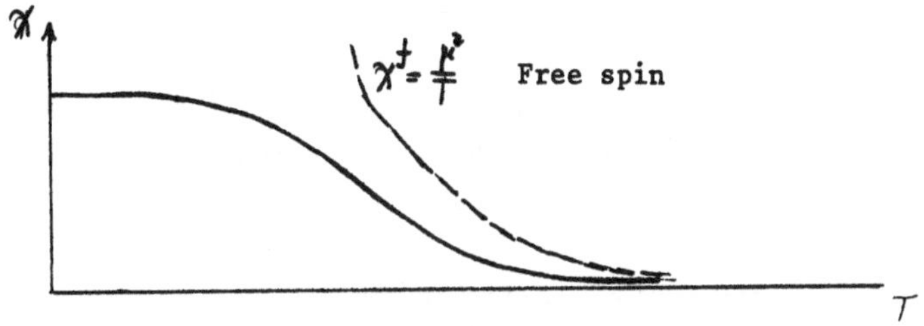

At high temperature the susceptibility tends towards the Curie law with logarithmic corrections typical of asymptotically free theories

$$\chi \to \frac{\mu^2}{T}\,[1+\frac{1}{\ell n \frac{T}{T_k}} + \ldots\,] \quad .$$

This indicates that the high temperature behaviors is controlled by the trivial fixed point of J=0. On the other hand, as we lower the temperature one gradually moves towards a strong coupling regime, and the susceptibility, instead of diverging, as is the case for free spins, tends towards a finite limit

$$\chi \to \frac{\mu^2}{\pi T_o} \quad .$$

This indicates a complete screening of the impurity through its interaction with the gas of electrons.

In terms of renormalization group flows we have a smooth flow of the interaction from the weak coupling point (J=0) controlling the high temperature/high field behavior towards the strong coupling point (J=∞) which controls the infrared behavior.

The introduction of flavor interaction changes this picture dramatically, although flavor is introduced diagonally and may look quite innocent at first sight.

Consider the stability of the trivial fixed points at J=0 and J=∞:

The weak coupling fixed point can be analyzed by perturbation theory and one immediately finds that it remains unstable, the contribution of flavor entering only to non-leading order. The coupling constant will flow away: 

To analyze the strong coupling fixed point imagine we consider the kinetic part as a perturbation on the interaction $H_I = J\,\psi^*_{am}(0)\sigma^i_{ab}\psi_{bm}(0)S^i$ where we introduce $\psi_{am}(x)$, the Fourier transform with respect to the momentum of $C_{qm;a}$. We also consider here S, the impurity spin to be in any representation of SU(2). We thus have two parameters. S, the dimension of spin representation, and $f = 2\ell_a+1$, the dimension of flavor representation.

We have, therefore, two cases to consider. 1) Under screening: $\frac{1}{2}f < S$. In this case the ground state of H_I is represented by:  , as a spin complex $S' = S - \frac{1}{2}f$. In other words: The interaction which is antiferromagnetic (J>0) is minimized by allowing for a maximum number of electrons at the site of the impurity x=0 to interact with it. The maximum number is limited by Pauli's principle to be f.

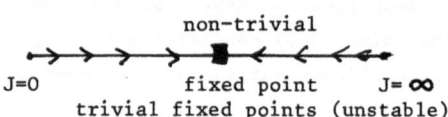

Now turn on the interaction. Allowing the electrons to interact with the complex we find that the effective interaction is ferromagnetic. The reason is that only electrons anti-parallel to the captured ones can interact. Their spin then, is parallel to the spin complex. Since the perturbation occurs to second order, the energy is lowered when the spins are parallel -- hence the interaction is effectively ferromagnetic. It is therefore stable and will flow into the fixed point at infinity. This yields then a situation similar to the one described by the canonical model.

2) Over screening: $\frac{1}{2}f > S$. In this case the ground state can be represented as 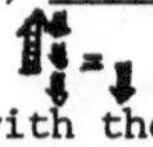 . The electrons which interact with the complex again must be anti-aligned with the electrons forming the complex, and therefore anti-aligned with the effective spin S'. As a result, the interaction between the Fermi sea and the effective spin is antiferromagnetic and we conclude that the fixed point at $J=\infty$ is unstable. Therefore a non-trivial fixed point must be present which will dominate the infrared behavior of the model leading to critical behavior and power laws as the temperature tends to zero. The flow diagram now becomes

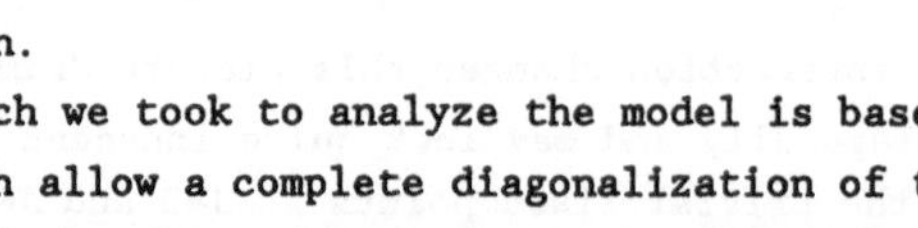

The analysis of Blandin and Nozières is compelling but of course heuristic. In particular it does not provide the means to calculate the critical exponents at the phase transition.

The approach we took to analyze the model is based on special properties it possesses, which allow a complete diagonalization of the hamiltonian and complete determination of the thermodynamics. It is the Bethe Ansatz approach. Since the subject is amply reviewed in the literature [see e.g. Andrei et al. 83] I will only present the line of thought rather than mathematical details.

Consider the hamiltonian in x-space

$$H = -i\int \psi^*_{am}(x)\partial_x\psi_{am}(x) + J\,\psi^*_{am}(0)\sigma^i_{am}\psi_{bm}(0)S^i \quad .$$

Since the number of electrons is conserved we can diagonalize it in the Hilbert space characterized by this number. A generic state would be of the form

$$|F\rangle = \int F_s(x,a,m)\Pi\psi^*_{am}(x)|0\rangle$$

where $|0\rangle$ is the state without any electrons, and $F_s(xam)$ is the impurity-electrons wave function. Acting on this state the second quantized hamiltonian is equivalent to the first quantized hamiltonian

$$h = \sum_j [-i\partial_j + 2J\delta(x_j)\sigma_j\cdot S] \quad .$$

Flavor interaction are not explicit in the hamiltonian. It may be surprising at first sight, but as the previous analysis indicated, flavor would make its appearance through the Pauli-principle. Indeed, in the first quantized picture the system can be represented as follows:

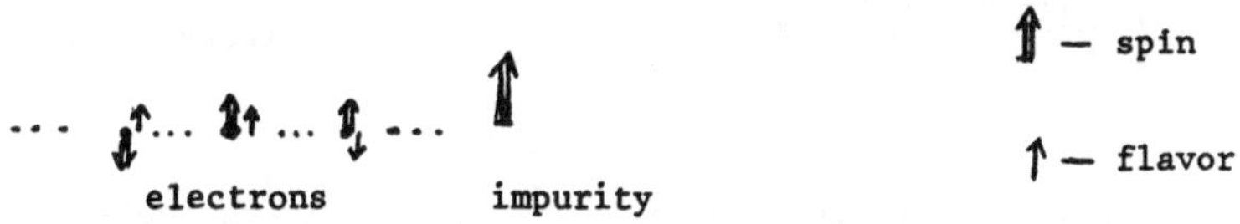

The impurity carries only spin degrees of freedom while the electrons carry both spin and flavor.

Obviously, flavor will not contribute if electrons cross the impurity one at a time. The only way flavor can enter is if it allows the formation of higher spin complexes pictorially represented as

which can interact with the impurity in a new way. Thus the role of flavor would be to overcome the Pauli-principle and lead to baryonic-like operators

$$\Psi_{a_1..a_n,m_1..m_n}(x) = \psi_{n_1a_1}(x)\ldots\psi_{m_na_n}(x) \quad , \qquad n < f$$

To see how this comes about we have to delve into the Bethe-Ansatz formalism. A Bethe-Ansatz soluble model has eigenstates that take the following form

$$F = \exp(i \sum_j k_j x_j) \sum_Q \theta(x_Q) S(Q) \Phi$$

where $\theta(x_Q)$ denote the ordering of the electrons according to a permutation Q:

$x_{Q_1} < x_{Q_2} < \ldots$, Φ is the spin-flavor amplitude in a reference region I: $x_1 < x_2 < x_3 < \ldots$ and $S(Q)\Phi$ is the amplitude in region Q. The Ansatz above is consistent if the S-matrix S(Q) can be uniquely expressed as a product along transpositions of elementary S-matrices S_{ij} which exchange two particles at a time. The condition for consistency are the Yang-Baxter relations

$$S^{ij}S^{jk}S^{jk} = S^{jk}S^{jk}S^{ij} \quad .$$

Since the S-matrix derives directly from the hamiltonian, its consistency indicates the integrability of the model. We find, indeed, in our case:

$$S^{ij} = \begin{cases} \dfrac{\lambda_j - \lambda_\ell - iJP_{j\ell}^{\text{spin}}}{\lambda_j - \lambda_\ell - iJ} \; \dfrac{\lambda_j - \lambda_\ell - iJP_{j\ell}^{\text{flavor}}}{\lambda_j - \lambda_\ell - iJ} \;, & \text{electron-electron} \\[2em] \dfrac{\lambda_j + 1 \; - iJ(\sigma_j \cdot S + \frac{1}{2})}{\lambda_j + 1 \; - iJ(S + \frac{1}{2})} \;, & \text{electron-impurity} \end{cases}$$

The quantities λ_j are given by $\lambda_j = \frac{k_j}{\Lambda}$ where Λ is the cut-off one needs to introduce into the problem, and which needs to be taken to infinity to obtain universal answers, in other words, answers which do not depend on the details of a particular cut-off scheme. P_{ij} is an exchange operator, be it for spin; P_{ij}^{spin} or flavor: P_{ij}^{flavor}. The matrices S^{ij} satisfy the consistency conditions and allow a complete determination of the spectrum.

The next step which I will omit and which can be found in our paper is the standard application of the technique to derive the equations which determine the momenta k_i appearing in the wave function and which yield the spectrum, through the expression

$$E = \sum_i k_i \left(1 + \frac{k_i}{\Lambda}\right) \; .$$

Now comes the main point: The Bethe-Ansatz equations thus obtained yield solutions for the momenta k_i which take complex values. More specifically, they are of the form

$$\lambda_j = P_j + \frac{i}{2} J(f-r+1) \qquad\qquad r=0,1 \; \ldots, \; f-1$$

These are so called "string" solutions, and describe a state where the momenta are grouped so that f of them have the same real part while their imaginary parts are separated by $iJ\Lambda$ and symmetrically distributed around the real axis. This leads to a new type of a "condensed" ground state where the wave function describes the "baryonic" operators. To see that consider the part of the wave-function corresponding to one particular string

$$F = \exp[-\frac{1}{2}\Lambda J \sum_{j<\ell} |x_j - x_\ell| + ip(x_1 + \ldots + x_f)] \; \{\ldots \text{ the part described by other strings } \ldots\}$$

In the limit $\Lambda \to \infty$ (the scaling limit), the only way to obtain a finite answer is if the particles described by momenta k_i belonging to the same string are squeezed to the same point. In this limit then, we find that the flavor interaction have introduced an attraction leading to the formation of these composites formed of f electrons coupled to give a flavor singlet and a maximal spin complex.

Having identified this mechanism of "dynamical Fusion", one may proceed in a standard fashion to determine the thermodynamic equation of the model.

The results corroborate the above present picture of a non-trivial fixed point appearing in the case $\frac{1}{2}$ f > S. We were able to calculate explicitly the critical exponents which characterize the fixed point and found for the specific heat exponent $\alpha = -\frac{4}{f+2}$ and for the susceptibility $\gamma = \frac{f-2}{f+2}$. It is interesting to note that they do not depend on the spin S.

In summary, the machinery of Bethe-Ansatz does apply to this new situation where a non-trivial fixed point makes its appearance. A new type of ground state is formed in the case of over screening where the impurity spin is screened but according to a power law with calculable exponents.

A different approach to the problem was presented by Tsvelick and Wiegman which is not based on the dynamical fusion picture but where the answer is derived from an Anderson model in the appropriate limit.

References

N. Andrei and C. Destri P.R.L. 52 p.364 (1984)

N. Andrei, K. Furuya, J.H. Lowenstein Rev. Mod. Phys. 55 p.331 (1983)

A.M. Tsvelick and P.B. Wiegman Z. fur Physik 54 p.201 (1974)

THE DIRECTED ANIMALS AND RELATED PROBLEMS

Deepak DHAR
Laboratoire de Physique Théorique et Hautes Energies
Université Pierre et Marie Curie
4, Place Jussieu - 75230 PARIS Cedex 05
and
Tata Institute of Fundamental Research
Homi Bhabha Road, Bombay, 400005, INDIA

The directed animals problem is one of the simplest of the many problems in statistical mechanics, characterized by an absence of reflection symmetry. Although this asymmetry, in the case of time-inversion, has been the central problem of non-equilibrium mechanics since the time of Boltzmann, the study of lattice models incorporating a preferred direction, and utilisation of the powerful techniques of equilibrium statistical mechanics to gain insight into non-equilibrium problems is much more recent. In this article, I shall describe the directed animals problem as a prototype of such models, and describe its relationship to crystal growth models, the Lee-Yang edge singularity, and to the Baxter's models of hard-square lattice gas with next-nearest neighbour interactions.

1) Directed Animals

Consider the two dimensional square lattice for simplicity. A directed animal of size n is defined as a connected set of n "occupied" sites, including the origin, such that for each occupied site (x,y) other than the origin, at least one of the sites (x-1, y) or (x, y-1) is also occupied. Generalization of this definition to other lattices and dimensions is straightforward. For large n, the total number distinct animals with n sites A_n, their average transverse and longitudinal diameters $\langle r_{\perp n}\rangle$ and $\langle r_{\parallel n}\rangle$ are expected to vary as

$$A_n \sim c\mu^n n^{-\theta} \tag{1.1}$$

$$\langle r_{\perp n}\rangle \sim n^{\nu_\perp} \tag{1.2}$$

$$\langle r_{\parallel n}\rangle \sim n^{\nu_\parallel} \tag{1.3}$$

We would like to calculate the exponents $\theta, \nu_\perp$ and $\nu_\parallel$ as these are expected to be universal in the sense that they depend on dimension but not on the detail of the lattice structure etc.

The problem can be solved exactly in d=1,2,3 and the corresponding values of θ are 0, 1/2, 5/6 respectively (1,2). For all dimensions $d \geq 7$ the mean field theory gives the exact results $\theta = 3/2$, $\nu_{\parallel} = 1/2$, $\nu_{\perp} = 1/4$ (3). For intermediate values of d, the exponents can be determined approximately by extrapolation of exact enumeration data (4), and to somewhat better reliability using rational approximants with the results of ϵ-expansion (5).

The fact that allowed configurations of occupied sites on the line x+y=T depend only on the configuration on the line x+y=T-1 in this problem (contrast with the case of undirected animals) leads to a very important simplification of the problem. If we identify T=x+y as the "time" coordinate of the site (x,y), this property may be called the Markovian property. Let C be the configuration of occupied sites on a line x+y=T. We define the generating function $A_C(x)$ as the sum of weights of all distinct configurations of animals at later times whose initial configuration is C, the weight of an animal of size n being x^n. Then the Markovian property leads to the recursion relations.

$$A_C(x) = x^{|C|} \left[1 + \sum_{C'} A_{C'}(x)\right] \qquad (1.4)$$

where $|C|$ is the number of sites in C, and the sum over C' is over all possible configurations of occupied sites at the time T+1, consistent with C. Using these recursions, the problem of explicit enumerations is made much more tractable, and fairly long series can be generated rather easily on the computer (6,7).

Alternatively, Eq(1.4) may be interpreted as the Chapman-Kolmogorov equation for a discrete time Markov process on a linear chain. Consider sequential occupation of sites in the plane by the following rule : At time τ=0, all sites below the line x+y = 0 are unoccupied. At time τ, sites on the line x+y = $-\tau$ are examined for occupancy. If both (x,y+1) and (x+1,y) are unoccupied, the the site (x,y) is occupied with probability p, otherwise left unoccupied.

Then, clearly, the probability that a site $A \equiv (x,y)$ is eventually occupied is p times the probability that both $A' \equiv (x+1,y)$ and $A'' \equiv (x,y+1)$ are empty. By the inclusion-exclusion principle, we get

$$\mathrm{Prob}(A) = p\left[1 - \mathrm{Prob}(A') - \mathrm{Prob}(A'') + \mathrm{Prob}(A'A'')\right] \qquad (1.5)$$

which is the same form as Eq.(1.4) when C consists of a single occupied site. In general we get

$$A_C(x = -p) = (-1)^{|C|}\, \mathrm{Prob}(C) \qquad (1.6)$$

This establishes the equivalence between the directed animals problem with a special case of a class of models, generally known as crystal growth models (CGM). The calculation of generating function of directed animals with a single point source is equivalent to that of determining the average density of occupied sites in the related CGM.

The CGM, in general, may have several states per site and other much more complicated transition rates. The general model, even in two dimensions shows a variety of complex behaviors (8,9), and can be solved exactly only for some special choices of transition rates (10-14). One such case is when the transition rates satisfy the detailed balance condition. Then the time invariant probability distribution is easily written down. The directed animals problems on a d-dimensional hypercubical lattice corresponds to time development of thermal relaxation of a lattice gas with nearest-neighbour exclusion on a (d-1) dimensional hypercubical lattice. The exact solution the a 1-dimensional hard core lattice gas with nearest neighbour exclusion then gives the complete enumeration of directed lattice animals on the square and triangular lattices (2). We find that

$$\theta = \nu_\perp = 1/2 \qquad \text{in} \quad d=2. \tag{1.7}$$

The exact expression for the density of the hard-hexagon gas as a function of its activity obtained by Baxter (15) can be used to determine the exponents for a 3-dimensional directed animals problem. This gives

$$\theta = 2\nu_\perp = 5/6 \qquad \text{in} \quad d=3. \tag{1.8}$$

For the exponent $\nu_\parallel$, no exact results are known in 2 or 3 dimensions. A finite size scaling analysis using transfer matrices by Nadal et al (16) gives a fairly precise estimate $\nu_\parallel \simeq .818$ in d=2.

The singularity of the density of a lattice gas as a function of its activity can be expressed in terms of the singularity of the density of zeroes of partition function in the complex activity plane (the Lee-Yang edge singularity). The edge singularity is conventionally studied for ferromagnetic interactions when the zeroes are confined to the unit circle. For antiferromagnetic interactions, the lines of zeroes can take fairly complicated shapes, but the singularity closest to the origin has the same exponent as the ferromagnetic case. The relation between the animal and Lee-Yang singularity exponents is

$$\theta(d) = \sigma(d-1) + 1 \tag{1.9}$$

where σ(d-1) is the Lee-Yang edge singularity exponent in (d-1) dimensions. In particular $\theta(3) = 5/6$ emplied that $\sigma(d=2) = -1/6$. Using the standard hyperscaling relation to the edge-singularity problem gives

$$\nu_\perp(d) \cdot (d-1) = \theta(d) \tag{1.10}$$

Parisi and Sourlas (17) have shown that the Lee-Yang edge singularity problem in d-dimensions describes an undirected animals problem in (d+2) dimensions, and

$$\theta_{\text{undirected}}(d+2) = \sigma(d) + 2 \tag{1.11}$$

Thus, knowing directed animal exponent in 3-dimensions, we are able to conclude that θ for undirected animals in 4 dimensions is 11/6.

2) Baxter's Hard-Square Lattice-Gas Model

This model describes a hard-core lattice gas with next-nearest neighbour interactions, and is defined by the Hamiltonian

$$H = -L \sum_{ij} n_{ij} n_{i+1,j+1} - M \sum_{ij} n_{ij} n_{i-1,j+1} - \mu \sum n_{ij} \tag{2.1}$$

where n_{ij} is the occupation number of the site (ij) on a square lattice taking values 0 and 1, and we impose the constraint that nearest neighbours cannot be simultaneously occupied.

For M=L=0, this describes a hard-square lattice gas with nearest-neighbour exclusion. For $M = -\infty$ and $L = 0$, this corresponds to the exactly soluble hard hexagon gas. Baxter's exact solution of this model (18-20) is restricted to the case when the three coupling constants are related by the condition

$$\exp(\mu) = (1-e^{L})(1-e^{M})/(e^{L+M}-e^{L}-e^{M}) \tag{2.2}$$

My discussion here will be restricted to the connection of this model with the directed animals problem. For very large negative values of μ, the free energy per site of this model may be explanded in powers of $z=e^{\mu}$, keeping L and M fixed. This Mayer series has a finite radius of convergence, and may be used to define the analytically continued value of free energy for negative z. The analytic continuation cannot be pushed to arbitrarity large negative values of z because of the occurence of a line of zeroes on the negative real axis in the complex z-plane. Let $-z_c(L,M)$ be the distance of the closest singularity to the origin on the negative real axis. Then in the neighbourhood of $z=z_c(L,M)$ the free energy per site of the model has a power law singularity

$$f(L,M,z) \sim [z-z_c(L,M)]^{\sigma+1} \tag{2.3}$$

where the exponent σ is the two dimensional Lee-Yang edge singularity, i.e $\sigma=-1/6$, at least so long as L and M are not too negative. This agrees with the known behavior of the free energy when Baxter's factorizability condition Eq.(2.2) holds.

However, the Hamiltonial (2.1) is a special case of the more general interactions round a face (IRF) model of Baxter. This general IRF model has special surfaces called disorder surfaces, in the parameters space of the coupling constants of the model for which the partition function per site reduces to a single algebraic expression. For the Hamiltonian (2.1), the equation of the disorder surface is (21)

$$z = (e^{-4L}-e^{-3L})/[1+e^{M}(e^{-L}-1)] \tag{2.4}$$

and on this surface, the density of the gas is given by

$$\rho = \tfrac{1}{2}[1-\{1+4e^{-2L}(e^{-L}-1)^2/z\}^{-1/2}] \tag{2.5}$$

This expression becomes singular and the density tends to $-\infty$ as the expression inside the curly brackets in Eq.(2.5) tends to zero. Thus we get the equation of line of intersection of the disorder surface Eq.(2.4) with the critical surface $z=z_c(L,M)$ as

$$z = -4e^{-2L}(e^{-L}-1)^2 = (e^{-4L}-e^{-3L})/[1+e^{M}(e^{-L}-1)] \tag{2.6}$$

And near this line, the singularity of free energy is given by $\sigma=-1/2$ from Eq.(2.5) for $L \neq M$. Note that $\sigma = -1/2$ corresponds to a one dimensional Lee-Yang edge singularity.

A different disorder surface is obtained by interchanging L and M in Eq. (2.4). The intersection of this surface with the critical surface $z=z_c(L,M)$ gives another line of critical points with $\sigma = -1/2$. The equation of this line is of cause, obtained by interchanging L and M in Eq.(2.6).

These two lines of critical points meet at the point $e^{-L} = e^{-M} = (-z)^{1/4} = 2/3$. It is easy to verify from the given expressions that if we approach this point along the line of intersection of the two disorder surfaces, then the free energy singularity corresponds to $\sigma = -1$.the value corresponding to the Lee-Yang edge singularity exponent in 0 dimensions.

Thus the critical surface $z=z_c(L,M)$ has on it areas corresponding to the two-dimensional critical behavior ($\sigma = -1/6$), lines corresponding to one dimensional critical behavior, and isolated points corresponding to zero-dimensional behavior. Ordinarily, models showing such dimensional crossovers involve Hamiltonians in which some parameter corresponding to interplanar coupling tending to zero. The present model is special, as no obvious decoupling between different directions is implied by the disorder condition. Further studies are needed to elucidate the nature and mechanism of this phenomenon of dimensional reduction.

References

(1) Dhar D, Phys. Rev. Lett. 49 959 (1982).
(2) Dhar D, Phys. Rev. Lett. 51 853 (1983).
(3) Lubensky T.C. and Vannimenus J, J.Physique Lett. 43, L377 (1982); Redner S. and Coniglio A, J. Phys. A, 15 L273 (1982).
(4) Stanley H.E, Redner S and Yang Z.R, J. Phys. A15 L569 (1982).
(5) Breuer N and Janssen H.K.,Z. Phys. B54, 175 (1984).
(6) Dhar D, Barma M and Phani M.K, J.Phys. A15, L279 (1982).
(7) Duarte JAMS, J. Phys. (Paris) 46 L523 (1985).
(8) Wolfram S, Rev. Mod. Phys. 55 601 (1983).
(9) Kinzel W, Z. Phys. B58 229 (1985).
(10) Verhagen A.M.W., J. Stat. Phys. 15, 219 (1976).
(11) Enting I.G., J.Phys. C10, 1379 (1977) ; J. Phys. A11, 555, 2001 (1978).
(12) Rujan P., J.Stat. Phys. 29 231, 247 (1982) ; 34 615 (1984).
(13) Domany E and Kinzel W. Phys. Rev. Lett. 53 311 (1984).
(14) Baxter R.J. J.Phys. A17 L911 (1984).
(15) Baxter R.J. J.Phys. A13 L61 (1980).
(16) Nadal J.P., Derrida B. and Vannimenus J., J.Phys. (Paris) 43, 1561 (1982).
(17) Parisi G. and Sourlas N, Phys. Rev. Lett. 46 871 (1981).
(18) Baxter R.J. and Pearce P.A., J. Phys. A15 897 (1982).
(19) Huse D.A., J.Phys. A16, 4357 (1983).
(20) Pearce P.A. and Baxter R.J., J. Phys. A17 2095 (1984).
(21) Baxter R.J., private communication.

INCOMMENSURATE STRUCTURES AND BREAKING OF ANALYTICITY

by

S. Aubry
Laboratoire Léon Brillouin
CEN-Saclay
91191 Gif-sur-Yvette, France

In this seminar, we essentially review our own results related to the concept of transition by breaking of analyticity[1,2] (TBA). This concept arises in incommensurate structures which unlike perfect crystals, are quasi-periodic structures with no translational invariance. In many cases, the structure can be viewed as a crystal with one or several superimposed static and periodic modulations, the periods of which are not rationally related to the periods of the initial crystal. These structures are degenerate with respect to the phase variations of the modulations. When the atomic coordinates vary analytically as a function of the phases, the structure is said to be analytic and exhibits a zero frequency mode (called Goldstone mode or phason). In the opposit case, the variation of the atomic coordinates exhibits infinitely many discontinuities which physically correspond to atomic jumps at the microscopic scale. There exists a finite gap in the phason spectrum.

The TBA corresponds to the transition from the analytic regime of the incommensurate structure to the non analytic one when some parameter of the model varies at fixed incommensurability ratio. Such a transition is rigorously proved to exist[3,4] in the Frenkel Kontorova model (FK model) this transition does not correspond to a standard symmetry breaking. But it can be shown that the TBA is associated to the breaking of a Kolmogorov Arnol'd Moser torus of an associated twist map into a Cantor set[5a,5b]. It is also a breaking of the Goldstone symmetry[5c]. It is also shown on the basis of numerical arguments that a TBA exists in other extended versions of the F.K. model including one-dimensional model with convex interactions between n^{th} neighbour atoms, with several sublattices[7] and also in two dimensional models[8]. (In that cases, the representation of the TBA by the breaking of an invariant curve into a Cantor set in the phase space of an associated Hamiltonian dynamical system is only formal because of the lack of theorems for these dynamical systems).

Numerical investigations of the TBA has determined many physical quantities which vanish at the critical point with non trivial exponents[9,10]. Unexpectedly their behaviors depend on the incommensurability ratio and is not universal. However, it seems that it does not depend on the details of the model providing that the incommensurability ratio stays the same. In fact, the renormalization approach shows that the critical behavior at the TBA is driven by the integer coefficients of the continued expansion of ζ.

A TBA has also been found for the propagation of the electrons (or waves) in a quasi-periodic lattice[11]. In that case, the non analyticity of the wave-function with respect to the phase of the potential is associated to the localization of the electrons (the wave functions of which become square summable). (Note that more complex situations may exist where the spectrum (eigen energies) becomes singular continuous with "intermittent" wave functions)[14].

Using arguments of self consistency in a system where the electrons and the lattice interact, we have conjectured[12] and next numerically checked[13] that a TBA also exists in models for Peierls one-dimensional deformable condutors. In that case the TBA is associated to a metal insulator transition by extinction of the Fröhlich superconductivity (in models at OK).

In many models for incommensurate structures (but not all), it is shown that the commensurability ratio varies as a Devil's staircase[2,5] (D.S.). Such a curve is continuous but exhibits a constant plateau at each rational value which is reached. We distinguish two kinds of DS : those which are incomplete where the total measure of the plateaus does not fill the full interval of variation of the parameter and those which are complete[12]. It is shown that incomplete DS are obtained when intermediate incommensurate structures are analytic while they are complete when all the incommensurate structures which are met during the parameter variation are non analytic.

Exact models with complete D.S. can be extended to more complex models[7-8]. Particularly a model introduced for describing an incommensurate structure submitted to an electric field[7], exhibits a polarization curve which has infinitely many plateaus and infinitely many discontinuities. This new kind of pathological curve has been called Manhattan profile.

When the convexity property for the atomic interaction is not fulfilled, we expect that the D.S. behavior may be drastically changed. We exhibited a model[22] where the variation curve of the commensurability ratio has unexpectedly no plateaus at the rational values while it has plateaus at a countable set of particular irrational values. Moreover it is probable that the variation curve also includes infinitely many discontinuities.

All the above results were obtained for classical model in one or several dimensions but always at OK. We have recently studied the thermodynamicl properties[17] of a one-dimensional incommensurate non-analytic structure at finite temperature. This study is done by a renormalization approach which is driven by the integer sequence of the continued fraction expansion of the incommensurability ratio ζ . It is shown that the density of state of the low energy excitation are hierarchically distributed according to rules determined by this continued fraction expansion.

The role of the continued fraction expansion has also been found to be essential in the dynamics of incommensurate structures. We found a sequence of dynamical instabilities in a model for a one-dimension analytic incommensurate charge density wave (CDW) submitted to an electric field[18]. However, the effect of the dimensionality of the model appears to be essential[19] and should introduce new kinds of instabilities. The investigation of a model for a non-analytic CDW in an electric field[20] also seems to reveals a role for the continued fraction expansion of the incommensurability ratio of the CDW. This new kind of approach for the dynamics of a CDW could be the basis for an explanation for many unexpected phenomena observed in the conductivity of CDW compounds[21] such as $NbSe_3$ or the blue bronze.

The matter of this seminar will be described in details in Ref. 23.

REFERENCES

1. **S. Aubry** in "Stochastic behavior in Classical and Quantum Systems" Lectures Notes in Physics **93**, 201-212 (1979) (Springer) ed. G. Cassati and J. Ford.
2. **S. Aubry** in "Soliton and Condensed Matter" Solid State Sciences, **8**, 264-278 (1978)(Springer) ed. A.R. Bishop and T. Schneider.
3. **S. Aubry, P.Y. Le Daëron, and G. André**, "Classical ground-states of one-dimensional model for incommensurate structures" 1982 unpublished.
4. **S. Aubry and P.Y. Le Daëron,** Physica **8D**, 38-422 (1983)
   S. Aubry Physica **7D**, 240-258 (1983)

5a. **S. Aubry,** in "Intrinsic stochasticity in plasmas" pp.63-83, ed. G. Laval and D. Gresillon, ed. de Physique (1979)

5b. Phys. Rep. **103**, 127-141 (1984)

5c. in "Symmetry and broken symmetries", Idset (Paris), p.313-322 (1981)) ed. N. Boccara

6. **M. Peyrard and S. Aubry,** unpublished (1983)
7. **S. Aubry, F. Axel and F. Vallet,** J. Phys. **C18**, 753-788 (1985)
8. **F. Vallet,** PHD, Dissertation (Paris 1986)
9. **M. Peyrard and S. Aubry,** J. Phys. **C16**, 1593-1608 (1983)
10. **L. de Sèze and S. Aubry** J. Physique **C17**, 389-403 (1984).
11. **S. Aubry and G. André,** Ann. of the Israël Phys. Soc. **3**, 133-164, ed. L.P. Horwitz and Y. Ne'eman (1980).

12. **S. Aubry** in "Bifurcation phenomena in Mathematical Physics and Related topics" ed. D. Bessis and C. Bardos (Riedel) 1980 p. 163-184.
13. **P.Y. Le Daëron and S. Aubry**, J. Physique **C16**, 4827-4838 (1983) J. Physique (Paris) **C3**, 1573-1577 (1983).
14. **M. Kohmoto**, Phys. Rev. Lett. **51**, 1198 (1983).
15. **S. Aubry** in "The Rieman problem, complete integrability and arithmetic applications", Lecture Notes in Math. **925**, 221-241 (1980).
16. **S. Aubry** in J. Physique **C16**, 2497-2508 (1983). **S. Aubry**, J. Physique Lett. **44**, L247-250 (1983).
17. **F. Vallet, R. Schilling and S. Aubry**, in preparation.
18. **S. Aubry and L. de Sèze**, in Festköperprobleme XXV (1985) in press. Proceedings of E.P.S. Conference (Berlin)
19. **S. Aubry, A. Bishop and P. Lomdhal**, in preparation.
20. **P. Quemerais and S. Aubry**, in preparation
21. See for example, Proc. of Budapest Conference, Charge Density Waves in Solid, Lectures Notes in Physics **217**, Springer (1984).
22. **S. Aubry, F. Fesser and A. Bishop**, J. of Phys. A, in press (1985).
23. Proceedings of "Structures et intabilities" Beg Rohu (france) (1985) To be published (in French) in Editions de Physique (1986).

LIST OF CONTRIBUTORS

N. ANDREI
Serin Physics Laboratory
Rutgers University
Piscataway, NJ 08854
U.S.A.

S. AUBRY
Laboratoire Léon Brillouin
CEN - Saclay
91191 Gif-sur-Yvette
France

J. AVAN
L.P.T.H.E. Jussieu
Tour 16, 1er étage
4 place Jussieu
75230 Paris Cedex 05
France

D. BERNARD
Groupe d'Astrophysique Relativiste
C.N.R.S. - Observatoire de Paris-Meudon
92195 Meudon Principal Cedex
France

E. CORRIGAN
Department of Mathematical Sciences
University of Durham
South Road
Durham DH1 3LE
United Kingdom.

C. DEVCHAND
Fackultat für Physik
Universität Freiburg
Hermann - Herder - Str. 3
D-7800 Freidburg
W - GERMANY

D. DHAR
Tata Institute of Fundamental Research
Homi Bhabha Road
400005 - Bombay
India

L.D. FADDEEV
Steklov Mathematical Institute
Fontanka 27
Leningrad D 11 - 191011
U.S.S.R.

M. FORGER
Theory Division, C.E.R.N.
1211 Geneva 23
Switzerland

B. GAFFET
Service d'Astrophysique
CEN - Saclay
91191 Gif-sur-Yvette Cedex
France

J.-L. GERVAIS
Physique Théorique, Ecole Normale Supérieure
24 rue Lhomond
75231 Paris Cedex 05
France

G.W. GIBBONS
Department of Applied Mathematics and Theoretical Physics,
University of Cambridge
Silver Street, Cambridge CB3 9EW
United Kingdom

M.B. GREEN
Physics Department, Queen Mary College
University of London
Mile End Road, London E1 4NS
United Kingdom

P. HAJICEK
Institute for Theoretical Physics
University of Bern
Sidlerstrasse 5, CH - 3012 Bern
Switzerland

S.W. HAWKING
Department of Applied Mathematics and Theoretical Physics
University of Cambridge
Silver Street, Cambridge CB3 9EW
United Kingdom

M. JIMBO
Research Institute for Mathematical Sciences
Kyoto University
Kyoto 606
Japan

G. JONA-LASINIO
Dipartamento di Fisica-Universita "La Sapienza"
GNSM and INFN - Roma
Piazzale Aldo Moro 2 - I-00185 Roma
Italy

N. KAMRAN
Centre de Recherches Mathématiques
Université de Montréal
Case Postale 6128, Succursale "A"
Montréal, Québec, H3C 3J7
Canada

B.G. KONOPELCHENKO
Institute of Nuclear Physics
Novosibirsk 90
630090 U.S.S.R.

A.I. NACHMAN
Department of Mathematics
University of Rochester
Rochester, NY 14627
U.S.A.

C.N. POPE
Blackett Laboratory, Imperial College
Prince Consort Road,
London SW7 2BZ
United Kingdom

N. SANCHEZ
ER 176, C.N.R.S., Departement d'Astrophysique Fondamentale,
Observatoire de Paris-Meudon
92195 Meudon Principal Cedex
France

W.F. SHADWICK
Department of Pure Mathematics
University of Waterloo
Waterloo, Ontario, N2L 3G1
Canada

A.A. STAROBINSKY
Landau Institute for Theoretical Physics
ul. Kosygina 2
Moscow 117 334
U.S.S.R.

LA.A TAKHTAJAN
Steklov Mathematical Institute
Fontanka 27,
Leningrad D11
191011 - U.S.S.R.

Lecture Notes in Physics

Vol. 214: H. Moraal, Classical, Discrete Spin Models. VII, 251 pages. 1984.

Vol. 215: Computing in Accelerator Design and Operation. Proceedings, 1983. Edited by W. Busse and R. Zelazny. XII, 574 pages. 1984.

Vol. 216: Applications of Field Theory to Statistical Mechanics. Proceedings, 1984. Edited by L. Garrido. VIII, 352 pages. 1985.

Vol. 217: Charge Density Waves in Solids. Proceedings, 1984. Edited by Gy. Hutiray and J. Sólyom. XIV, 541 pages. 1985.

Vol. 218: Ninth International Conference on Numerical Methods in Fluid Dynamics. Edited by Soubbaramayer and J.P. Boujot. X, 612 pages. 1985.

Vol. 219: Fusion Reactions Below the Coulomb Barrier. Proceedings, 1984. Edited by S.G. Steadman. VII, 351 pages. 1985.

Vol. 220: W. Dittrich, M. Reuter, Effective Lagrangians in Quantum Electrodynamics. V, 244 pages. 1985.

Vol. 221: Quark Matter '84. Proceedings, 1984. Edited by K. Kajantie. VI, 305 pages. 1985.

Vol. 222: A. García, P. Kielanowski, The Beta Decay of Hyperons. Edited by A. Bohm. VIII, 173 pages. 1985.

Vol. 223: H. Saller, Vereinheitlichte Feldtheorien der Elementarteilchen. IX, 157 Seiten. 1985.

Vol. 224: Supernovae as Distance Indicators. Proceedings, 1984. Edited by N. Bartel. VI, 226 pages. 1985.

Vol. 225: B. Müller, The Physics of the Quark-Gluon Plasma. VII, 142 pages. 1985.

Vol. 226: Non-Linear Equations in Classical and Quantum Field Theory. Proceedings, 1983/84. Edited by N. Sanchez. VII, 400 pages. 1985.

Vol. 227: J.-P. Eckmann, P. Wittwer, Computer Methods and Borel Summability Applied to Feigenbaum's Equation. XIV, 297 pages. 1985.

Vol. 228: Thermodynamics and Constitutive Equations. Proceedings, 1982. Edited by G. Grioli. V, 257 pages. 1985.

Vol. 229: Fundamentals of Laser Interactions. Proceedings, 1985. Edited by F. Ehlotzky. IX, 314 pages. 1985.

Vol. 230: Macroscopic Modelling of Turbulent FLows. Proceedings, 1984. Edited by U. Frisch, J.B. Keller, G. Papanicolaou and O. Pironneau. X, 360 pages. 1985.

Vol. 231: Hadrons and Heavy Ions. Proceedings, 1984. Edited by W.D. Heiss. VII, 458 pages. 1985.

Vol. 232: New Aspects of Galaxy Photometry. Proceedings, 1984. Edited by J.-L. Nieto. XIII, 350 pages. 1985.

Vol. 233: High Resolution in Solar Physics. Proceedings, 1984. Edited by R. Muller. VII, 320 pages. 1985.

Vol. 234: Electron and Photon Interactions at Intermediate Energies. Proceedings, 1984. Edited by D. Menze, W. Pfeil and W. J. Schwille. VII, 481 pages. 1985.

Vol. 235: G.E.A. Meier, F. Obermeier (Eds.), Flow of Real Fluids. VIII, 348 pages. 1985.

Vol. 236: Advanced Methods in the Evaluation of Nuclear Scattering Data. Proceedings, 1985. Edited by H.J. Krappe and R. Lipperheide. VI, 364 pages. 1985.

Vol. 237: Nearby Molecular Clouds. Proceedings, 1984. Edited by G. Serra. IX, 242 pages. 1985.

Vol. 238: The Free-Lagrange Method. Proceedings, 1985. Edited by M.J. Fritts, W.P. Crowley and H. Trease. IX, 313 pages. 1985.

Vol. 239: Geometrics Aspects of the Einstein Equations and Integrable Systems. Proceedings, 1984. Edited by R. Martini. V, 344 pages. 1985.

Vol. 240: Monte-Carlo Methods and Applications in Neutronics, Photonics and Statistical Physics. Proceedings, 1985. Edited by R. Alcouffe, R. Dautray, A. Forster, G. Ledanois and B. Mercier. VIII, 483 pages. 1985.

Vol. 241: Numerical Simulation of Combustion Phenomena. Proceedings, 1985. Edited by R. Glowinski, B. Larrouturou and R. Temam. IX, 404 pages. 1985.

Vol. 242: Exactly Solvable Problems in Condensed Matter and Relativistic Field Theory. Proceedings, 1985. Edited by B.S. Shastry, S.S. Jha and V. Singh. V, 318 pages. 1985.

Vol. 243: Medium Energy Nucleon and Antinucleon Scattering. Proceedings, 1985. Edited by H.V. von Geramb. IX, 576 pages. 1985.

Vol. 244: W. Dittrich, M. Reuter, Selected Topics in Gauge Theories. V, 315 pages. 1986.

Vol. 245: R.Kh. Zeytounian, Les Modèles Asymptotiques de la Mécanique des Fluides I. IX, 260 pages. 1986.

Vol. 246: Field Theory, Quantum Gravity and Strings. Proceedings, 1984/85. Edited by H.J. de Vega and N. Sánchez. VI, 381 pages. 1986.

Selected Issues from

Lecture Notes in Mathematics